The Salton Sea Centennial Symposium

Developments in Hydrobiology 201

Series editor

K. Martens

The Salton Sea Centennial Symposium

Proceedings of a Symposium Celebrating a Century of Symbiosis Among Agriculture, Wildlife and People, 1905–2005, held in San Diego, California, USA, March 2005

Edited by

Stuart H. Hurlbert

San Diego State University, San Diego, California, USA

Reprinted from Hydrobiologia, Volume 604 (2008)

 Springer

Library of Congress Cataloging-in-Publication Data

A C.I.P. Catalogue record for this book is available from the Library of Congress.

ISBN-13: 978-1-4020-8805-6

Published by Springer,
P.O. Box 17, 3300 AA Dordrecht, The Netherlands

Cite this publication as Hydrobiologia vol. 604 (2008).

Cover illustration: The Salton Sea in January 2008, as seen from North Shore, California. Photo credit: Doug Barnum, US Geological Survey.

TABLE OF CONTENTS

Hydrobiologia (2008) 604:1–3
DOI 10.1007/s10750-008-9314-3

SALTON SEA

Preface

Stuart H. Hurlbert

The Salton Sea is the largest lake in California and occupies a below sea level depression in the desert just north of the border with Mexico. The current lake formed accidentally in 1905 as a result of a breaching of diversion structures on the Colorado River. Ever since then it has persisted, despite the hot desertic climate, as 1.6 billion cubic meters of agricultural wastewaters flow into it every year. From its earliest days, this lake was prime habitat for fish and wildlife. Over time, residential communities, marinas, a wildlife refuge, and a state park developed along its shoreline, and it became a mecca for sport fishing, boating, water sports, birdwatching, and camping. Over the last three decades, however, fluctuating lake levels, rising salinity and continued eutrophication have caused increasing problems for both wildlife and man.

In the late 1990s, in response mainly to large fish kills and heavy bird mortalities at the Salton Sea, significant funding finally became available for scientific analysis of the ecology of the Sea and of possible solutions to its problems. Much of the resultant new scientific information on the Salton Sea was published in four earlier volumes of

---

Guest editor: S. H. Hurlbert
The Salton Sea Centennial Symposium. Proceedings of a Symposium Celebrating a Century of Symbiosis Among Agriculture, Wildlife and People, 1905–2005, held in San Diego, California, USA, March 2005

Hydrobiologia (Zheng et al., 1998; Melack et al., 2001; Barnum et al., 2002; Melack, 2007).

In order to stimulate synthesis and publication of new findings on the Salton Sea ecosystem, the Salton Sea Centennial Symposium was convened in San Diego in March 2005, under the auspices of the U.S. Geological Survey (USGS) Salton Sea Science Office and San Diego State University's Center for Inland Waters, with support from the Water Education Foundation, the SDSU President's Leadership Fund, and the California Department of Water Resources. Additional support for publishing the symposium papers was provided by the Salton Sea Authority. Planning of the scientific program was carried out by Douglas Barnum (USGS Salton Sea Science Office) and the editor. The present volume contains papers based on 14 of the 35 oral presentations made at that symposium. A companion set of symposium papers with new biological findings on the Salton Sea is being published simultaneously as a special issue of the journal *Lake and Reservoir Management* (23(5), 2007).

Several major speakers and events at the symposium are not represented in this collection of papers. Experts on three other large, saline, aquatic ecosystems were keynote speakers at the symposium. Enrique Bucher (Professor of Ecology, National University of Cordoba, Cordoba, Argentina) talked on Mar Chiquita, a 5,000 km^2 lake in northern Argentina. Philip Micklin (Professor of Geography, Western Michigan University, Kalamazoo, Michigan) presented a talk on the Aral Sea. Jose Campoy Favela

 Springer

(Director, Northern Gulf of California and Colorado River Delta Biosphere Reserve, San Luis Rio Colorado, Sonora, Mexico) gave a talk on wetlands in the lower portion of the Colorado River delta in Mexico. We were also honored by a brief visit and discussions with a delegation of 16 scientists and officials from Uzbekistan, Turkmenistan, Kazakhstan, and Tajikistan, countries bordering the Aral Sea. Finally, Rita Schmidt Sudman (Executive Director, Water Education Foundation) organized and moderated a lively panel discussion on the topic, *The Link Between Water Supply, Science, Restoration and The Law.*

The future of the Salton Sea ecosystem is uncertain, but it certainly is headed toward uncharted waters. The ecosystem analyzed in these studies already has changed. Water inflows are declining. The fish are essentially gone except for the hardy tilapia. Salinity is presently 47–48 g l^{-1}, matching the previous historic high of the mid-1930s, and continues to rise. We may intervene or not, but the Salton Sea ecosystem of the last half century is no more, and just as we understood it better than ever before!

Much of what we have learned about it has proved useful in at least developing plans for a brighter future. After many years of study and discussion involving large numbers of stakeholders, the California Resources Agency has put forward its preferred alternative for a "Salton Sea Ecosystem Restoration Program" (available at http://www.saltonsea.water.ca.gov/). This ambitious and complex plan defies concise description. It envisions use of dikes, berms, canals, and other elements to create: a narrow, 182 km², horseshoe-shaped salt lake, stabilized at 30–40 g l^{-1}, around the perimeter of the northern two-thirds of the present lake; a 251 km² complex of tiered, shallow, saline (20–200 g l^{-1}) wetlands, mostly around the southern end of the present lake; and a large central area that eventually will consist of exposed lakebed or playa (429 km²) and two very shallow brine lakes (69 km²). Aquatic habitat diversity will be greatly increased in the region, though total area of aquatic habitat will be about half that of the present lake (930 km²). Project capital costs are estimated at USD 8.9 billion, with post-construction operating and maintenance costs estimated at USD142 million per year. Many large technical and financial issues concerning the proposed project are not yet resolved. The state of California is in a fiscal crisis at this time; its population continues to grow, water demand is up, and climate models predict increasing aridity over large

portions of the American Southwest. Whatever will be the consequences of these colliding plans and forces for the Salton Sea, they will—at a minimum—be "interesting" over the next decades.

Acknowledgments

The sponsoring and funding agencies are acknowledged above. Their moral support and financial contributions were critical to the success of the symposium and were obtained primarily through the efforts of Douglas Barnum of the USGS Salton Sea Science Office. Rita Schmidt Sudman, Judy Maben, and Sue McClurg of the Water Education Foundation did superb jobs of overseeing the registration process and all arrangements for facilities and services at the symposium site, moderating the panel discussion, and producing a summary of the abstracts of all talks given (WEF, 2005). Joan Dainer and Jim Zimmer helped with various logistical matters. A Cooperative Agreement between the USGS Salton Sea Science Office and the SDSU Research Foundation supported the editor's involvement in planning of the scientific program and the early editorial work on this volume and its companion in *Lake and Reservoir Management.*

Each manuscript submitted for this volume was reviewed by 2–4 referees plus the editor, and most manuscripts underwent extensive revision before being accepted. Many thanks to all those referees whose constructive suggestions helped authors to "put their best foot forward": K.K. Bertine, J. Bloesch, M. Bowes, C. Brauner, M.T. Brett, M.J. Cohen, D.M. Dexter, M. Dittrich, J. Fram, C. Franson, E. Furlong, R. Gersberg, M. Gurol, K. Hoef-Emden, G.C. Holdren, K.A. Hovel, A.H. Hurlbert, R. Jellison, S. Kelly, L.Y. Lewis, T.W. Lyons, L. Majewski, A.K. Miles, S. McIntyre, E. McNaughton, M.F. Moreau, C.R. Phillips, T.S. Presser, K.M. Reifel, D.M. Robertson, T.E. Rocke, D. Schlenk, R.A. Schroeder, J.L. Scott, J.G. Setmire, P. Smith, V.H. Smith, B.K. Swan, M.A. Tiffany, D. Trolle, D.L. Valentine, J.M. Watts, E. Welch, W.A. Wurtsbaugh. A final thanks to the authors themselves for their contributions, their patience, and their stoic courtesy in putting up with unending editorial suggestions.

Stuart H. Hurlbert
San Diego, CA
USA

References

Barnum, D. A., J. F. Elder, D. Stephens & M. Friend (eds), 2002. The Salton Sea: Proceedings of the Salton Sea Symposium, 2000. Hydrobiologia 473: xii + 306 pp. (Developments in Hydrobiology 161: xii + 306 pp).

Melack, J. M., R. Jellison & D. B. Herbst (eds), 2001. Saline Lakes: Publications from the 7th International Conference on Salt Lakes. Hydrobiologia 466: ix + 347 pp. (Developments in Hydrobiology 162: ix + 347 pp).

Melack, J. M. (ed.), 2007. Saline waters and their biota. Hydrobiologia 576: iii + 203 pp.

WEF, 2005. The Salton Sea Centennial Symposium: Summary of Abstracts. Water Education Foundation, Sacramento, California, 15 pp.

Zheng, M., S. H. Hurlbert & W. D. Williams (eds), 1998. Saline Lakes VI: Proceedings of the VIth International Symposium on Inland Salt Lakes. Hydrobiologia 381: x + 200 pp.

Hydrobiologia (2008) 604:5–19
DOI 10.1007/s10750-008-9321-4

Response in the water quality of the Salton Sea, California, to changes in phosphorus loading: an empirical modeling approach

Dale M. Robertson · S. Geoffrey Schladow

Abstract Salton Sea, California, like many other lakes, has become eutrophic because of excessive nutrient loading, primarily phosphorus (P). A Total Maximum Daily Load (TMDL) is being prepared for P to reduce the input of P to the Sea. In order to better understand how P-load reductions should affect the average annual water quality of this terminal saline lake, three different eutrophication programs (BATH-TUB, WiLMS, and the Seepage Lake Model) were applied. After verifying that specific empirical models within these programs were applicable to this saline lake, each model was calibrated using water-quality and nutrient-loading data for 1999 and then used to simulate the effects of specific P-load reductions. Model simulations indicate that a 50% decrease in external P loading would decrease near-surface total phosphorus concentrations (TP) by 25–50%. Application of other empirical models demonstrated that this decrease in loading should decrease near-surface chlorophyll a concentrations (Chl a) by 17–63% and increase Secchi depths (SD) by 38–97%. The wide range in estimated responses in Chl a and SD were primarily caused by uncertainty in how non-algal turbidity would respond to P-load reductions. If only the models most applicable to the Salton Sea are considered, a 70–90% P-load reduction is required for the Sea to be classified as moderately eutrophic (trophic state index of 55). These models simulate steady-state conditions in the Sea; therefore, it is difficult to ascertain how long it would take for the simulated changes to occur after load reductions.

Keywords Eutrophication · TMDL · Chlorophyll · Secchi depth · Saline lake

Guest editor: S. H. Hurlbert
The Salton Sea Centennial Symposium. Proceedings of a Symposium Celebrating a Century of Symbiosis Among Agriculture, Wildlife and People, 1905–2005, held in San Diego, California, USA, March 2005.

D. M. Robertson (✉)
U.S. Geological Survey, 8505 Research Way, Middleton, WI 53562, USA
e-mail: dzrobert@usgs.gov

S. G. Schladow
Department of Civil and Environmental Engineering, University of California-Davis, Davis, CA 95616, USA
e-mail: gschladow@ucdavis.edu

Introduction

The Salton Sea, the largest lake in California, USA, is located in the southeastern desert region (Fig. 1). The current Salton Sea was formed during a 17-month period from October 1905 to February 1907 following a high-flow event in the Colorado River that resulted in the river breaching a levee of an early irrigation diversion channel and flooding the low-lying desert of Imperial and Riverside counties. The Salton Sea is a terminal water body (lakes with inlets

Fig. 1 Salton Sea, California, with major tributaries and U.S. Geological Survey gaging stations identified

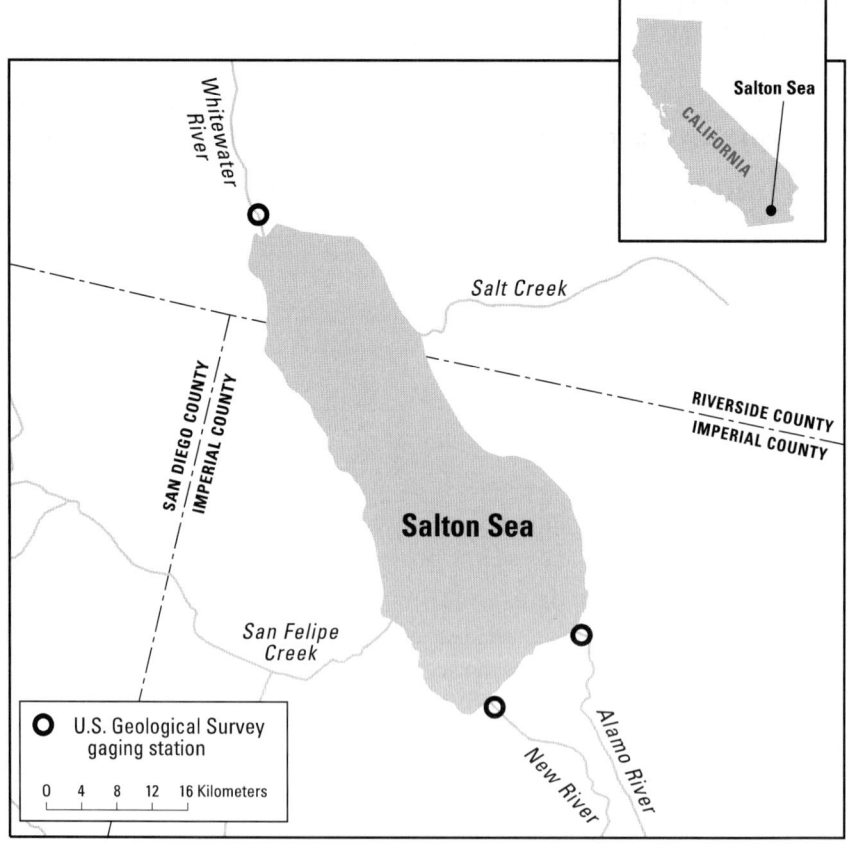

but no outlets) that has no natural tributaries and would have completely evaporated if it were not for receiving agricultural discharges from the Coachella (Whitewater River), Imperial (Alamo and New rivers), and Mexicali valleys along with municipal and industrial effluent from Mexicali, Mexico (New River). One of the major functions of the Sea, since the mid 1920s, is to serve as a sump for agricultural wastewater for the Imperial and Coachella Valleys (Redlands Institute, 2002). As a result of the Salton Sea being a terminal water body in an area with very high evaporation, the salinity of the sea has increased to over 44 g/l in 2000 (Schroeder et al., 2002).

As a result of the high external loading of nutrients, the Salton Sea has become a eutrophic to hypereutrophic waterbody characterized by high nutrient concentrations, high algal biomass as demonstrated by high chlorophyll a concentrations (Chl a), high fish productivity, low clarity, frequently very low dissolved oxygen concentrations, massive fish kills, and noxious odors (Holdren & Montaño, 2002a). Unlike many other saline systems, its

eutrophic condition is believed to be controlled or limited by the phosphorus (P) concentrations in the Sea rather than nitrogen (N) (Holdren & Montaño, 2002a; Robertson et al., 2008, this issue). Most of the P input to the Salton Sea on an annual basis is from tributary loading (Robertson et al., 2008, this issue). Nutrient and sediment concentrations and loads in the tributaries to the Sea are very high because they carry agricultural discharge and, in addition, the New River carries municipal and industrial effluent from Mexicali, Mexico. Although tributary loading is believed to be the major source of nutrients, it is uncertain how important the release of P from the sediments of the Sea is to driving short-term algal blooms and anoxic conditions. In many lakes, P is released from the sediments in deep areas of the lake during low redox conditions or released during with the resuspension of sediment in shallow areas of the lake associated with strong winds and waves. In order to reduce the P loading to the Sea, a P Total Maximum Daily Load (TMDL; a detailed plan to reduce P loading to a specified value; USEPA, 2000) is being prepared by

the Colorado River Regional Water Quality Control Board (CRRWQCB, 2006), that, if implemented, is intended to improve the water quality and reduce the problems associated with eutrophication.

In order to aid in developing the TMDL for the Salton Sea, a better understanding is needed of how nutrient-load reductions will affect the Sea's water quality. One way to determine how much of the P that enters a waterbody needs to be eliminated to improve water quality or to determine how increases in P loading may degrade water quality, is through the use of empirical or semi-empirical models that relate P loading with measured water quality (USEPA, 2000; National Research Council, 2001; Minnesota Pollution Control Agency, 2006). These types of models have been developed on the basis of comparisons between hydrologic and nutrient loads determined for many different waterbodies and specific measures describing water quality, such as near-surface total phosphorus (TP), Chl *a*, and Secchi depths (Walker, 1986; Chapra, 1997). Several of these empirical models are contained within the Wisconsin Lakes Modeling Suite (WiLMS; Panuska & Kreider, 2002) and the semi-empirical models within BATHTUB (Walker, 1996). In BATHTUB, steady-state water- and nutrient-balance calculations are performed in a spatially segmented hydraulic network to account for advective and diffusive transport, and nutrient sedimentation. Changes in water-quality conditions are then predicted using empirical relationships derived from assessments of data in various lakes and reservoirs (Walker, 1985, 1986). Most of these empirical models were developed with data from non-terminal, freshwater lakes, and reservoirs, which typically differ both chemically and biologically from their saline counterparts. No empirical eutrophication models have been specifically developed from data on terminal saline lakes. Therefore, it is important to determine if the existing empirical models are applicable to saline lakes, specifically the Salton Sea, prior to using them to determine how a saline lake should respond to changes in nutrient loading.

Although dynamic or deterministic models are useful for understanding processes that occur in a waterbody during the year(s) examined, specific processes could dramatically change in response to changes in nutrient concentrations, such as those associated with changes in specific species or populations of fish. As these changes cannot usually be foreseen, the altered processes that would not be incorporated into the dynamic model make long-term predictions questionable. Therefore, it is advantageous to use empirical models, based on the differences that have been documented in a wide range of systems, to predict the long-term effects of nutrient-load reductions, rather than using more complicated dynamic models.

Two empirical modeling studies have been previously conducted on the Salton Sea to determine if its water quality responds to nutrient loading like other lakes and reservoirs (Setmire et al., U.S. Geological Survey (USGS), pers. comm.) and to determine how the Sea should respond to incremental changes in P loading (Anderson, 2003). Both these studies were primarily based on the water quality measured in the Sea and the estimated tributary P loading in 1999 by Holdren & Montaño (2002a, b). Setmire et al. (USGS, pers. comm.) applied several of the empirical models contained in WiLMS to the Salton Sea and concluded that the TP in the Sea was lower than what would be expected for a lake with the Sea's morphometry and 1999 P-loading rates.

Anderson (2003) applied several of the P (first- and second-order P-settling algorithms), Chl *a*, and Secchi depth models within BATHTUB, and a coupled sediment/water-column model that was developed for the Salton Sea to determine how the Sea's water quality should respond to incremental changes in P loading. The coupled sediment/water-column model separately apportioned loading to external sources and internal recycling, and allowed for P removal via settling. In all the P models, internal loading of P from the sediments was included at a rate of 3.2 mg/m^2/day (for the 1999 base-case scenario), based on results of sediment core incubations in the laboratory. This resulted in the internal P load from the sediment being slightly less than the external P load from the watershed. Anderson found that the coupled sediment/water-column model and BATHTUB models accurately simulated the TP measured in the Salton Sea in 1999, but to do so the models required large calibration factors. Based on the results of these models, a 50% decrease in external P loading should cause TP in the Sea to decrease by ~30% (second-order model) to 50% (first-order and coupled sediment/water-column model). Based on the coupled sediment/water-column model, it was concluded that a 25% reduction in external TP loading would reduce the total P loading to the lake by 50% and that complete

Hydrobiologia (2008) 604:5–19

responses to changes in external loading should occur in as little as 3 years.

There are several problems with the way the empirical models were applied and the conclusions that were made in the previous studies. First, several of the models used by Setmire et al. (USGS, pers. comm.) were not applicable to terminal lakes. Second, based on the water-quality data measured in this saline lake, there is little or no net internal P loading from the deep sediments (Holdren & Montaño, 2002a; Robertson et al., 2008, this issue); however, there may be substantial input of P associated with the resuspension of shallow sediment. Even if there was internal P loading, as assumed by Anderson (2003), a 25% reduction in external P loading would reduce external loading by 25% and indirectly reduce part of the internal loading, and thus would only reduce the total P loading by ≤25%. Finally, if extensive internal P loading was present in the Sea, then internal P loading would persist much longer than 3 years and possibly reduce the immediate effects of external P-load reductions. Internal P loading has been found in many lakes to delay the effects of P-load reductions (for example, Shagawa Lake, Minnesota; Larsen et al., 1981).

Methods

In order to predict how the average whole-lake water quality of the Salton Sea should respond to changes in P loading, models in three eutrophication programs were applied to the Sea with the conditions measured in 1999 (Holdren & Montaño, 2002a), which was the base case for calibration and the case with which to compare other simulations. The eutrophication programs included BATHTUB (Walker, 1996), WiLMS (Panuska & Kreider, 2002), and the Wisconsin Department of Natural Resource's (WDNR's) Seepage Lake Model (J. Panuska, WDNR, pers. comm.). Models within each of these programs were calibrated, if necessary, and then applied with changes in P loading ranging from reductions in P loadings from tributaries and drainage canals by 10–100% and to increases in loading from these sources by 10–100%. 1999 was used as a base case and for calibration because it is the only year that had sufficient coinciding information on external loading and in-Sea water quality. Since the Sea is believed to be potentially limited by P (Holdren & Montaño, 2002a;

Robertson et al., 2008, this issue), only the effects of alterations to P loads (TP and orthophosphate-P (OP) concentrations) were simulated; loadings of all other constituents were kept similar to those measured in 1999. Although there is considerable temporal and spatial variability in the water quality of the Salton Sea, these models were used to predict changes in the average whole-lake near-surface water quality.

Model input

Four types of data are required as input for these empirical eutrophication models: morphometric and physical characteristics, hydrologic, nutrient-loading, and water-quality data (Tables 1 and 2). The time period for the hydrologic and nutrient-loading data used in BATHTUB is dependent on the P-turnover ratio or the number of times the P mass in the lake is displaced during the averaging period. BATHTUB should be run for a period that results in a P-turnover ratio >2. Annual simulations resulted in the P-turnover ratio of 2.02. Annual water-loading rates and volumetrically weighted mean concentrations for 1999 were obtained from Robertson et al. (2008, this issue) except those for the Whitewater River. Volumetrically weighted concentrations for the Whitewater River for 1999, based on the long-term calibration of a loading model, appeared to overestimate the concentrations for 1999. Given that only 2 of 18 TP concentrations measured in the Whitewater River by Reclamation in 1999 exceeded 1.0 mg/l (Holdren & Montaño, 2002b), an average concentration of 1.004 mg/l appeared to be too high. Therefore, a P load for the Whitewater River for 1999 was obtained by multiplying the monthly or biweekly TP concentrations (measured by Reclamation) by the coinciding total streamflows (measured by the USGS), and then a volumetrically weighted concentration was obtained by dividing the total P load by the total annual flow. This resulted in an average TP concentration of 0.88 mg/l for the Whitewater River. Average TP, OP, total nitrogen (TN), and inorganic N in the tributaries to the Salton Sea for 1999 are listed in Table 1. The total annual P load to the Salton Sea in 1999 was ~ 1,440,000 kg, which equates to a unit-area yield from the basin of ~ 67 kg/km^2 and an areal loading to the Sea of ~ 1.5 g/m^2/year.

Even though loading data are summarized for the entire year, BATHTUB typically estimates water quality only for the growing season (May through

Table 1 Morphometric, hydrologic, and tributatry inputs for the empirical models for the Salton Sea, California, for 1999 (water quality for the base-case scenario)

Salton Sea morphometric and physical characteristics data

Average water level[a]	69.4 m below sea level	Width	24 km
Area[b]	980.8 km^2	Volume[b]	9.257 × 10^9 m^3
Mean depth[b]	9.44 m	Mixed layer depth[c]	6.6 m
Length	56 km	Non-algal turbidity factor	0.45

Precipitation, evaporation, and hydraulic flushing

Precipitation[d]	0.076 m	13.5 kg of total phosphorus per km^2/year	
Evaporation[d]	1.663 m		
Hydraulic flushing rate	0.18 year^{-1}		

Tributary inflow volumes and nutrient concentrations: loading information

Parameter	Alamo River	New River	Whitewater River	Direct Drains[g]
Flow (× 10^6 m^3)[e]	761	603	65	128
Total phosphorus (mg/l)[e]	0.716	1.218	0.880	0.716
Orthophosphate-P (mg/l)[f]	0.409	0.697	0.710	0.409
Total nitrogen (mg/l)[f]	9.263	8.289	16.367	15.072
Inorganic nitrogen (mg/l)[f]	7.713	7.267	15.072	7.713

[a] Based on elevation data obtained from the Bureau of Reclamation (P. Weghorst, Bureau of Reclamation, unpubl. raw data)

[b] Based on bathymetric data obtained from the Bureau of Reclamation (P. Weghorst, Bureau of Reclamation, unpubl. raw data)

[c] Based on algorithms in BATHTUB (Walker, 1996) using the morphometry and water quality measured in the Salton Sea

[d] Based on estimated precipitation and evaporation from Robertson et al. (2008, this issue)

[e] From loading estimated by Robertson et al. (2008, this issue) using U.S. Geological Survey flow data for the gages identified in Fig. 1 and data collected by various agencies, except the Whitewater River that is described in the text

[f] Based on average monthly concentrations for 1999 (Holdren and Montaño, 2002b)

September in temperate lakes and reservoirs). The models within WiLMS and the Seepage Lake model also use annual hydrologic and P loadings, but

Table 2 Monthly average near-surface water quality for the Salton Sea in 1999

Constituent	Concentration/ depth	Trophic state index value
Total phosphorus	0.077 mg/l	67
Chlorophyll *a*	33 μg/l	65
Secchi depth	0.81 m	63
Orthophosphate-P	0.022 mg/l	–
Total nitrogen	3.80 mg/l	–
Organic nitrogen	2.58 mg/l	–
Total suspended sediment	45.7 mg/l	–

All values were based on data collected by the Bureau of Reclamation (Holdren and Montaño, 2002b), except chlorophyll *a* concentrations that were obtained from San Diego State University (Tiffany et al., in press) and represent a whole-lake average value

typically simulate the water quality for different discrete seasons. Since algal growth occurs throughout the year in the Salton Sea, output from all the models were compared with the average monthly water quality for the entire 1999 year (Table 2). All these models estimate the water quality of the upper mixed layer of the water column; therefore, results of model simulations were compared with measured near-surface water-quality data. Additional P from internal loading (chemical diffusion of P from the deep sediments) was not included as a P source for BATHTUB, WiLMS, and the Seepage Lake Model because: (1) based on the water quality measured in the Sea, there was minimal net internal P loading (increased TP immediately above the bottom was not detected even during anoxic periods) and (2) even if there was internal P loading that was typical of other lakes and reservoirs (as suggested by Anderson and Amrhein, 2002), most empirical models (such as those used in this study) inherently incorporate this source and, therefore, additional P from internal

loading should only be added when a lake/reservoir has abnormally high internal P-loading rates (Walker, 1996, p. 4-24).

Algorithms and calibration of models

Phosphorus

Most empirical models used to estimate TP, including those in this study, are based on the Vollenweider eutrophication modeling approach (Vollenweider, 1975). With this approach, the average TP in the lake or reservoir is estimated as a function of the annual areal loading rate of P (L), the mean depth of the lake (z), the P-settling coefficient (σ), and the hydraulic flushing rate (ρ) (Eq. 1). The main difference in most of the algorithms used to estimate average TP is in how σ is estimated.

$$TP = L/(z\,(\sigma + \rho)). \quad (1)$$

When applying BATHTUB, specific models (algorithms) must be selected to simulate each water-quality constituent. The algorithms chosen to simulate TP, Chl a, and Secchi depth (SD) are given in Table 3. For TP, three different algorithms were chosen: one first-order P-settling algorithm and two second-order P-settling algorithms (available-P and decay-rate algorithms). The first-order algorithm

assumes that changes in particulate concentrations over time are directly proportional to the concentration of the particulates in the water column (σ is set equal to 1.0). The second-order algorithms assume that changes in concentrations over time are proportional to the square of the particulate concentration (TP is computed with Eq. 2). The second-order, available-P algorithm (σ computed with Eq. 3) is the default algorithm in BATHTUB and the one most generally applicable to lakes and reservoirs (Walker, 1996). This algorithm performs mass-balance calculations on "available P," a weighted sum of OP and non-OP, and places a higher emphasis on OP (the more biologically available component) (Walker, 1996). Walker (1996) states that the second-order, decay-rate algorithm (σ computed with Eq. 4) may be most applicable to lakes with long residence times, such as the Salton Sea. The first-order and second-order decay-rate algorithms were used by Anderson (2003) to simulate changes in TP in the Salton Sea in response to various changes in P loading.

$$TP = \{-1 + [1 + 4\,\sigma L/(z\,\rho^2)]^{0.5}\}/(2\sigma/\rho) \quad (2)$$

where

$$\sigma = 0.17\,Q_s/(Q_s + 13.3) \quad (3)$$

or

Table 3 Models used to estimate changes in the water quality of the Salton Sea, California

Eutrophication program	Model (algorithm) description	Pre-calibration concentration, μg/l	Post-calibration concentration, μg/l	Calibration factor
Total phosphorus—measured 0.077 mg/l				
BATHTUB	First order P-settling	0.156	0.078	2.00
BATHTUB	Second order P-settling, available P	0.078	0.078	1.00
BATHTUB	Second order P-settling, decay rate	0.084	0.077	1.18
WiLMS	Canfield & Bachman (1981)	0.095	0.077	−23%
Seepage Lake model				$V_s = 19.0$
Chlorophyll a—measured 33 μg/l				
BATHTUB	Based on P and N concentrations, light, and flushing rate	19	32	1.65
WiLMS/Seepage Lake model	Carlson TSI equations	40	33	−20%
Secchi depth—measured 0.81 m				
BATHTUB	Based on P concentration and turbidity	0.70	0.80	1.25
WiLMS/Seepage Lake model	Carlson (1977) TSI equations	0.60	0.81	−25%

The pre- and post-calibration values and calibration factors used for each empirical model are listed. V_s is a settling velocity

$$\sigma = 0.056\ Q_s/[\text{FOP}\ (Q_s + 13.3) \qquad (4)$$

where Q_s is the maximum of the total depth $\times\ \rho$ or 4; FOP is fraction of tributary OP loading to TP loading.

Of the 13 empirical models contained within WiLMS, only three of the models were relatively insensitive to the residence time of water in the lake; most of the models are not capable of simulating water quality in terminal lakes, lakes with no outlets and therefore very long residence times. Simulated TP from the 10 other models increased dramatically as the residence time increased. Of the three models relatively insensitive to residence time, only the Canfield & Bachman (1981) natural-lake model (σ for Eq. 1 is computed with Eq. 5) was applicable to the hydrology, loading rates, and TP concentrations of the Salton Sea,

$$\sigma = 0.162\ (L/z)^{0.458}. \qquad (5)$$

The Seepage Lake Model (J. Panuska, WDNR, pers. comm.), specifically developed for terminal lakes, estimates a lake's average TP as a function of areal P loading (L), areal water loading (Q_A), and an apparent settling velocity (V_s) for P (Eq. 6). The value of V_s required to accurately estimate the measured TP in a lake can be used to infer processes occurring in a lake that control TP. V_s values less than 0 m/year are indicative of high internal P loading, values between 0 and 6 m/year are indicative of moderate internal P loading, values between 7 and 12 m/year are indicative of low internal P loading, values between 12 and 20 m/year are indicative of no internal P loading and that marling (co-precipitation with calcium carbonate) or other removal mechanisms occur for P, and values more than 20 m/year are indicative of very high marling or extensive P removal (J. Panuska, WDNR, pers. comm.). A V_s value of 19 m/year was required to estimate the measured TP of 0.077 mg/l in the Salton Sea. Therefore, the Seepage Lake Model indicates that there is no net internal P loading and suggests that there are removal mechanisms occurring in the Sea.

$$\text{TP} = L/(Q_A + V_s). \qquad (6)$$

None of these P algorithms was specifically developed using data for terminal saline lakes, such as the Salton Sea; however, they were developed using data from lakes and reservoirs with a wide range of environmental and hydraulic conditions. Therefore, their applicability to the Salton Sea can be evaluated by the calibration coefficients required to accurately simulate the water quality measured in the Sea in 1999 (Table 3). Predictions made with the models prior to and following calibration are also shown in Table 3. All these coefficients are within the typical range applied for most lakes (Walker, 1996). By not including the internal P loading (as included in the models by Anderson, 2003), only the first-order model required much, if any, calibration for predicting TP. Therefore, these empirical P models appear to be able to simulate changes in TP in this terminal saline lake.

Chlorophyll a *and Secchi depth*

The TP estimated with BATHTUB, WiLMS, and the Seepage Lake Model, were then used to estimate Chl *a* and SD. Chl *a* was estimated with BATHTUB using an algorithm that is a function of TP, TN, light, and flushing rate (similar to that used by Anderson, 2003). SDs were predicted with BATHTUB using the algorithm that is a function of TP and a non-algal turbidity factor (0.45 m^{-1}, computed within BATHTUB). Non-algal turbidity is the portion of light extinction that is due to factors other than algae, such as inorganic suspended solids, dissolved organic matter, and color (Walker, 1996). The non-algal turbidity factor was held constant in all simulations. The calibration coefficients applied to accurately simulate the Chl *a* and SD for 1999 with BATHTUB are given in Table 3. These calibration coefficients are within the typical range applied for most lakes (Walker, 1996), which again indicates that these empirical models appear to be able to simulate changes in Chl *a* and SD in this terminal saline lake.

The TP concentrations predicted with WiLMS (Canfield & Bachman 1981 model) and the Seepage Lake Model were used to estimate Chl *a* and SD through the use of Carlson's (1977) trophic state index (TSI) equations. In other words, the Chl *a* and SD were computed that yielded similar TSI values as the predicted TP. There are no calibration factors when the Carlson's TSI equations are used to estimate Chl *a* and SD; however, the output can be adjusted to account for model biases by only interpreting the results as a percentage of change from present conditions. In other words, the percent bias found in predicting concentrations or depths for 1999 (Table 3) is removed from all the other

predictions. Changes in Chl *a* were also estimated with algorithms by Jones & Bachman (1976) and Dillon & Rigler (1974), but the results (simulated Chl *a* concentrations of 37 and 40 µg/l, respectively) were similar to those of the Carlson TSI equations and therefore are not presented. The uncalibrated models were within 25% of the observed Chl *a* and SD, which is well within the variability found for non-saline lakes, which again suggests that the Salton Sea responds to changes in TP similar to other lakes.

In a previous study, Chl *a* was found to be unusually low in prairie saline lakes in Canada compared to other lakes with similar TP and lower than predicted with empirical models (Campbell & Prepas, 1986). However, the Chl *a* in the Salton Sea are not unusually low and are predicted well by the empirical models used in this study (Table 3). The unusually low concentrations measured in the prairie saline lakes may have been caused by those lakes being N limited compared to the Salton Sea which is usually P limited; therefore, the empirical models that are based on P limitation and used for the prairie saline lakes in Canada were not appropriate for that application.

Results

In order to simulate the effects of changes in P loading, the average TP and OP concentrations (or loads) for each of the tributaries in 1999 (Table 1) were decreased by 10–100% and increased by 10–100%; N concentrations were not modified (average monthly concentrations for 1999 were used in the models). Since the Salton Sea is P limited (Holdren & Montaño, 2002a; Robertson et al., 2008, this issue), altering N concentrations should and had little effect on the simulated results (these results are not presented).

Response in phosphorus concentration

On the basis of results of the first-order P-settling BATHTUB model and the Seepage Lake Model, TP in Salton Sea should have a linear response to a linear change in P loading (Fig. 2). On a percentage basis, results from both models indicate that the percent changes in TP are the same as the percent changes in P loadings, i.e., a 50% decrease in P loading should cause a 50% decrease in TP in the Sea.

On the basis of results of simulations with both second-order P-settling models in BATHTUB and the Canfield & Bachman (1981) model within WiLMS, TP in the Salton Sea should have a relatively linear response to a linear change in P loading, except with very large decreases in P loading for which the changes in TP in the Sea are predicted to be larger (Fig. 2). On a percentage basis, results from these three models indicate that the changes in TP are ∼30–40% of the changes in P loadings, except with very large decreases in loading (>80% decreases) for which the decreases in TP in the Sea become non-linear and

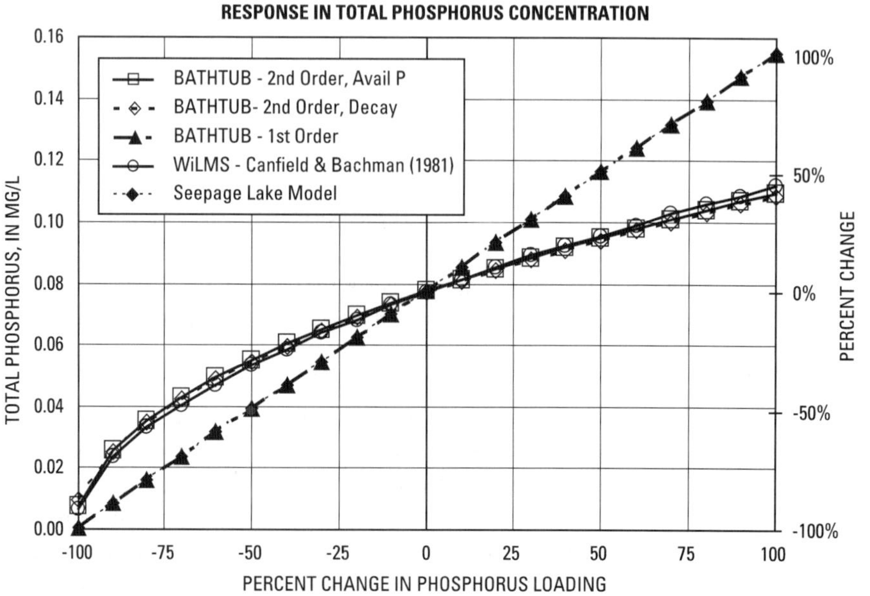

Fig. 2 Modeled changes in near-surface total phosphorus concentrations in response to changes in external phosphorus loading to the Salton Sea

should be larger. A 50% decrease in P loading should cause a ∼30% decrease in TP in the Sea.

With a 50% decrease in P loading, results of these models indicate TP in the Salton Sea should decrease from 0.077 to ∼0.039–0.055 mg/l (30–50% decrease). The second-order P-settling models in BATHTUB and the Canfield & Bachman model should be the most applicable models to the Salton Sea; therefore, a 50% decrease in P loading, should decrease TP from 0.077 to ∼0.055 mg/l (a 30% decrease).

Response in chlorophyll *a* concentration

On the basis of the results of the first-order and the second-order P-settling models in BATHTUB (all three using the same Chl *a* algorithm), Chl *a* should have a larger response to decreases in P loading than to increases in P loading (Fig. 3). The differences in the results from these three models are caused only by the differences in the predicted TP. The largest relative response in Chl *a* occurs with P-load reductions >80%. Results from the second-order P-settling models indicate a smaller response in Chl *a* to changes in P loading than with the first-order P-settling model. Based on results from the BATHTUB models, a 50% decrease in P loading should cause a 17–34% decrease in Chl *a* in the Sea.

Results from the Canfield & Bachman (1981) model within WiLMS and the Carlson TSI equations, indicate a response similar to the first-order P-settling decay model in BATHTUB for decreases in P loading, but larger responses than any of the BATHTUB models for increases in P loading (Fig. 3). A 50% decrease in P loading should cause a ∼40% decrease in Chl *a* in the Sea.

Results from the Seepage Lake Model and the Carlson TSI equation indicate a much larger response in Chl *a* than indicated from the other models (Fig. 3). The larger response in Chl *a* from this combination of models was primarily caused by the larger predicted changes in TP. With a 50% decrease in P loading, results from these models indicate a 63% decrease in Chl *a*. Results for 50% increase in P loading indicate an 80% increase in Chl *a*.

With a 50% decrease in P loading, results of these models indicate Chl *a* in the Salton Sea should decrease from 33 to ∼12–26 μg/l (a 17–63% decrease). Since the second-order P-settling models and the Canfield & Bachman model appear to be the most applicable models for the Salton Sea, it is expected that a 50% decrease in P loading should decrease average Chl *a* from 33 to ∼19–26 μg/l (17–43% decrease). The changes in response to increased P loading based on the Carlson TSI equation may be larger than the other models because these equations

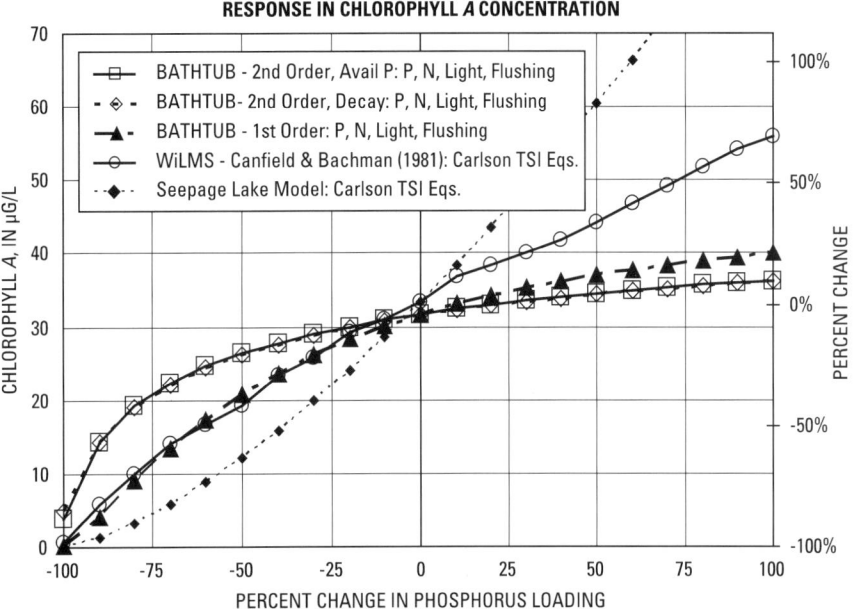

Fig. 3 Modeled changes in near-surface chlorophyll *a* concentrations in response to changes in external phosphorus loading to the Salton Sea

do not incorporate the effects of the high non-algal turbidity that occurs in the Salton Sea.

Response in Secchi depth

Results of all the models indicate that SD (water clarity) is much more responsive to decreases in P loading (especially to decreases in loading by >50%) than to increases in P loading (Fig. 4). Results of all the models indicate that with a 50% increase in P loading, the average SD should decrease by ~0.1–0.3 m. However, the responses are quite variable to decreases in P loading. Results of the BATHTUB models indicate the smallest response in SD, and results of the Seepage Lake Model indicate the largest response.

With a 50% decrease in P loading, results of these models indicate that SD in the Salton Sea should increase from ~0.8 to ~1.1–1.6 m (a 38–97% increase). Since the second-order P-settling models and the Canfield & Bachman model should be the most applicable models for the Salton Sea, it is expected that a 50% decrease in P loading should increase the average SD from ~0.8 to ~1.1–1.2 m (38–50% increase). The main reason for the wide range in the responses in SD with decreased P loading is that the algorithms in BATHTUB assume that the non-algal turbidity will remain constant, whereas, the

Carlson TSI equations assume that the turbidity is primarily a function of the amount of algae in the water column.

Response in trophic status

TSI values indicate that Salton Sea is presently eutrophic to hypereutrophic (TSI values ranging from 63 to 67; Table 2). TSI values were computed from the results of the models for changes in P loading. On the basis of simulation results from BATHTUB, WiLMS, and the Seepage Lake Model, any increase in P loading should result in the Sea remaining or becoming more hypereutrophic (Fig. 5). Results from the Seepage Lake Model and Canfield & Bachman model indicate higher TSI values than do the results from the models in BATHTUB. Results of all the models indicate that decreases in P loading can convert the Salton Sea to a eutrophic system (TSI values from 50 to 60). The results indicate that P-load reductions of ~50% (Seepage Lake Model) to 80% (Canfield & Bachman model in WiLMS) are required for the Sea to be classified as moderately eutrophic (a TSI value of 55) with respect to TP; load reductions of ~50% (Seepage Lake Model) to 90% (second-order P-settling BATHTUB models) to be classified as moderately eutrophic with respect to Chl a; and P-load reductions of ~50% (Seepage Lake Model) to

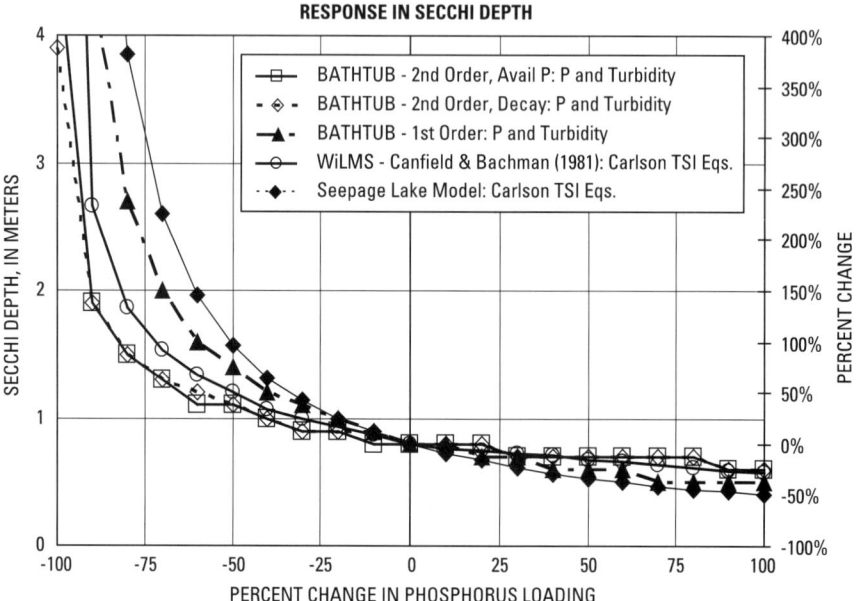

Fig. 4 Modeled changes in Secchi depth in response to changes in external phosphorus loading to the Salton Sea

Fig. 5 Modeled changes in the trophic state of the Salton Sea in response to changes in external phosphorus loading

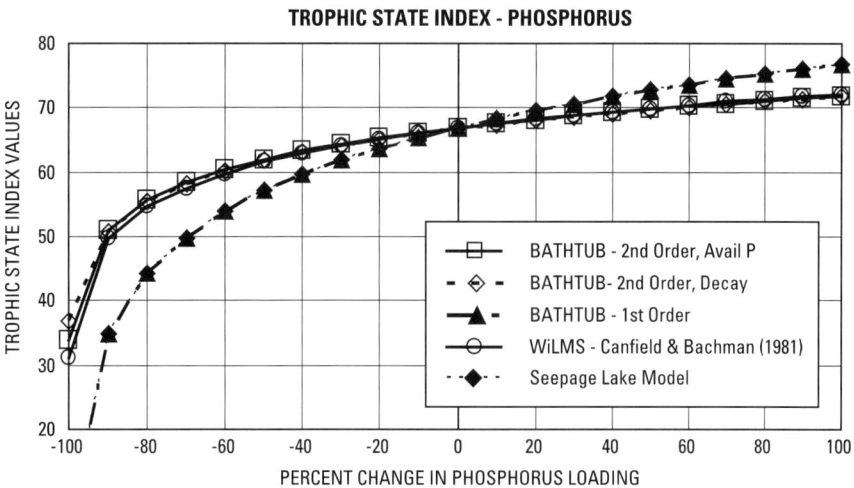

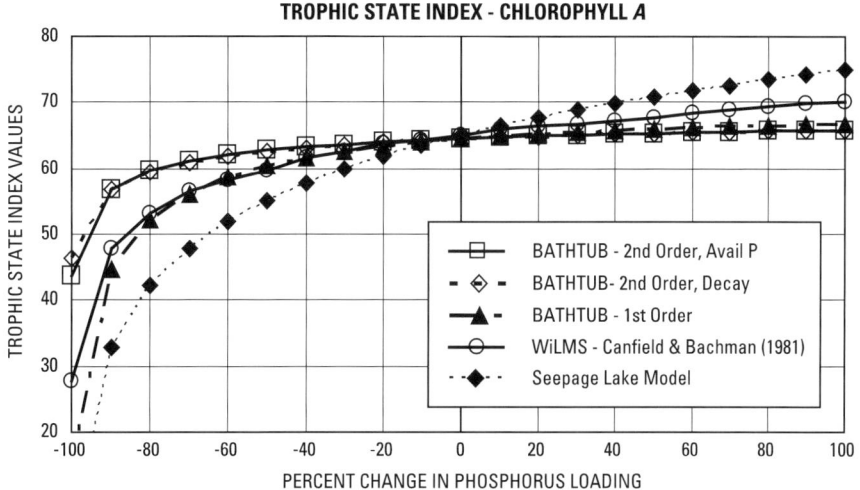

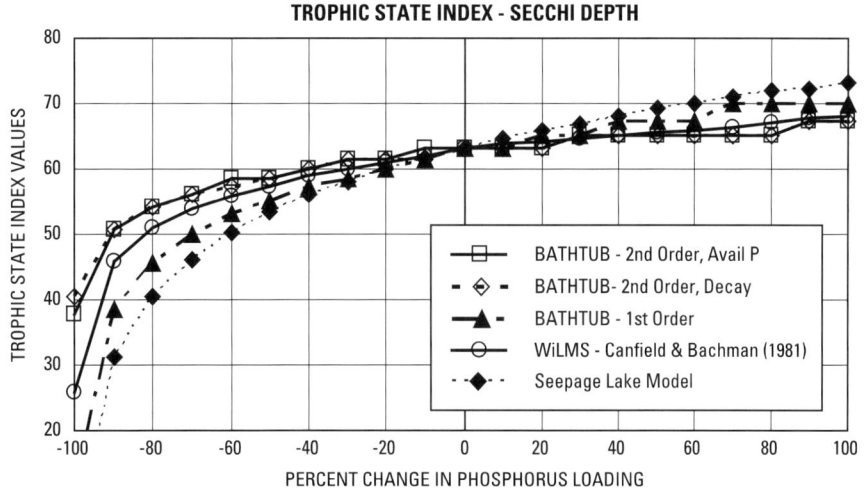

80% (second-order P-settling BATHTUB models) to be classified as moderately eutrophic with respect to water clarity (SD). If just the results of the second-order P-settling models and Canfield & Bachman model, which should be the most applicable models for the Salton Sea, are considered, a P-load reduction of ~70–90% is required for the Salton Sea to be classified as moderately eutrophic (for TP, Chl *a*, and SD).

Algal-bloom response

All the results presented above describe changes in average annual water quality. However, what is often

perceived as the biggest problem in lakes is the extremes in water quality, for example, algal blooms. A Chl *a* concentration of 10 µg/l represents a relatively minor algal bloom (the typical concentration in a eutrophic lake), whereas a concentration of 60 µg/l represents an extremely severe algal bloom. A Chl *a* concentration of 30 µg/l represents what many would consider a moderate algal bloom. The predicted percentages of days with Chl *a* exceeding specified concentrations from BATHTUB and WiLMS are compared to the percentages based on measured Chl *a* data collected from 1997 to 2000 (Tiffany et al., in press) in Fig. 6. Both eutrophication programs estimate the percentage of days above

Fig. 6 Measured and modeled percentage of days with algal blooms (chlorophyll *a* concentrations exceeding a specified value), and changes in percentage of days with algal blooms in response to a 50% reduction in total phosphorus loading. BATHTUB models used the chlorophyll *a* algorithm that is a function of total phosphorus, total nitrogen, light, and flushing rate, and WiLMS model used Carlson (1977) trophic state index equations

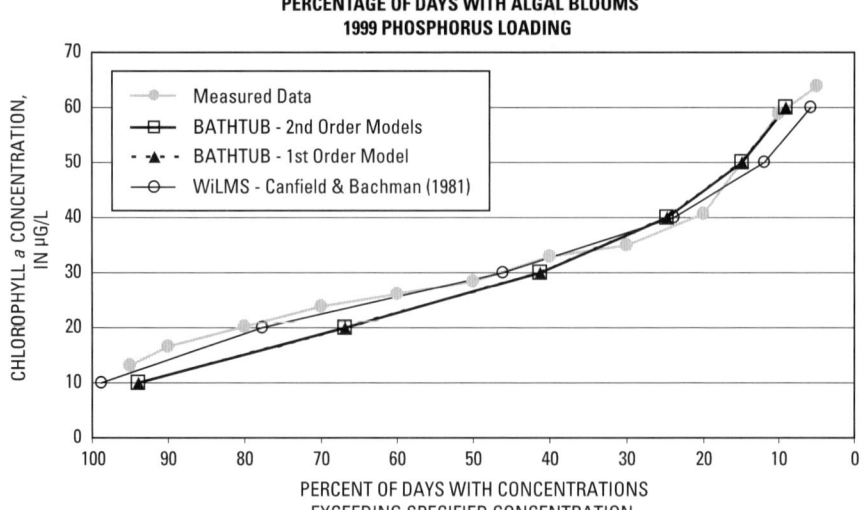

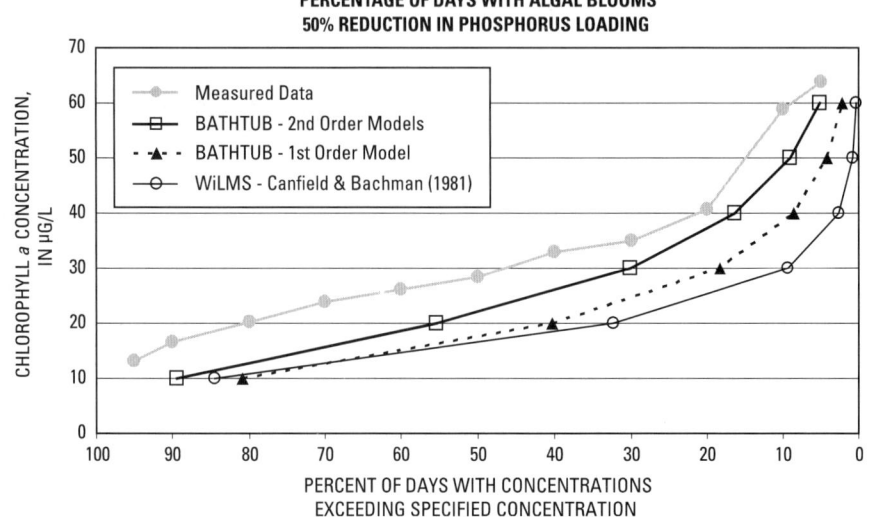

specific concentrations based on an algorithm from Walker (1984). All the models simulated the percentage of days above specified concentrations during 1997–2000 very well, except for possibly the percentage of days above relatively low Chl *a* concentrations, which the models within BATHTUB slightly overestimated.

Results from all the models indicate that as the P loading to the Sea increases the percentage of days described as having algal blooms (by any definition) and as the P loading to the Sea decreases percentage of days described as having algal blooms decreases (Fig. 6). In other words, decreases in P loading causes the curves in Fig. 6 to be shifted to the right. Results for both second-order P-settling models in BATH-TUB gave essentially the same response in the percentage of days described as having algal blooms for 1999 and for all P-load scenarios. Results of the models indicate that a 50% reduction in P loading should cause the percentage of days described as having algal blooms defined as Chl *a* >10 µg/l to decrease from currently ~95% of the time to ~80% (first-order P-settling BATHTUB model) to 90% (second-order P-settling BATHTUB models) of the time. If the percentage of days described as having algal blooms is defined as Chl *a* >30 µg/l, then the percentage of days should decrease from currently ~40% of the time to ~10% (WiLMS) to 30% (second-order P-settling BATHTUB models) of the time. If the percentage of days described as having algal blooms is defined as Chl *a* >60 µg/l, then the percentage of days should decrease from currently ~8% of the time to ~0.3% (WiLMS) to 5% (second-order P-settling BATHTUB models) of the time.

Discussion

Comparison of results with those of other studies

After applying several of the empirical eutrophication models in WiLMS, Setmire et al. (USGS, pers. comm.) concluded that the TP in the Sea was lower than what would be expected for a lake with the Salton Sea's morphometry and 1999 P-loading rates. However, most of the empirical models in WiLMS are not applicable to terminal lakes like the Salton

Sea, which has no outlet. Simulated TP based on most of the models in WiLMS increased dramatically as the residence time in the lake increased; therefore, their results are questionable for terminal lakes. Results from the one model in WiLMS that was applicable to the Salton Sea (Canfield & Bachman 1981 model) indicate that the TP in the Sea was slightly lower than predicted by the model. However, the predicted TP was only ~23% more than what was measured in the Sea, which is well within the expected accuracy of this model. All the TP, Chl *a*, and SD algorithms estimated the water-quality conditions measured in the Salton Sea in 1999 with only minor calibration; therefore, these models should be capable of simulating changes in the water quality of this terminal saline lake in response to changes in P loading.

Results from this study indicate that a 50% decrease in P loading should decrease in-Sea TP by 25–50%, similar to that estimated by Anderson (2003). A 50% decrease in P loading should decrease the average Chl *a* by 17–63%; whereas, Anderson (2003) concluded this reduction would decrease Chl *a* by 15–32%. A 50% decrease in P loading should increase the average SD from ~0.8 to ~1.1–1.6 m (a 38–97% increase); whereas, Anderson (2003) concluded this reduction would only increase the average SD from ~0.8 to ~0.9–1.1 m. The main reason for these differences in the responses of Chl *a* and SD in these two studies is that it was assumed that non-algal turbidity would remain constant in the BATHTUB models, whereas, the Carlson TSI equations assume that the turbidity is only a function of the amount of algae in the water column and that it will be reduced with reductions in P loading. If non-algal turbidity does not improve with reduced P loading, a 50% decrease in P loading should only decrease the Chl *a* by 17–43% and increase the average SD to ~1.1 m, which is similar to that found by Anderson. However, it is uncertain how the turbidity will change with a reduction in external P loading.

Annual P loading to the Salton Sea has been estimated from 1965 to 2002 by Robertson et al. (2008, this issue). They estimated that the loading has increased by ~55% over that period. However, because of the limited water-quality data collected in the Sea over this time period, it is not possible to use these data to evaluate the ability of these models to

predict in-Sea water quality with reduced P loading. Therefore, it is necessary to rely on models, such as those applied in this study, to determine how the water quality of the Salton Sea should respond to changes in nutrient loading in the future and how much of the P that enters the Salton Sea needs to be eliminated to improve its water quality.

TMDL for the Salton Sea

Preliminary numeric targets (specific water-quality goals) for the P TMDL for the Salton Sea were defined based on Salton Sea Science Committee recommendations, results of previous modeling studies, and the scientific literature (CRRWQCB, 2006). The goals were to have the water quality of Salton Sea similar to that of a moderately eutrophic lake or a Carlson TSI value of 55. By definition, eutrophic lakes have an average TP of 0.024–0.049 mg/l, with a mean of 0.035 mg/l; average Chl *a* of 7–20 µg/l, with a mean of 12 µg/l; and SD of 1–2 m, with a mean of 1.4 m (Carlson, 1977). If TP and Chl *a* could be improved, it is assumed that water clarity would increase, and the other problems associated with eutrophication such as low dissolved oxygen concentrations and fish kills would be reduced.

Based on the results of the empirical modeling conducted in this study, it is estimated that to attain the targeted values for all three of these parameters and have the Salton Sea be classified as a moderately eutrophic waterbody would require that the total annual P load be reduced by 55–90%, with a best estimate of 70–90%. In order to achieve attainment of the respective upper ranges of these parameters (a TSI value of 60) and have the Salton Sea be classified as an upper eutrophic waterbody would require the total annual P load be reduced from 35 to 80%, with a best estimate from 50 to 80%.

Response to phosphorus-load reductions

The empirical models used in this study simulate steady-state conditions that should occur as a result of the hydrology and nutrient-loading input into the models; however, the simulated changes may be expected to take place several years after load reductions occur because of the P stored in the sediments of a lake. Exactly how long it would take

for these changes to occur is unknown. The internal release of P from the deep sediment of a lake associated with low redox potential at the sediment surface has been found in many lakes to delay the effects of improvements made in the watershed (for example, Shagawa Lake, Minnesota; Larsen et al., 1981), but this release does not appear to occur in the Salton Sea (Robertson et al., 2008, this issue). However, particulate and dissolved P stored in the near-shore sediment and released during short-term resuspension events may drive the short-term algal blooms and anoxic conditions. If external P loading is reduced, it is uncertain how long sediment resuspension can continue to influence P concentrations in the Sea and short-term algal blooms. Therefore, results of the empirical models indicate the potential long-term changes in the water quality of the Salton Sea associated with P-load reductions, with which there may be much variability because of events like sediment and P resuspension. Chung et al. (2008, this issue) have used dynamic models to describe and quantify these short-term effects of resuspension events.

Conclusion

Application of three different empirical or semi-empirical eutrophication programs (BATHTUB, WiLMS, and the Seepage Lake Model) demonstrated that specific models in these programs that were developed primarily based on freshwater lakes and reservoirs are applicable to the Salton Sea, a terminal saline lake. The applicable models were then used to quantify how specific changes in P loading to the Salton Sea should affect its average annual water quality. Results of this effort indicate that a 50% decrease in external P loading would decrease near-surface TP by 25–50%, decrease near-surface Chl *a* by 17–63%, and increase average SD by 38–97%. Results also indicate that P-load reductions of ∼50–90% are required for the Sea to be classified as moderately eutrophic (TSI value of 55). If only the models that should be most applicable to the Salton Sea (second-order P-settling models and Canfield & Bachman 1981 model) are considered, a P-load reduction of ∼70–90% is required for the Salton Sea to be classified as moderately eutrophic. The wide range in estimated responses was primarily

caused by uncertainty in how non-algal turbidity in the Sea would respond to P-load reductions. All these models simulate steady-state conditions in the Sea; therefore, the simulated changes would be expected to take place several years after P-load reductions occur.

References

Anderson, M. A., 2003. Bioavailability, resuspension and control of sediment-borne nutrients in the Salton Sea. Final report to the California River Basin Regional Water Quality Control Board. Department of Environmental Sciences, University of California, Riverside.

Anderson, M. A. & C. Amrhein, 2002. Nutrient cycling in the Salton Sea. Final report to the Salton Sea authority. Department of Environmental Sciences, University of California, Riverside.

Campbell, C. E. & E. E. Prepas, 1986. Evaluation of factors related to unusually low chlorophyll levels in prairie saline lakes. Canadian Journal of Fisheries Aquatic Sciences 43: 846–854.

Canfield, D. E. & R. W. Bachmann, 1981. Prediction of total phosphorus concentrations, chlorophyll-a, and Secchi depths in natural and artificial lakes. Canadian Journal of Fisheries Aquatic Sciences 38: 414–423.

Carlson, R. E., 1977. A trophic state index for lakes. Limnology & Oceanography 22: 361–369.

Chapra, S., 1997. Surface Water-Quality Modeling. McGraw-Hill Publishers, Inc.

Chung, E., S. G. Schladow, J. Perez-Losada & D. M. Robertson, 2008. A linked hydrodynamic and water quality model for the Salton Sea. Hydrobiologia (this issue). doi: 10.1007/s10750-008-9311-6.

CRRWQCB (Colorado River Regional Water Quality Control Board), 2006. Total maximum daily load program. Accessed at http://www.waterboards.ca.gov/coloradoriver/tmdl.html.

Dillon, P. J. & F. H. Rigler, 1974. The phosphorus-chlorophyll relationship in lakes. Limnology & Oceanography 19: 776–773.

Holdren, G. C. & A. Montaño, 2002a. Chemical and physical characteristics of the Salton Sea, California. Hydrobiologia 473: 1–21.

Holdren, G. C. & A. Montaño, 2002b. Chemical and physical limnology of the Salton Sea, California—1999. Technical Memorandum No.8220-03-02, U.S. Department of the Interior, Bureau of Reclamation, Denver, CO.

Jones, J. R. & R. Bachman, 1976. Prediction of phosphorus and chlorophyll levels in lakes. Journal of the Water Pollution Control Federation 48: 2176–2184.

Larsen, D. P., D. W. Schultz & K. W. Malueg, 1981. Summer internal phosphorus supplies in Shagawa Lake, Minnesota. Limnology & Oceanography 26: 740–753.

Minnesota Pollution Control Agency, 2006. Lake Pepin Watershed TMDL eutrophication and turbidity impairments project overview. Water Quality/Impaired Waters #9.01a, 2 pp.

National Research Council, 2001. Assessing the TMDL approach to water quality management. National Academy Press, Washington, DC, 109 pp.

Panuska, J. C. & J. C. Kreider, 2002. Wisconsin lake modeling suite program documentation and user's manual, Version 3.3 for Windows. Wisconsin Dept. Natural Resources, PUBL-WR-363-94, 32 p. Accessible at http://www.dnr.state.wi.us/org/water/fhp/lakes/laketool.htm.

Redlands Institute, 2002. Salton Sea Atlas. The Redlands Institute for Environmental Design, Management, and Policy, University of Redlands, Redlands, CA, 127 pp.

Robertson, D. M., S. G. Schladow & G. C. Holdren, 2008. Long-term changes in the phosphorus loading to and trophic state of the Salton Sea. Hydrobiologia (this issue). doi:10.1007/s10750-008-9310-4.

Schroeder, R. A., W. H. Orem & Y. K. Kharaka, 2002. Chemical evolution of the Salton Sea, California: nutrient and selenium dynamics. Hydrobiologia 473: 23–45.

Tiffany, M. A., M. R. Gonzalez, B. K. Swan, K. M. Reifel, J. M. Watts & S. H. Hurlbert (in press). Phytoplankton dynamics in the Salton Sea, California, 1997–1999. Lake and Reservoir Management.

USEPA (United States Environmental Protection Agency), 2000. Nutrient criteria technical guidance manual: lakes and reservoirs, Report no. EPA-822-B00-001, Washington, DC, variously paginated.

Vollenweider, R. A., 1975. Input-output models with special reference to phosphorus loading concept in limnology. Schweizerische Zeitschrift Fur Hydrologie 37: 53–84.

Walker, W. W., 1984. Statistical bases for mean chlorophyll a criteria. Lake and Reservoir Management 2: 57–62.

Walker, W. W., 1985. Empirical methods for predicting eutrophication in impoundments; Report 3, Phase III: Model refinements. Technical report E-81-9, U.S. Army Engineer Waterways Experiment Station, Vicksburg, MS.

Walker, W. W., Jr., 1986. Empirical methods for predicting eutrophication in impoundments; Report 3, Phase III: Applications manual. Technical report E-81-9, U.S. Army Engineer Waterways Experiment Station, Vicksburg, MS. Accessible at http://www.wes.army.mil/el/elmodels/emiinfo.html.

Walker, W. W., Jr., 1996. Simplified procedures for eutrophication assessment and prediction: U.S Army Corps of Engineers, Instruction report W-96-2, variously paginated.

Hydrobiologia (2008) 604:21–36
DOI 10.1007/s10750-008-9312-5

Long-term changes in the phosphorus loading to and trophic state of the Salton Sea, California

Dale M. Robertson · S. Geoffrey Schladow ·
G. Chris Holdren

Abstract The Salton Sea (Sea) is a eutrophic to hypereutrophic lake characterized by high nutrient concentrations, low water clarity, and high biological productivity. Based on dissolved phosphorus (P) and nitrogen (N) concentrations and N:P ratios, P is typically the limiting nutrient in the Sea and, therefore, should be the primary nutrient of concern when considering management efforts. Flows in the major tributaries to the Sea have been measured since 1965, whereas total P (TP) concentrations were only measured intermittently by various agencies since 1968. These data were used to estimate annual P loading from 1965 to 2002. Annual loads have increased steadily from $\sim$940,000 kg around 1968 to $\sim$1,450,000 kg in 2002 ($\sim$55% increase), primarily a result of increased TP concentrations and loads in the New River. Although the eutrophic condition of the Salton Sea is of great concern, only limited nutrient data are available for the Sea. It is difficult to determine whether the eutrophic state of the Sea has degraded or possibly even improved slightly in response to the change in P loading because of variability in the data and changes in the sampling and analytical methodologies.

Keywords Load · Eutrophication · Saline · Water quality

Guest editor: S. H. Hurlbert
The Salton Sea Centennial Symposium. Proceedings of a Symposium Celebrating a Century of Symbiosis Among Agriculture, Wildlife and People, 1905–2005, held in San Diego, California, USA, March 2005

Electronic supplementary material The online version of this article (doi:10.1007/s10750-008-9312-5) contains supplementary material, which is available to authorized users.

D. M. Robertson (✉)
U.S. Geological Survey, 8505 Research Way, Middleton, WI 53562, USA
e-mail: dzrobert@usgs.gov

S. G. Schladow
Department of Civil and Environmental Engineering, University of California, Davis, CA 95616, USA
e-mail: gschladow@ucdavis.edu

G. C. Holdren
Denver Federal Center, Bureau of Reclamation, Denver, CO 80225, USA
e-mail: choldren@do.usbr.gov

Introduction

The Salton Sea is the largest lake in California, USA, located in the southeastern desert region (Fig. 1). The current Salton Sea was formed during 17 months from October 1905 to February 1907 following a high-flow event in the Colorado River that resulted in the river breaching a levee of an early irrigation diversion channel and flooding the low-lying desert of Imperial and Riverside counties. After repairing the diversion structure and the levee, the Salton Sea

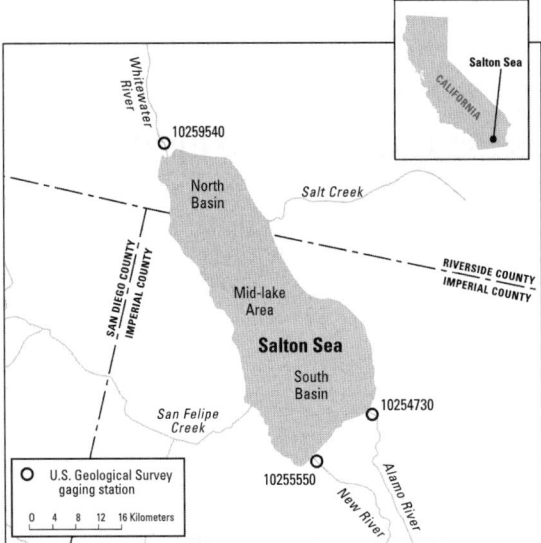

Fig. 1 Salton Sea, California, with tributaries and U.S. Geological Survey gaging stations identified. Discharge measurements and water samples were collected at these sites

began to quickly evaporate because the area around the Sea is extremely arid, receiving only ~75 mm of annual precipitation (NCDC, 2002) while evaporating at a rate of ~1.65 m/yr (Farnsworth et al., 1982). The Salton Sea is a terminal water body that has no natural tributaries and would have completely evaporated if it were not for receiving agricultural discharges from the Imperial (Alamo and New Rivers), Coachella (Whitewater River), and Mexicali valleys along with municipal and industrial effluent from Mexacali, Mexico (New River). One of the major functions of the Sea, since the mid 1920s, is to serve as a sump for agricultural wastewater for the Imperial and Coachella Valleys (Redlands Institute, 2002).

Since the Salton Sea's formation, it has faced and continues to face many of the problems typical of a terminal lake: large fluctuations in water level (Schroeder et al., 2002), increased salinity (Schroeder et al., 2002), and increased concentrations of other constituents, such as nutrients (Holdren & Montaño, 2002a). The water level in the Sea has changed dramatically since its formation; however, the water level has been relatively constant from 1965 to 2002. In 1965, the water level was ~71 m below sea level, and in 2002 it was ~69 m below sea level (Schroeder et al., 2002). The high nutrient concentrations in the Sea have resulted from high nutrient

loads from its tributaries because they carry agricultural discharge and municipal and industrial effluent, primarily from Mexicali, Mexico (Bain et al., 1970). These high loadings have resulted in the eutrophication of the Salton Sea, characterized by high nutrient concentrations, high algal biomass as demonstrated by extensive algal blooms, high fish productivity, low water clarity, frequent low dissolved oxygen concentrations (DO), massive fish kills, and noxious odors (Holdren & Montaño, 2002a). Redlands Institute, California (2002) has assembled a detailed historical chronology of changes and problems that occurred in the Salton Sea and surrounding area.

One method of classifying the water quality of a waterbody is with trophic state index (TSI; Carlson, 1977) values based on near-surface total phosphorus (TP) concentrations (source of the problem), chlorophyll a (Chl a) concentrations (biological response), and Secchi depths (SD; physical response). The indices were developed to place these three characteristics on similar scales to allow comparisons among water bodies. Oligotrophic water bodies (TSI values < 40) have a limited supply of nutrients, typically are clear with low nutrient concentrations and low algal populations, and generally contain DO throughout the year in their deepest zones. Mesotrophic water bodies (TSI values between 40 and 50) have a moderate supply of nutrients, moderate clarity, and are prone to moderate algal blooms; and occasional DO depletions in the deepest zones are possible. Eutrophic water bodies (TSI values between 50 and 60) are nutrient rich with correspondingly severe water-quality problems, such as frequent seasonal algal blooms and poor clarity; DO depletion is common in the deeper zones. Water bodies with TSI values > 60 are often considered hypereutrophic and are extremely nutrient rich and usually experience extensive algal blooms.

A detailed study of the water quality of the Salton Sea and its tributaries was conducted in 1999 by Holdren & Montaño (2002a, b). Based on average TSI values computed from these data (TSI for TP was 67 and for SD was 63), the Salton Sea was classified as eutrophic to hypereutrophic, and based on dissolved P and nitrogen (N) concentrations and N:P ratios (discussed in detail later) P is typically the limiting nutrient and therefore, should be the primary nutrient of concern when considering management efforts. Most of the P input to the Salton Sea on an

annual basis is from tributary loading (Holdren & Montaño, 2002a); therefore, reduction in P loading from the tributaries to the Sea would be a logical management goal. In order to reduce the P loading, a P Total Maximum Daily Load (TMDL; a detailed plan to reduce P loading to a specified value; USEPA, 2000) is being prepared that, if implemented, is intended to improve the water quality and eliminate or, at least, reduce the problems associated with eutrophication.

In order to better understand the sources of P to the Salton Sea, detailed P budgets have previously been assembled by Bain et al. (1970), Watts (in Holdren & Montaño, 2002a), Cagle (1998), Holdren & Montaño (2002a), and Anderson & Amrhein (2002). Bain et al. (1970) used data collected by the Colorado River Regional Water Quality Control Board (CRRWQCB) from 1968 to 1969 to estimate that the annual P load to the Sea was 807,000 kg, which equates to an areal loading of ~ 0.8 g/m^2 (1,000,000 kg equates to an areal loading of 1.02 g/m^2). Watts (Holdren & Montaño, 2002a) used average water-quality data from Bain et al. (1970) to estimate that the annual P load in the late 1960s was $\sim 663,000$ kg. Cagle (1998) used data collected by the CRRWQCB from 1980 to 1992 to estimate that the average annual P load was 1,135,000 kg. Watts (Holdren & Montaño, 2002a) also used data collected by the Imperial Irrigation District and Coachella Valley Water District to estimate that the annual P load in 1996–1997 was 1,515,000 kg. The other two studies estimated the P load for 1999 based on TP measured in the tributaries by Holdren & Montaño (2002a). These two studies estimated a similar annual load of $\sim 1,400,000$ kg for 1999. Therefore, based on the historical studies, P loading to the Sea appears to have approximately doubled from the late 1960s to 1999; however, various different methodologies were used to compute the loads in these studies.

Although most of the problems in Salton Sea are associated with its water quality (high salinity and excessive productivity), no long-term monitoring program exists for the Salton Sea and only a few detailed water-quality studies dealing with nutrients have been conducted. During 1906–1911, a water-quality study was conducted by Ross (1914). In 1955, a study was conducted by the State of California Department of Fish and Game (Walker, 1961). In 1968–1969, a study was conducted by the Federal Water Quality Administration, Pacific Southwest Region (FWQA; Bain et al., 1970). During 1997–2000, a study was conducted by the San Diego State University (SDSU; Hurlbert, written commun., 2001), and in 1999, a study was conducted by the U.S. Bureau of Reclamation (Holdren & Montaño, 2002a, b). Nutrient data from the earliest studies (prior to 1968) are of limited use because of unknown analytical accuracy, limited constituents were examined, and often only summary statistics were reported. Based on the data from 1968 to 1969 and 1999, Holdren & Montaño (2002a) concluded that TP concentrations in the Sea apparently decreased, while P loadings to the Sea approximately doubled.

In order to aid in developing a P TMDL for the Salton Sea, a collaborative study by the University of California-Davis (UC-Davis), the U.S. Geological Survey (USGS), and the National Aeronautic and Space Administration (NASA) was performed to better understand the P dynamics in the Salton Sea and their relation to internal and external P loading. In this article, we summarize the available nutrient, Chl a, and SD data for the Salton Sea and its tributaries, use a consistent methodology to estimate the sources of P to the Sea in terms of P budgets from 1965 to 2002, describe how P loading has changed through time, and describe how the water quality (TP, Chl a, and SD) of the Sea has responded to the changes in P loading. In separate papers in this issue, Robertson & Schladow (2008, this issue) use empirical water-quality models to describe how reductions in P loading should affect the long-term changes in the water quality of the Salton Sea and Chung et al. (2008, this issue) use dynamic models to describe the effects of short-term events such as sediment and P resuspension. Most of the modeling in these other two papers is based on data from 1999; therefore, annual average values for 1999 are presented for annual loads and in-lake water quality.

Study area

At an elevation of 69.4 m below sea level (average elevation in 1999), the Salton Sea has a length of ~ 56 km, width of ~ 24 km, area of 980.8 km^2, volume of 9.26×10^9 m^3, and mean depth of 9.44 m (Weghorst, U.S. Bureau of Reclamation, unpublished

raw data). The overall drainage area of the Salton Sea is ~22,000 km^2 (Agajanian et al., 2005); however, the drainage area is difficult to quantify because the Sea receives much of its water from diversions from the Colorado River and from inputs from Mexico.

Methods

Most of the P enters the Salton Sea associated with the input of water. Therefore, to determine where the water and P originate and how this loading has changed, we quantify the water and P budgets of the Sea. Holdren & Montaño (2002a) estimated that loading from the three major tributaries to the Salton Sea (Alamo, New, and Whitewater Rivers; Fig. 1) contributed ~93% of the total external P loading; therefore, in this study, we concentrated on re-estimating this part of the load and used assumptions similar to those used in previous studies to estimate the loading from the other minor sources. Daily average discharges in the three major tributaries have been continually monitored since 1965 by the USGS (Fig. 1), while water-quality data for the tributaries have been collected only for short periods of time by various agencies. To take advantage of all the discontinuous in-stream data, tributary loading was estimated by use of a regression approach (described below). Although eutrophication of the Salton Sea is of great concern, very limited in-lake water-quality data are available. In this article, in-lake data describing the trophic status of the Sea (TP, Chl *a*, and SD) from 1968 to 1969 (Bain et al., 1970) and 1997 to 1999 (data from Holdren & Montaño, 2002a, b supplemented with Chl *a* from Tiffany, et al. in press and SD data from Hurlbert et al., San Diego State University, unpubl. raw data) are presented and discussed with respect to the calculated changes in external P loading. Detailed water-quality data collected in 1999 by Holdren & Montaño (2002a, b) are used to describe seasonal changes in tributary and in-lake nutrient concentrations and demonstrate that P is typically the limiting nutrient in the Salton Sea.

Field sampling

Water samples were collected from the Alamo, New, and Whitewater Rivers (Fig. 1) in 1968 and 1969 by

FWQA (Bain et al., 1970); although many samples were collected, only average concentrations were available. From 1969 to 1981, the USGS collected water-quality data in the Alamo and New Rivers; these data were obtained from the USGS National Water Information System (NWIS; USGS, 1998). From 1980 to 1993, the CRRWQCB collected water-quality data for the three major tributaries (CRRWQCB staff, written commun., 2003). From 1996 to 1998, the Imperial Irrigation District collected water-quality data in the Alamo and New Rivers, and the Coachella Valley Water District collected water-quality data in the Whitewater River (CRRWQCB staff, written commun., 2003). During 1999, Holdren & Montaño (2002a, b) collected water-quality data in the three major tributaries. During 2002–2003, a few samples were analyzed for nutrient concentrations in the Alamo and New Rivers by the CRRWQCB (CRRWQCB staff, written commun., 2003). Water samples were collected from each tributary as either surface grabs (all agencies except the USGS) or by use of the Equal-Width-Increment (EWI) method using a depth integrated sampler (USGS).

Bain et al. (1970) measured SD and collected water samples to be analyzed for nutrients and Chl *a* from multiple depths at ten locations throughout the Salton Sea (mostly around the periphery of the Sea) from July 1968 to May 1969, although only lake-wide seasonally averaged data were available. Holdren & Montaño (2002b) measured SD and collected water samples on a monthly (January–March) or biweekly (October–December) basis in 1999 to analyze nutrients from three stations in the lake (middle of the North Basin, mid-lake, and middle of the South Basin) from multiple depths by use of a Kemmerer sampler; however, only the data from the surface and near-bottom samples were examined in this study. For each sampling date, the near-surface concentrations and SDs were used to compute lake-wide averages. Tiffany et al. (in press) collected near-surface water samples on approximately a monthly basis from 1997 to 2000 from three stations in the lake (middle of the North Basin, mid-lake, and middle of the South Basin) to be analyzed for Chl *a* using a 3 m long, 4 cm diameter PVC tube sampler. These concentrations were used to obtain average near-surface lake-wide concentrations. Hurlbert et al. (San Diego State University, unpubl. raw data)

measured SD at multiple locations from 1997 to 2000; however, only the data from the North Basin were examined in this study.

Analytical techniques

Holdren & Montaño (2002a, b) analyzed water samples for nutrient species with a Perstorp autoanalyzer. TP was determined using the ascorbic acid method following persulfate digestion (EPA Method 365.4, USEPA, 1983). Orthophosphate-P (OP) was determined with EPA Method 365.1, ammonia with EPA Method 350.1, nitrate plus nitrate with EPA Method 353.2, and Kjeldahl nitrogen with EPA Method 351.2 (USEPA, 1983). Each sample batch included spikes, duplicates, and blanks. The quality-assurance project plan required spike recoveries of 75–125% and a relative difference between duplicates of <20% before data were accepted. These criteria were met for all sample batches, including samples from late August and September 1999 when TP results were inexplicably below the detection limit of 0.005 mg/l. These sampling events in late August and September followed fish kills associated with "green-water" events caused by the formation and precipitation of gypsum crystals (Watts et al., 2001). It is possible that some unknown factor related to these events contributed to the low analytical results.

The USGS also analyzed for TP with an autoanalyzer using the ascorbic acid method following persulfate digestion (EPA Method 365.4, USEPA, 1983; Skougstad et al., 1979). Each sample batch included blanks, replicates, and standards. The relative deviation of this method is estimated to be <12% at 0.20 mg/l and <3% at 0.67 mg/l.

The exact methods used by the other agencies for TP were not specified; however, by the early 1970s, most laboratories were using either automated or manual versions of the ascorbic acid method (EPA 365.1 and EPA 365.2, respectively; USEPA, 1983). Several different methods were available for digesting samples for TP analysis; however, no information is available on the digestion methods used by the CRRWQCB, or Imperial and Coachella Valley Irrigation Districts. Bain et al. (1970) did report using the ascorbic acid method with persulfate digestion for data from 1968 to 1969, but indicated

a possible underestimation of TP due to incomplete digestion and a potential interference from the $HgCl_2$ that was used as a preservative.

The method used in 1968–1969 by Bain et al. (1970) to analyze for Chl a was not specified. Chl a concentrations were determined in 1997–1999 by Tiffany et al. (in press) by filtering 500 ml of water through 200-μm GF/F filters within 24 h of collection. The filters were then stored at −80°C until standard extraction and spectrophotometric analysis could be carried out with the trichromatic method (APHA, 1998).

Load computations

A regression approach was used to estimate the P loads in each of the major tributaries from 1965 to 2002. This method enabled loads to be estimated for years in which little or no water-quality data were collected and took full advantage of all the available in-stream data. The regression method uses a relation between concentration (or load) and daily average flow (and other independent variables) to estimate daily concentrations (or loads) of the constituent. Annual P loads (calculated by summing daily loads) were estimated for each tributary using the Estimator program (Cohn et al., 1989) based on relations between constituent load and variables describing flow (Q), time of the year (T, in radians), and year (DY, decimal year). The general form of the model was

$$\ln(L) = a + b\,[\ln(Q) - c] + d\,[\sin(T)] + e\,[\cos(T)] + f\,[DY] \tag{1}$$

Values for the coefficients (a, b, c, d, e, and f) were computed for each site with multiple regression analyses between daily loads (daily average Qs multiplied by instantaneous measured concentrations) and daily average Q, T, and DY. The sine and cosine terms were included in the general model to account for possible seasonality in the relations and the decimal year (DY) term was included to account for possible long-term trends in the relations (in other words, trends in the flow-adjusted concentrations). Water-quality data from Bain et al. (1970) and the most recent data from CCRWQCB (2002 and 2003) were not included in the calibration process. Data from Bain et al. (1970) were not included because only average concentrations were available and

daily concentrations are needed for the regression approach. In each regression, only the terms in Eq. 1 that were significant at $P < 0.05$ were included. Daily loads were then estimated with the calibrated equations using daily streamflows, day of the year, and decimal year. Since a logarithmic transformation was used in these relations, daily loads were adjusted to account for a retransformation bias by use of the minimum variance unbiased estimate procedure (see Cohn et al., 1989, for a complete discussion). Annual loads were obtained by summing the daily estimated loads.

Computations of TSI values

TSI values based on TP (TSI_P), Chl a (TSI_C), and SD (TSI_{SD}) measured in 1999 were computed for each sampling by the use of Eqs. 2–4 (Carlson, 1977) and were then used to compute lake-wide average TSI values:

$$TSI_P = 4.15 + 14.42 \ [\ln TP \ (\text{in } \mu g/l)] \qquad (2)$$

$$TSI_C = 30.6 + 9.81 \ [\ln \text{Chl } a \ (\text{in } \mu g/l)] \qquad (3)$$

$$TSI_{SD} = 60.0 - 14.41 \ [\ln SD \ (\text{in m})] \qquad (4)$$

Results

Tributary water quality and loading

Water budget

The hydrology of the Salton Sea can be described in terms of components of its water budget. The water budget of the Salton Sea (a terminal lake) can be represented by:

$$\Delta S = PPT + SW_{In} + GW_{In} - Evap \qquad (5)$$

where ΔS is the change in the volume of water stored in the Sea during the period of interest. Water enters the Sea as precipitation (PPT), surface-water inflow (SW_{In}), and ground-water inflow (GW_{In}) and leaves the waterbody through evaporation (Evap). The water level of the Salton Sea has been relatively stable since 1965 (Schroeder et al., 2002) and therefore, on an

annual basis ΔS was assumed to be zero, and Eq. 1 was simplified to:

$$SW_{In} + PPT + GW_{In} = Evap \qquad (6)$$

Daily streamflows from 1965 through 2002 for the three main tributaries to the Sea (Alamo River— USGS site 10254730, New River—USGS site 10255550, and Whitewater River—USGS site 10259540; Fig. 1) were retrieved from the USGS Automated Data-Processing System (ADAPS) within NWIS (USGS, 1998). The average annual flow (1965–2002) from the three main tributaries was $1,391 \times 10^6 \ m^3$. Annual inputs from small tributaries and direct drains (collectively referred to as direct drains) were assumed to contribute a constant annual volume of $130 \times 10^6 \ m^3$ of water ($\sim 8\%$ of the total inflow; Holdren & Montaño, 2002a). The area near the Salton Sea is one of the most arid areas of the United States. The average annual input from direct rainfall to the surface of the Sea was assumed to be a constant 76 mm (NCDC, 2002); with a surface area of 980.8 km^2 this equates to $75 \times 10^6 \ m^3$. Annual ground water input has been estimated to be $\sim 61 \times 10^6 \ m^3$ ($\sim 3.7\%$ of the total inflow; Holdren & Montaño, 2002a). Annual evaporation from the Salton Sea is approximately 1.65 m (Farnsworth et al., 1982) which equates to $1,618 \times 10^6 \ m^3$. In order to validate the above assumptions, the water budget for 1999, which had little change in storage, was examined in more detail. The total input of water to the Sea was $1,695 \times 10^6 \ m^3$ and the total output was 1.65 m per year or $1,618 \times 10^6 \ m^3$ (see Electronic Supplementary Material—Appendix 1); therefore, the assumptions made above appear to be valid.

The total input from the three major tributaries fluctuated only slightly from year to year, and their relative contributions remained very similar (Fig. 2). Only small variations occurred among years because most of the input of water is associated with agricultural drainage from the Imperial and Coachella Valleys and municipal wastewater discharges from Mexico. The maximum annual input of water occurred in 1975 ($1,765 \times 10^6 \ m^3$) and minimum annual input occurred in 1965 ($1,476 \times 10^6 \ m^3$; see Electronic Supplementary Material—Appendix 1). On average, the Alamo River contributed $\sim 45\%$ of the total input, the New River contributed $\sim 33\%$, and the Whitewater River contributed $\sim 6\%$. Other

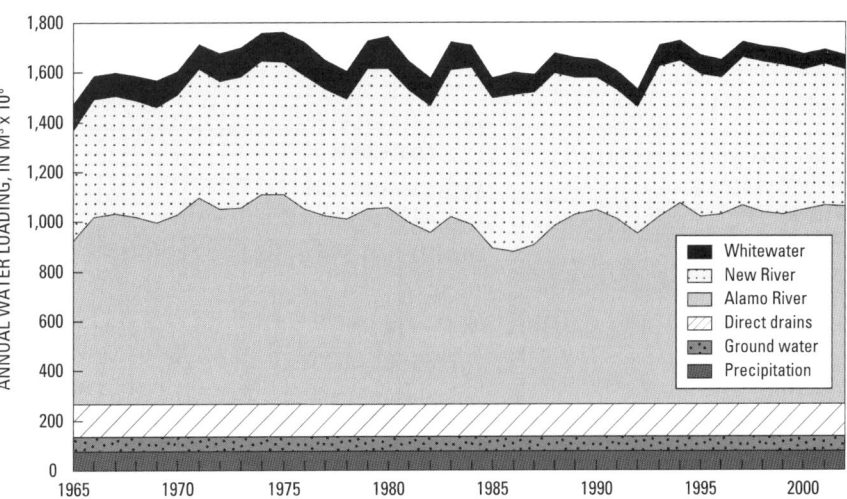

Fig. 2 Annual water loading to the Salton Sea by source from 1965 to 2002. Values for direct drains, ground water, and precipitation to the surface of the Sea were assumed to be constant. Annual data are provided in Electronic Supplementary Material—Appendix 1

direct drains, ground water, and precipitation contributed $\sim 8\%$, $\sim 4\%$, and $\sim 5\%$, respectively. Direct precipitation on the surface of the Sea and the contributions lumped as direct drains could be slightly more important during wet years than shown here. The relative contributions during 1999 are similar to the long-term average contributions. No long-term trend in water inputs to the Salton Sea is apparent.

Nutrient concentrations in the tributaries

TP concentrations in the Alamo River have ranged from 0.1 to 2.6 mg/l, but there was no long-term trend from 1969 to 1999; the average concentration was ~ 0.7 mg/l (Fig. 3). The earliest data from Bain et al. (1970) had a lower average concentration (0.33 mg/l) than the rest of the data; however, they stated that their method of analysis may have been biased low because it did not digest all the organic matter present and some analytical difficulties arose due to interference with the preservative used. These biases not only affected concentrations, but also would affect any load computations based solely on these data. Recent data by the CRRWQCB (2002–2003), although quite variable, may indicate an increase in concentrations in the Alamo River. During 1999, the average monthly concentration in the Alamo River was 0.70 mg/l (Table 1). The highest concentrations were measured in January through May and lowest concentrations were measured in August.

TP concentrations in the New River (Fig. 3) significantly increased from 1969 to the present ($P \ll 0.001$). During 1969–1981, the mean concentration was ~ 0.75 mg/l, slightly higher than was measured in the Alamo River during this period. The average concentrations, however, steadily increased to ~ 0.92 mg/l (1980–1992), to ~ 1.26 mg/l in 1996–1998, decreased slightly to ~ 1.11 in 1999, and increased again to ~ 1.38 mg/l in 2002–2003. During 1999, the average monthly concentration in the New River was 1.11 mg/l (Table 1). Concentrations were slightly higher in winter than in summer.

Less TP data were available for the Whitewater River than the other two main tributaries (Fig. 3). The average TP concentration in the Whitewater River was relatively low (0.23 mg/l) during the 1980s and early 1990s and then significantly increased to ~ 0.85 mg/l in the mid 1990s ($P \ll 0.001$). During 1999, the average monthly TP in the Whitewater River was 0.88 mg/l (Table 1). Little seasonal variability was observed in TP concentrations.

Concentrations of OP in the tributaries are discussed only based on the data collected in 1999 (Table 1; Holdren & Montaño, 2002a). Average monthly OP concentrations were highest in the Whitewater and New Rivers (~ 0.7 mg/l) and lowest in the Alamo River (~ 0.4 mg/l). No distinct seasonality was observed in OP; however, highest concentrations were observed in February in all three tributaries. On average, OP represented ~ 60 to 80% of the TP in these tributaries; however, during some months this dropped to less than 30% and increased to 100% in other months (in February the OP in the

Hydrobiologia (2008) 604:21–36

Fig. 3 Total phosphorus concentrations with average concentrations for each study for the Alamo, New, and Whitewater Rivers. The single data points at the end of 1968 represent an average reported value from Bain et al. (1970)

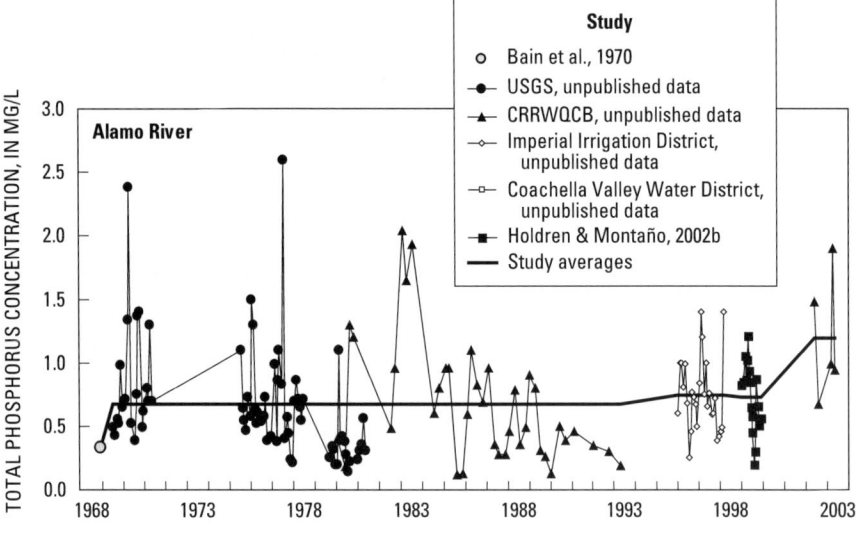

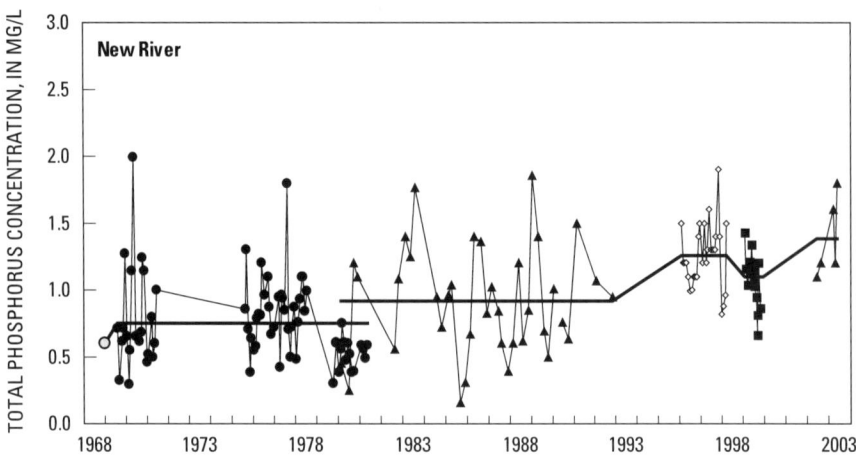

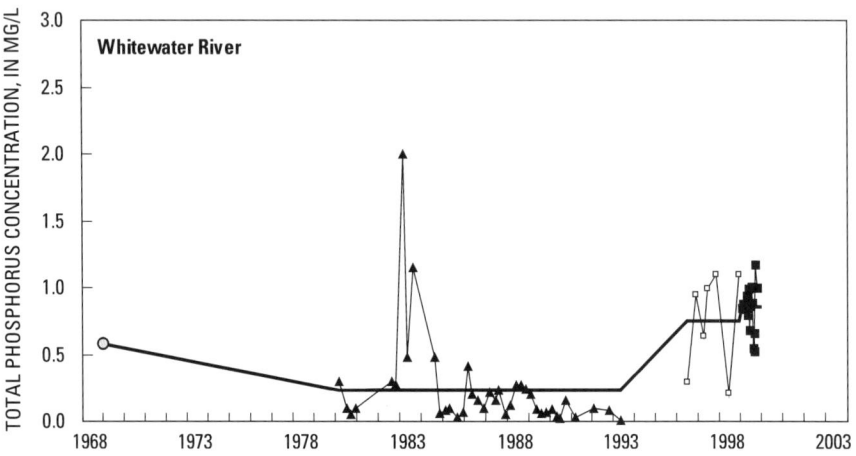

Table 1 Monthly average water quality in the Alamo, New, and Whitewater Rivers in 1999 based on data from Holdren & Montaño (2002b)

Month	Total phosphorus (mg/l)			Orthophosphate-P (mg/l)			Percent orthophosphate		
	Alamo	New	Whitewater	Alamo	New	Whitewater	Alamo	New	Whitewater
January	0.82	1.43	0.85	0.37	0.99	0.77	44.5%	69.3%	90.7%
February	0.85	1.16	0.88	0.97	1.03	1.17	100%[a]	88.8%	100%[a]
March	1.05	1.04	0.84	0.48	0.44	0.58	46.1%	42.4%	69.2%
April	0.94	1.18	0.93	0.46	0.60	0.72	48.4%	50.3%	77.7%
May	1.07	1.24	0.90	0.42	0.57	0.72	39.3%	46.4%	80.3%
June	0.52	1.09	0.95	0.31	0.56	0.70	60.6%	51.2%	74.0%
July	0.52	1.09	0.95	0.18	0.52	0.77	34.3%	47.9%	81.2%
August	0.33	0.81	0.54	0.21	0.56	0.43	63.9%	69.8%	79.2%
September	0.59	1.00	0.91	0.52	0.91	0.79	88.1%	90.9%	86.2%
October	0.66	0.87	1.00	0.58	0.76	0.83	88.6%	87.8%	83.2%
November	0.51	1.47	0.85	0.27	1.11	0.51	53.3%	75.5%	60.1%
December	0.56	0.90	0.97	0.14	0.30	0.53	25.3%	33.1%	54.6%
Mean	0.70	1.11	0.88	0.41	0.70	0.71	57.7%	62.8%	78.0%
Coefficient of variation	0.34	0.19	0.13	0.55	0.37	0.27	0.407	0.319	0.160

[a] Orthophosphate concentrations exceeded total phosphorus concentrations

Alamo and Whitewater Rivers exceeded TP indicating some error in the analytical process).

Concentrations of N species (nitrite plus nitrate, ammonia, and Kjeldahl nitrogen) in the tributaries were also measured in 1999 (Holdren & Montaño, 2002a). Average monthly total nitrogen (TN) concentrations ranged from 8.3 mg/l in the New River, to 9.3 mg/l in the Alamo River, to 16.4 mg/l in the Whitewater River. This resulted in average N:P ratios ranging from ~8 in the New River, to ~13 in the Alamo River, to ~19 in the Whitewater River. Most of the N (80–90%) was in inorganic form. Ammonia concentrations were highest in the New River (3.7 mg/l average) and represented ~45% of the N. Ammonia concentrations in the other tributaries were lower (0.7–1.2 mg/l averages) and represented a smaller part of the TN.

Phosphorus loading

Long-term flow and P data for each of the tributaries (Figs. 2 and 3) were used to estimate the P loading to the Sea from 1965 to 2002 using regression models (Eq. 1; Fig. 4; and see Electronic Supplementary Material—Appendix 2). The final load model for the Alamo River included daily flow and seasonality (day of the year) and explained ~50% of the total variability in measured daily loads. The final model for the New River included daily flow and decimal year and explained ~45% of the total variability in measured daily loads. The final model for the Whitewater River also included daily flow and decimal year, but only explained ~22% of the total variability in measured daily loads. The significance of the decimal-year terms demonstrates a significant increase in flow-adjusted TP concentrations in the New and Whitewater Rivers.

Annual P loadings from direct drains (93,000 kg; Fig. 4; and see Electronic Supplementary Material—Appendix 2) were computed assuming they contributed ~130 × 10^6 m^3 of water (Holdren & Montaño, 2002a) with a TP concentration similar to that measured in the Alamo River in 1999 (volumetrically weighted average of 0.72 mg/l). The TP concentration in the Alamo River was chosen for the direct drains because both areas drain similar types of agricultural land. Annual loading from precipitation was computed assuming a constant deposition of 14.8 kg/km^2 (Jassby et al., 1994) or an annual deposition of 14,500 kg. This deposition rate was obtained from Lake Tahoe, California, ~725 km north of the Salton Sea. However, if P input similar to that found from other desert areas (southeastern

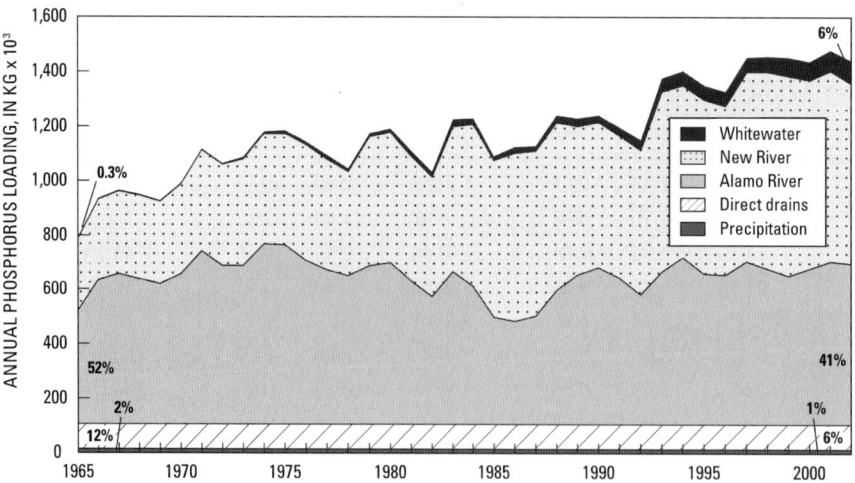

Fig. 4 Annual total phosphorus load by source for the Salton Sea from 1965 to 2002. Annual tributary loads were estimated with regression models to account for periods when data were unavailable. Annual loads from direct drains and atmospheric deposition were assumed to be constant. Percentages of the total annual phosphorus loading are given for 1965 and 2002. Annual data are provided in Electronic Supplementary Material—Appendix 2

Mediterranean) were used (~ 40 kg/km^2; Herut et al., 2002), this could increase to $\sim 40,000$ kg, which is about half of that estimated from the direct drains. The lower estimate from precipitation was used in computing the annual P budgets. Contributions from ground water were assumed to be negligible.

The long-term average annual P load (1965–2002) to the Salton Sea was $\sim 1,184,000$ kg, which equates to an areal loading to the Sea of ~ 1.2 g/m^2. Annual loads have ranged from $\sim 790,000$ kg in 1965 to $\sim 1,480,000$ kg in 2001 (see Electronic Supplementary Material—Appendix 2). The upward trend in total loading from 1965 to 2002 was statistically significant ($P \ll 0.001$) and equates to an increase in annual P loading of $\sim 14,130$ kg/yr or an increase of $\sim 55\%$ since the late 1960s. The P load to the Salton Sea in 1999 was $\sim 1,452,000$ kg, which equates to a unit-area yield from the basin of 67.1 kg/km^2 and an area loading to the Sea of ~ 1.5 g/m^2.

The Alamo River contributed $\sim 46\%$ of the long-term average annual P load to the Salton Sea. Annual loading from the Alamo River ranged from $\sim 375,000$ kg in 1986 to >650,000 kg in 1974 (see Electronic Supplementary Material—Appendix 2). No long-term trend in loading was apparent; a result of no trends in flows and little if any trend in concentrations prior to the data collected after 2000. In 1999, it is estimated that $\sim 544,000$ kg of P was transported to the Sea from the Alamo River, and the average volumetrically weighted concentration was 0.72 mg/l.

The New River contributed $\sim 43\%$ of the long-term average annual P load to the Salton Sea. Annual loading from the New River ranged from $\sim 266,000$ kg in 1965 to $\sim 734,000$ kg in 1999 (see Electronic Supplementary Material—Appendix 2). There was a statistically significant trend ($P \ll 0.001$) in loading from the New River as a result of the dramatic increases in TP concentrations. The P load in 1999 was $\sim 734,000$ kg, and average volumetrically weighted concentration was 1.22 mg/l.

Over the long-term, the Whitewater River has contributed only $\sim 2\%$ of the P loading to the Salton Sea; however, the proportion has increased from $\sim 0.3\%$ in 1965 to $\sim 6\%$ in 2002 (see Electronic Supplementary Material—Appendix 2). There was a very strong trend in loading from the Whitewater River as a result of the dramatic increases in TP concentration in the late 1990s and limited data prior to 1980. The P load in 1999 was $\sim 65,000$ kg, and average volumetrically weighted concentration was 1.00 mg/l. The drainage area of the Whitewater River is 3,872 km^2 (Agajanian et al., 2005); therefore, the unit-area yield from this basin is 16.8 kg/km^2. This is the only tributary with a defined drainage basin.

The relative contributions of P to the Salton Sea have changed dramatically since 1965 (Fig. 4). In 1965, the Alamo River contributed 52% of the total

load; however, in 2002, it contributed only 41%. This was primarily caused by the increased loading by the New and Whitewater Rivers, rather than a reduction in loading in the Alamo River. In 1965, the New and Whitewater Rivers contributed 34% and 0.3%, respectively (however, very little data were available for the New River prior to 1980). In 2002, the New and Whitewater Rivers contributed 46% and 6%, respectively. The large increases in P loading from the New River may have been caused by increased loading from municipal and industrial effluent from Mexicali, Mexico. Mexicali almost doubled in population between 1970 and 2000 (from ~400,000 to ~750,000; INEGI, 2000).

Water quality of the Salton Sea

Phosphorus concentrations

Since 1968, the reported lake-wide average TP concentrations in the Salton Sea have ranged from near detection to >0.2 mg/l, with very large variations occurring between consecutive samplings (Fig. 5A). The average near-surface concentration for all the data collected in 1968–1969 was 0.095 mg/l and in 1999 was 0.077 mg/l. TP concentrations in the Sea are about an order of magnitude less than those in its tributaries.

In order to determine whether the slightly higher values measured in 1968–1969 were an artifact of averaging the concentrations over depth, the near-surface TP were compared with those measured near the bottom in 1999. Although TP (and OP) concentrations were often quite uniform throughout the water column, near-bottom concentrations were frequently lower than near-surface concentrations (Table 2). Lake-wide average TP concentrations that include lower TP near the bottom, such as included in 1968–1969, should have resulted in the lake-wide average concentrations being slightly lower than those based only on the near-surface concentrations, rather than the higher reported values. The data from 1968 to 1969, however, did incorporate many samples collected near the periphery of the Sea, which could have resulted in the higher estimated concentrations. Therefore, on average it appears that the TP concentrations in the Salton Sea may have decreased slightly from 1968–1969 to 1999; however, given the

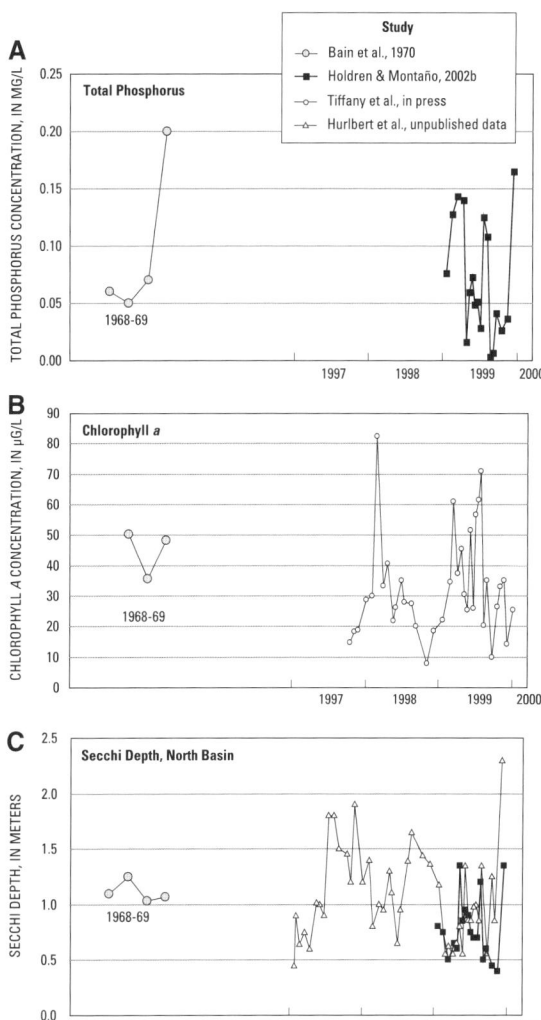

Fig. 5 Water quality measured in the Salton Sea from 1968 to 2000: (**A**) Near-surface total phosphorus concentration, (**B**) near-surface chlorophyll *a* concentration, and (**C**) Secchi depth. Only seasonal data were available for 1968–1969 (Bain et al., 1970)

variability in TP measured in the Sea and changes in the data-collection and analytical methodologies significant changes are difficult to ascertain.

The detailed water chemistry measured in 1999 (Holdren & Montaño, 2002b) was used to describe spatial and temporal changes in TP throughout the Sea. Small differences in TP were typically observed between sampling sites, with the highest concentrations in the North Basin, slightly lower concentrations in the narrow middle area, and lowest concentrations in the South Basin. On average the North Basin had TP ~0.015 mg/l higher than the South Basin.

Table 2 Monthly average surface and near-bottom lake-wide water quality in the Salton Sea in 1999

Month	Total phosphorus (mg/l)		Orthophosphate-P (mg/l)		Total nitrogen (mg/l)		N:P ratio	Chlorophyll *a* (µg/l)	Secchi depth (m)
	Surface	Bottom	Surface	Bottom	Surface	Bottom	Surface	Surface	Surface
January	0.076	0.059	0.003	0.003	2.16	1.77	28.3	22.2	0.88
February	0.128	0.066	0.034	0.034	2.34	1.59	18.3	34.6	0.72
March	0.143	0.108	0.023	0.025	2.24	1.56	15.6	60.9	0.68
April	0.078	0.067	0.003	0.003	5.01	5.10	64.1	41.5	0.65
May	0.066	0.069	0.008	0.009	4.33	4.49	65.4	27.9	0.92
June	0.050	0.035	0.006	0.004	4.97	4.88	99.6	38.8	0.96
July	0.077	0.081	0.003	0.005	3.65	5.37	47.6	63.0	0.75
August	0.056	0.054	0.032	0.014	4.33	5.37	77.6	27.7	1.02
September	0.024	0.019	0.020	0.019	4.37	4.42	179.7	9.8	0.78
October	0.027	0.013	0.027	0.012	4.30	4.35	161.2	26.5	0.72
November	0.036	0.038	0.018	0.036	4.20	3.65	115.5	34.1	0.67
December	0.165	0.123	0.090	0.088	3.73	2.90	22.6	14.3	1.02
Mean	0.077	0.061	0.022	0.021	3.80	3.79	74.6	33.4	0.81
Coefficient of variation	0.59	0.54	1.10	1.15	0.27	0.39	0.74	0.48	0.17

All data were from Holdren & Montaño (2002b), except chlorophyll *a* data which were from Tiffany et al. (in press)

The average monthly OP concentration in 1999 was ~0.02 mg/l near the surface and near the bottom (Table 2). From April through July, concentrations were <0.01 mg/l and very close to detection limits. Only during November were concentrations near the bottom higher than at the surface. OP concentrations near detection (when dissolved N species were abundant, described later) suggest that P was the limiting nutrient, and OP concentrations near detection just above the bottom during anoxic periods indicate a little or no net internal loading of P from the deep sediments.

In addition to the data shown in Fig. 5A, OP concentrations were measured in the Salton Sea annually in May or June from 1906 to 1911 (Ross, 1914). The average OP concentration during that period was ~0.03 mg/l. During 1999, the average OP concentration was 0.007 mg/l. This would indicate that OP concentrations may have been much higher shortly after the Salton Sea formed than they are presently. However, the exact methods used in the early study and the accuracy of the methods are unknown.

Nitrogen concentrations

Changes in TN are discussed based only on the data collected in 1999 (Table 2; Holdren & Montaño,

2002b). The average monthly near-surface and near-bottom TN concentrations in 1999 were ~3.8 mg/l. Monthly concentrations ranged from 1.6 to 2.2 mg/l in February to ~5.0 mg/l in April, and July and August near the bottom. On an annual basis, 65% of the N was in organic form and 35% was in inorganic forms (mostly as ammonia). Ammonia concentrations typically were higher near the bottom of the Sea than near the surface; whereas, organic forms of N typically were quite uniform throughout the water column.

Nitrogen to phosphorus ratios

The N:P ratio is often used to determine whether N or P should potentially limit algal productivity in a lake. The specific value of this ratio that determines which nutrient is potentially limiting, varies under different conditions such as water temperature, light intensity, and nutrient deficiencies (Corell, 1998); however, a ratio greater than ~12:1, by weight, usually indicates that P is the potentially limiting nutrient. Average monthly N:P ratios in the Salton Sea ranged from ~16:1 to 180:1 (Table 2); therefore, P should usually be the potential limiting nutrient for algal growth in the Sea. Concentrations of OP were often near

Fig. 6 Trophic state indices (Carlson, 1977) for the Salton Sea, based on 1999 data for total phosphorus and Secchi depth (Holdren & Montaño, 2002b) and chlorophyll *a* (Tiffany et al., in press)

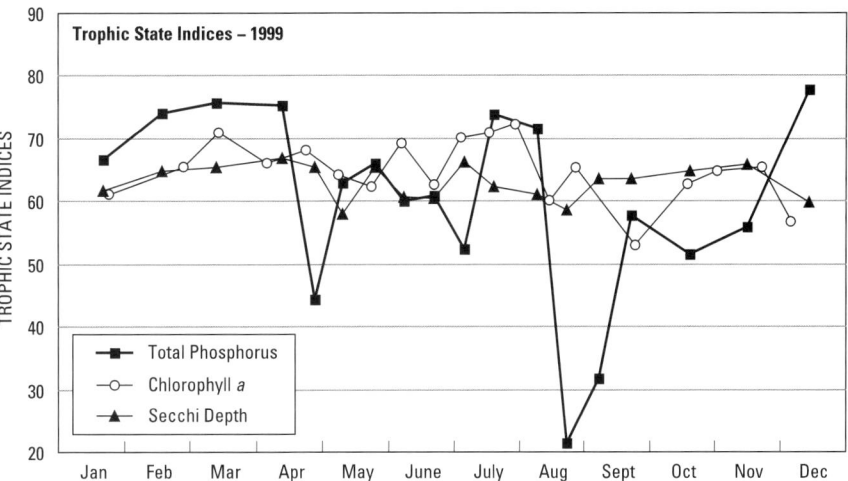

detection limits; whereas, dissolved nitrate and ammonia were often quite high (Holdren & Montaño, 2002b), which also indicates that P is the nutrient that is typically limiting or potentially limiting algal growth. Therefore, P should be the nutrient to primarily focus on when considering management efforts to improve water quality.

Chlorophyll a concentrations

Chl *a* in the Salton Sea ranged from ∼6 µg/l to over 80 µg/l (Fig. 5B). The average surface concentration for the data collected in 1968–1969 was 45 µg/l, and the average for 1997–1999 was 33 µg/l. Given the variability in Chl *a* measured in the Sea and changes in the data-collection methodologies and potential changes in analytical techniques, it is difficult to determine whether Chl *a* concentrations have changed. Concentrations generally were highest (>30 µg/l) between February and August (Table 2), indicative of algal blooms. The average monthly lake-wide near-surface Chl *a* concentration measured in 1999 was 33.4 µg/l.

Water clarity

Since 1968, SD ranged from <0.5 m to >2 m, with an average depth of ∼1 m (Fig. 5C). Given the variability in SD and the limited data, no trend in clarity is apparent. A slight difference in clarity was observed among sampling sites, with the clarity in the

North Basin being slightly better than the narrow middle area of the Sea, which in turn was slightly better than the South Basin. Based on data from 1999 (Holdren & Montaño, 2002b), the average SD in the North Basin was ∼0.09 m greater than in the South Basin. The average annual SD for 1999 (based only on Holdren & Montaño data) was 0.81 m (Table 2).

Trophic state indices

Average annual TSI values, computed based on TP and Chl *a* concentrations and SD, indicate that the Salton Sea was usually eutrophic (TSI values 50–60) to hypereutrophic (TSI values > 60) during 1999 (Fig. 6). The average monthly TSI value based on TP was 62, SD was 63, and Chl *a* was 64. On a few occasions, however, the TSI values based on TP indicated the Sea was mesotrophic (TSI values between 40 and 50) and even oligotrophic (TSI values < 40) in late August and early September, whereas TSI values for Chl *a* and SD during these periods indicated the Sea was eutrophic to hypereutrophic. It is not unusual for TSI values to diverge during major algal blooms, but the low TSI values based on TP during August and September were the result of very low, and possibly erroneous, analytical results (see discussion under Analytical Techniques). The good agreement among TSIs based on TP, Chl *a*, and SD for the rest of the year clearly indicate that the Salton Sea is usually eutrophic to hypereutrophic.

Discussion

Comparison of water quality in the Salton Sea with other lakes

The lower TP and OP concentrations near the bottom of the Salton Sea are the opposite of that found in most eutrophic lakes during stratified periods (Wetzel, 2001). In most lakes, P is released from the bottom sediments when the water above the sediments goes anoxic. However, average monthly near-bottom TP concentration was 0.061 mg/l compared to 0.077 mg/l near the surface, and the average near-bottom OP concentration was 0.021 mg/l compared to 0.022 mg/l near the surface. The similar or slightly lower concentrations near bottom in the Salton Sea typically occurred during periods of mixing and during stratified periods when the deep water was anoxic, except in November when higher OP concentrations were measured near the bottom. The Salton Sea is a polymictic lake; however, extensive periods of weak thermal stratification have been documented (for example in 1997 by Cook et al., 2002 and in 1999 by Holdren & Montaño, 2002a). Lower TP and similar OP near the bottom during stratified periods, especially during anoxic periods, indicate little or no net internal loading of P from the deep sediments in the Salton Sea. Rodriguez et al. (2008, this issue) have shown that this lack of internal loading of P in the Salton Sea is the result of P coprecipitating with calcite. This removal process has been documented to also occur in fresh hardwater lakes (Robertson et al., 2007).

Comparison of phosphorus loading with previous estimates

Holdren & Montaño (2002a) concluded that the total P loading to the Salton Sea had doubled from the late 1960s to 1999. This conclusion was based on an annual loading estimate for 1967–1968 of 663,000 kg, which was based on average TP concentration data from Bain et al. (1970). Bain et al. (1970), however, used the individual TP concentration data to estimate a total annual P load for 1968–1969 of ~807,000 kg. Bain et al. (1970) stated that the estimated TP concentrations in the tributaries may have been biased low because their analytical

technique did not digest all the organic matter and some analytical difficulties arose due to interference with the preservative used. The more extensive data collected by the USGS from 1969 to 1981 which followed strict quality controls indicated that TP concentrations in the tributaries were higher than those estimated by Bain et al. (Fig. 3) and resulted in higher loading estimates in the late 1960s. In this study, we estimated the annual P loading for 1968–1969 was ~940,000 kg (see Electronic Supplementary Material—Appendix 2), which is ~15% more than Bain et al. (1970) and ~40% more than Holdren & Montaño (2002a). Cagle (1998) estimated the total annual loading for the period 1980–1992 to be ~1,135,000 kg, which is similar to that estimated in this study for those years (1,218,000 kg). Holdren & Montaño (2002a) estimated the total annual load for 1999 to be 1,385,000 kg. In this study, we estimated that the total annual load for 1999 was ~1,450,000 kg. The slightly higher estimated load in 1999 from this study was because of slightly higher estimates from the New and Whitewater Rivers. Our estimates indicated that annual P loads have increased from ~940,000 kg (late 1960s) to ~1,450,000 (around 2000), an increase of ~55%.

Response in the Salton Sea to changes in phosphorus loading

In order to determine how the water quality of the Sea has changed in response to the changes in P loading, extensive efforts were made to acquire data from past studies on the Salton Sea. Based on results from these studies, the average near-surface TP concentration in the late 1960s was 0.095 mg/l compared to 0.077 mg/l measured in 1999. The average surface Chl a was 45 µg/l in the late 1960s compared to 33 µg/l measured in 1997–1999. The average SD was 1.1 m in the late 1960s compared to 0.8–1.0 m measured in 1999. All the data indicate that the Salton Sea has been eutrophic to hypereutrophic since the 1960s and has responded very little or may have improved slightly to the documented increase (~55%) in P loading. However, given the variability in the water quality measured in the Sea and the changes in data-collection and analytical methodologies, significant trends or responses to the changes in P loading could not be quantified. The lack of response in the

eutrophic state of the Salton Sea to increased P loadings may have been caused by an increased rate of coprecipitation of P with calcite that may have occurred with the increased salinity of the Sea.

Most lakes continue to experience degraded water quality for several years after reduction in P loading because of the continued release of P from the sediments of the lake, especially during anoxic periods (e.g., Lake Sammamish, Wash.; Welch et al., 1986). However, it appears that P is not released from the deep sediments of the Salton Sea because of coprecipitation with calcite (Rodriguez et al., 2008, this issue), as demonstrated by lower TP concentrations and similar OP concentrations being measured near the bottom during stratified periods, especially during anoxic periods. Therefore, reductions in external P loading may be expected to reduce TP concentrations in the Salton Sea more quickly than in other non-saline lakes.

Conclusions

Although most of the major problems in Salton Sea are associated with its water quality (high salinity and excessive productivity), there has been no long-term monitoring program and only a few detailed water-quality studies have been conducted. Therefore, extensive efforts were made in this study to acquire original reports and data compilations from past water-quality studies on the Salton Sea and its tributaries to enable accurate load estimates to be made and to evaluate how the water quality in the Sea has responded to changes in P loading.

Based on N:P ratios and concentrations of dissolved nutrients from these past studies, P is the limiting or potentially limiting nutrient in the Salton Sea. Therefore, a reduction in P loading from the tributaries to the Sea through the implementation of a P TMDL would be a logical management goal to eliminate or at least reduce the problems associated with eutrophication. To aid in developing a P TMDL for the Salton Sea, we quantified the sources of P to the Sea from 1965 to 2002. Annual P loads have increased steadily from $\sim$940,000 kg (late 1960s) to $\sim$1,450,000 kg (around 2000), or an increase of $\sim$55%. The major reason for the increase in P loading was the increased loading from the New River, which may have been caused by increased

contributions from municipal and industrial effluent from Mexicali, Mexico, which almost doubled in population during this period. The overall increase in P loading to the Sea is much less than that estimated by the Holdren & Montaño (2002a), who concluded that the loading has more than doubled during this period. The overestimate in the overall change in P loading was primarily caused by the underestimate in the P loading in the early 1960s.

It is often difficult to document subtle changes in the trophic state of most water bodies with long-term, consistent monitoring programs; however, it is even more difficult to document changes when only intermittent monitoring data from various short-term studies are available. Based on the limited data available from the 1960s, it appears that the water quality in the Salton Sea has responded very little or may have even slightly improved in response to the documented increase in external P loading ($\sim$55%) that has occurred since the late 1960s. However, given the variability in the water quality measured in the Sea and the changes in data-collection and analytical methodologies, we were not able to determine whether the trophic state of the Salton Sea has significantly changed in response to the increased P loading. Therefore, it is necessary to rely on various types of water-quality models to determine how the water quality of the Salton Sea should respond to changes in nutrient loading in the future and how much of the P that enters the Salton Sea needs to be eliminated to improve its water quality. In separate papers in this issue, Robertson & Schladow (2008, this issue) use empirical water-quality models to describe how reductions in P loading should affect the long-term changes in the water quality of the Salton Sea and Chung et al. (2008, this issue) use dynamic models to describe how short-term events such as sediment and P resuspension should affect the short-term changes in water quality.

References

Agajanian, J., L. A. Caldwell, G. L. Rockwell & G. L. Pope, 2005. Water resources data California: Water year 2004, U.S. Geol. Survey Water-Data Rep. CA–04–1, 549 pp.

Anderson, M. A. & C. Amrhein, 2002. Nutrient cycling in the Salton Sea. Final Report to the Salton Sea Authority. Dept. of Envir. Sciences, Univ. of California Riverside.

APHA (American Public Health Association), 1998. Standard Methods for the Examination of Water and Wastewater, 20th ed. Washington, D.C., USA, 1220 pp.

Bain, R. C., A. M. Caldwell, R. H. Clawson, H. L. Scotten & R. G. Wills, 1970. U.S. Department of Interior 1970 Report. Based on Salton Sea Project, California. Federal-State Reconnaissance Report. Dept. of the Interior and Resources Agency of California. Also accessible at http://www.sci.sdsu.edu/USDI1970.html

Cagle, F., 1998. Geomedicine in the Salton Basin. In Lowell, L. (ed.) Geology and Geothermal Resources of the Imperial and Mexicali Valleys. San Diego Assoc. of Geologists, 1169–1187.

Carlson, R. E., 1977. A trophic state index for lakes. Limnology & Oceanography 22: 361–369.

Chung, E., S. G. Schladow, J. Perez-Losada & D. M. Robertson, 2008. A linked hydrodynamic and water quality model for the Salton Sea. Hydrobiologia (this issue). doi: 10.1007/s10750-008-9311-6.

Cohn, T. A., L. L. DeLong, E. J. Gilroy, R. M. Hirsch & D. K. Wells, 1989. Estimating constituent loads. Water Resources Research 25: 937–942.

Cook, C. B., G. T. Orlob & D. W. Huston, 2002. Simulation of wind-driven circulation in the Salton Sea: implications for indigenous ecosystems. Hydrobiologia 473: 59–75.

Corell, D. L., 1998. The role of phosphorus in the eutrophication of receiving waters—a review. Journal of Environmental Quality 27: 261–266.

Farnsworth, R. K., E. S. Thompson & E. L. Peck, 1982. Evaporation atlas for the contiguous 48 United States: National Oceanographic and Atmos. Adm. Technical Rep. NWS 33. Washington, D.C. National Weather Service, 26.

Herut, B., R. Collier & M. D. Krom, 2002. The role of dust in supplying nitrogen and phosphorus to the Southeast Mediterranean. Limnology & Oceanography 47: 870–878.

Holdren, G. C. & A. Montaño, 2002a. Chemical and physical characteristics of the Salton Sea, California. Hydrobiologia 473: 1–21.

Holdren, G. C. & A. Montaño, 2002b. Chemical and physical limnology of the Salton Sea, California—1999. Technical Memorandum No. 8220-03-02, U.S. Dept. of the Interior, Bureau of Reclamation, Denver, Colorado.

INEGI (Instituto Nacional De Estadistica Geografica e Informatica), 2000. Censo de poblacion y vivienda del ano 2000.

Jassby, A. D., J. E. Reuter, R. P. Axler, C. R. Goldman & S. H. Hackley, 1994. Atmospheric deposition of nitrogen and phosphorus in the annual nutrient load of Lake Tahoe (California-Nevada). Water Resources Research 30: 2207–2216.

NCDC (National Climatic Data Center), 2002. Climatography of the U.S.—Monthly Station Normals of Temperature, Precipitation, and Heating and Cooling Degree Days, 1971–2002. National Oceanic and Atmospheric Adm., Asheville, North Carolina.

Redlands Institute, 2002. Salton Sea Atlas. The Redlands Institute for Environmental Design, Management, and Policy, University of Redlands, Redlands, California, 127 pp.

Robertson, D. M., H. S. Garn & W. J. Rose, 2007. Response of Nagawicka Lake, a calcareous lake in Wisconsin, to changes in phosphorus loading. Lake and Reservoir Management 23: 298–312.

Robertson, D. M. & S. G. Schladow, 2008. Response in the water quality in the Salton Sea to changes in phosphorus loading: an empirical modeling approach. Hydrobiologia (this issue). doi:10.1007/s10750-008-9321-4.

Rodriguez, I. R., C. Amrhein & M. A. Anderson, 2008. Laboratory studies of coprecipitation of phosphate with calcium carbonate in the Salton Sea, California. Hydrobiologia (this issue). doi:10.1007/s10750-008-9310-7.

Ross, W. H., 1914. Chemical composition of the water of Salton Sea, and its annual variation in concentration 1906–1911. In the Salton Sea: study of the geography, the geology, the floristics, and the ecology of a desert basin, by D. T. MacDougal. Carnegie Inst. Washington, Publ., No. 193, 182 pp.

Schroeder, R. A., W. H. Orem & Y. K. Kharaka, 2002. Chemical evolution of the Salton Sea, California: nutrient and selenium dynamics. Hydrobiologia 473: 23–45.

Skougstad, M. W., M. J. Fishman, L. C. Friedman, D. E. Erdman & S. S. Duncan (eds), 1979. Methods for the determination of inorganic substances in water and fluvial sediments: Techniques of Water-Resources Investigations of the U.S. Geological Survey, Book 5, Chap. A1, 626.

Tiffany, M. A., M. R. Gonzalez, B. K. Swan, K. M. Reifel, J. M. Watts & S. H. Hurlbert, in press. Phytoplankton dynamics in the Salton Sea, California, 1997–1999. Lake and Reservoir Management.

USEPA (United States Environmental Protection Agency), 1983. Methods for chemical analysis of water and wastes. Report No. EPA-600/4-079-020. U.S. EPA, Cincinnati, Ohio.

USEPA (United States Environmental Protection Agency), 2000. Nutrient criteria technical guidance manual: lakes and reservoirs. Report No. EPA-822-B00-001, Washington, D.C., variously paginated.

USGS (U.S. Geological Survey), 1998. National Water Information System (NWIS): U.S. Geol. Survey Fact Sheet FS-027-98. Also accessible at http://waterdata.usgs.gov/nwis

Walker, B. W., 1961. The ecology of the Salton Sea, California, in relation to the sportfishery. State of California Dept. of Fish and Game. Fish Bulletin No. 113.

Watts, J. M., B. K. Swan, M. A. Tiffany & S. H. Hurlbert, 2001. Thermal, mixing, and oxygen regimes in the Salton Sea, California, 1997–1999. Hydrobiologia 466: 159–176.

Welch, E. B., D. E. Spyridakis, J. I. Shuster & R. R. Horner, 1986. Declining lake sediment phosphorus release and oxygen deficit following wastewater diversion. Journal of Water Pollution Control Federation 58: 92–96.

Wetzel, R. G., 2001. Limnology. Lake and River Ecosystems, 3rd ed. Academic Press, San Diego, 1006 pp.

Hydrobiologia (2008) 604:37–44
DOI 10.1007/s10750-008-9313-4

Reducing dissolved phosphorus loading to the Salton Sea with aluminum sulfate

I. R. Rodriguez · C. Amrhein · M. A. Anderson

Abstract The primary productivity of the Salton Sea, California is excessively high, leading to low-oxygen conditions, low clarity, and odors associated with algal decomposition. Treating the inflow water with aluminum sulfate (alum) to remove soluble phosphorus (P), the limiting nutrient, is being considered to improve water quality. The objective of this study was to evaluate the use of alum to remove dissolved phosphorus from New River water, and the potential for the Al-bound P to be released into the Salton Sea. The New River is dominated by agricultural wastewater and has a salinity somewhat higher than normally encountered for alum treatment (total dissolved solids = 2,300 mg l^{-1}), thus, evaluation of alum's effectiveness is needed. In addition, alum may be dosed directly into the New River and the floc allowed to flow into the Salton Sea if the precipitated P is stable in Salton Sea water. In this study, we evaluated the potential for floc-bound P to be desorbed in Salton Sea water, which has an unusually high salinity (46 g l^{-1}). Aluminum at a 5- mg l^{-1} dose was effective in removing over 90% of the soluble phosphorus from the New River water. However, when the alum floc was added to Salton Sea water, up to 100% of the Al-bound P was released into the Sea water due to desorption, dissolution, and recrystallization of the alum floc. These results indicate that treatment of agricultural drainage water to reduce P-loading can be effective if the alum floc is settled and not allowed to enter the saline Salton Sea. In addition to alum costs, estimated at US$13 million year^{-1}, settling basin construction and maintenance for floc removal would be required.

Keywords Phosphate · Eutrophication · External loading · Nutrients · Alum · Silicate

Guest editor: S. H. Hurlbert
The Salton Sea Centennial Symposium. Proceedings of a Symposium Celebrating a Century of Symbiosis Among Agriculture, Wildlife and People, 1905–2005, held in San Diego, California, USA, March 2005

I. R. Rodriguez · C. Amrhein (✉) · M. A. Anderson
Department of Environmental Sciences, University of California, Riverside, CA 92521, USA
e-mail: amrhein@ucr.edu

Introduction

The Salton Sea is the largest lake in California (95,000 ha), with a total volume of 9.3 km^3. The inflow to the Sea is predominantly agricultural and municipal wastewater, with the Alamo and New Rivers comprising ~80% of the total inflow, and the balance coming from the Whitewater River at the north end, and miscellaneous drains (Holdren & Montaño, 2002). The inflows are high in plant nutrients that can produce eutrophic conditions, resulting in high chlorophyll a concentrations and high turbidity. Decomposition of the algal material leads to low dissolved oxygen concentrations, fish kills, and H_2S due to sulfate reduction (Watts et al., 2001). Increasingly, there is environmental and political pressure to improve the water quality of the Sea by reducing nutrient loading.

It is generally recognized that primary production in the Salton Sea is limited by phosphorus (Holdren & Montaño, 2002). In 1999, the average dissolved ortho-P concentration of the Salton Sea was 0.02 mg l^{-1} and the total P was 0.07 mg l^{-1}, while the nitrogen to phosphorus ratios averaged 200:1 in the summer months and 25:1 in the winter months (Holdren & Montaño, 2002). Although N:P ratios are not always indicative of nutrient limitations for algal growth, especially at low concentrations (Tank and Dodds, 2003), the high concentrations of ammonium in the Salton Sea (>1 mg l^{-1}) suggests P is the limiting nutrient. In addition, we conducted laboratory enrichment studies in which we added P and N to Salton Sea water in varying amounts and measured chlorophyll a with time. In winter-sampled water, the chlorophyll a concentrations increased up to 1.7-fold due to P fertilization, while there was no response to added nitrate or ammonium (authors' unpublished data). Thus, a reduction in phosphorus loading should reduce the algal blooms and the low dissolved oxygen conditions and odors that accompany algal decomposition.

Recent studies on the sediments of the Salton Sea have shown that internal loading of P from the sediments averaged 1 mg m^{-2} d^{-1} in the winter and increased to 5–7 mg m^{-2} d^{-1} during the spring, depending upon pileworm (*Neanthes succinea* Frey and Leuckart) activity, and dropped to 4 mg m^{-2} d^{-1} during the summer when the hypolimnion is anaerobic (Swan et al., 2007). Unpublished studies from our lab found internal P loading rates of 1 mg m^{-2} d^{-1} in the winter and 5–10 mg m^{-2} d^{-1} during the summer. Calcite precipitation within the sediments and the coprecipitation of P with the calcite has been shown to be an important mechanism controlling P release from the sediments (Rodriguez et al., 2008). Based on these studies and the external loading reported by Holdren & Montaño (2002) we estimate that the external P load comprises above 55% of the total phosphorus loading to the Sea. Thus, controlling external loading of P should have a significant impact on improving the water quality of the Salton Sea.

Dissolved phosphorus removal from water is often accomplished by treatment with alum (aluminum

sulfate). Alum has been used to treat municipal wastewater prior to discharge, and has been added directly to eutrophic lakes to remove phosphate on a long-term basis (Welch & Cooke, 1999). It has been used for three decades to capture phosphorus migrating from lake sediments regardless of redox status (Rydin et al., 2000). Pilgrim & Brezonik (2005a) recently reported that the treatment of lake inflows with 6 mg Al l^{-1} was effective in reducing algal growth if internal loading was also controlled.

It is generally recognized that two mechanisms are responsible for phosphorus removal by alum (Jenkins & Hermanowicz, 1991; Cooke et al., 1993). First, alum dissolves quickly and reacts with phosphate to form aluminum phosphate precipitates. The precipitation can be described as:

$$Al^{3+} + H_2PO_4^- + 2OH^- \rightarrow Al(H_2PO_4)(OH)_2(s)$$

$$Al^{3+} + HPO_4^{2-} + OH^- \rightarrow Al(HPO_4)(OH)(s)$$

Second, alum reacts with water leading to the formation of an amorphous aluminum hydroxide solid, $Al(OH)_3$ or "floc," as summarized by the following reaction:

$$Al^{3+} + 3H_2O \rightarrow Al(OH)_3(s) + 3H^+$$

The solid amorphous hydrous aluminum oxide is able to incorporate soluble phosphorus into its structure via coprecipitation or adsorption onto the surface (Omoike & vanLoon, 1999; Berkowitz et al., 2006). Aluminum in excess of phosphate will react to form the aluminum hydroxide floc (Jenkins & Hermanowicz, 1991), which transforms over time from an amorphous, high surface area phase to a more ordered, lower surface area gibbsite phase (Berkowitz et al., 2005). $Al(OH)_3(s)$ is useful for phosphorus removal because of its low toxicity to lake biota under normal treatment conditions (pH > 6.0), and its high adsorption capacity for P, with essentially irreversible binding except at high pH (Birchall et al., 1989; Cooke et al., 1993).

Alum is frequently added directly to lakes where internal loading of P from the sediments is significant and external loading of P can be controlled (Kennedy & Cooke, 1982; James et al., 1991; Boers et al., 1993). When alum is added directly to a lake, the floc settles to the bottom forming a reactive blanket over the sediments. The effectiveness and longevity of alum treatment depends on the continuing adsorption

capacity and the distribution of the floc (Rydin & Welch, 1998). Treatment effectiveness usually persists for 5 years or more in lakes or lake basins in which macrophytes do not interfere with the floc on the bottom of the lake (Welch & Schrieve, 1994), while others have seen treatment effectiveness averaging about 10 years (Welch & Cooke, 1999).

There has been a report of benthic invertebrate mortality due to the heavy accumulation of alum floc during treatment at 8 mg Al L^{-1} to the inflow water of a freshwater lake in Minnesota (Pilgrim & Brezonik, 2005b). In this case, the aquatic insects (mostly caddisfly family), amphipods, snails, ostracods, and leaches were smothered in the floc-settling basin. There were no measurable effects on invertebrates at a treatment concentration of 1 mg Al l^{-1}. Depending on the treatment dose and the distribution of the alum floc as it enters the Salton Sea, there may be some impact on *Neanthes succinea*, the dominant benthic organism.

Since the internal loading of P in the Salton Sea is controlled to some extent by calcite precipitation and phosphate coprecipitation reactions (Rodriguez et al., 2008), direct alum treatment of the Salton Sea is not necessarily needed, but treatment of the inflow to remove externally loaded P could significantly reduce eutrophication.

Addition of alum directly to the rivers flowing into the Salton Sea is being considered to remove dissolved P. The resultant floc could be allowed to enter and settle within the Sea. This system eliminates the need for settling basins and periodic dredging to remove accumulated floc. However, there have been no studies on the chemistry of alum or aluminum hydroxide floc in saline water, and desorption of P from the floc may occur. This phosphorus could then ultimately become bioavailable, which would cancel the initial beneficial effect of the alum treatments in removing the phosphorus from the river water.

The objective of this study was to evaluate the use of aluminum sulfate to remove soluble phosphorus from the New River water, and the potential for the P adsorbed to the floc to be released once it enters the Salton Sea. The river and Sea waters were filtered prior to use to eliminate suspended solids (algae, mineral matter, diatoms, and organic detritus) that would have confounded the study. It is recognized that alum treatment of the unfiltered river water will

be different because the clays and silt material will adsorb some of the Al^{3+}, and the alum-floc/suspended solids agglomerate will have different settling properties compared to alum floc alone. However, this study was designed to isolate the reactions of dissolved phosphorus with alum and to evaluate the stability of the alum floc in Salton Sea water.

The New River was chosen because it is the tributary with the highest soluble phosphorus concentrations, averaging ∼1 mg l^{-1}, and contributing nearly half of the total external P load (Holdren & Montaño, 2002). Although alum has been frequently used to remove soluble P, the rivers flowing into the Sea are much saltier than waters where alum has been used previously. Thus, the general rule of thumb of adding an equal weight of alum for phosphate to be removed may not apply (Jenkins & Hermanowicz, 1991). The first part of the study determined the relationship between alum dose and P removal. The second part of the study determined if the floc-adsorbed P would desorb in the Salton Sea water. These studies together established the quantity of aluminum sulfate required to remove soluble reactive phosphorus from the New River, any changes the floc may undergo in a saline system, and the cost-effectiveness of treating the tributaries (New, Alamo, and Whitewater Rivers) with aluminum sulfate.

Material and methods

A 50-l sample of New River water was collected at the river gauging station where Vail Road meets the New River (33° 06′ 12.2″ N, 115° 39′ 32.7″ W) and a 30-l sample of Salton Sea surface water was collected at the center of the north basin (33° 25′ N, 115° 57′ W) on September 12, 2001. The river gauging station on the New River is approximately 4 km from the Salton Sea. The river water and Sea water were pre-filtered through Whatman #5 filter paper and then through a 0.45-μm pore-size membrane filter (Fisherbrand MSE) using vacuum filtration within 24 h of sampling. The filtered waters were placed in acid-washed carboys and refrigerated at 9°C, but allowed to come up to room temperature before use. A 1,000-mg/l Al^{3+} stock solution was prepared using Al$_2$(SO$_4$)$_3$·18H$_2$O (Mallinckrodt analytical reagent grade alum).

Experiment 1: phosphorus removal from New River water

The effect of alum dose on phosphorus removal was evaluated using batch experiments conducted in 1 L polyethylene bottles that had been acid washed with 10% HCl and rinsed well with deionized water. Filtered New River water (1.0 l) and a stir bar were added to each of 45 bottles and maintained at room temperature ($\sim 22°C$). The river water was spiked with varying amounts of 1,000 mg l^{-1} Al stock solution to yield final concentrations of 0, 0.1, 0.3, 1.0, 3.0, 5.0, 7.5, 10.0, and 30 mg l^{-1} Al. These concentrations had been determined from preliminary experiments. All reactions were replicated five-fold. The solutions were continuously stirred and allowed to react for 2 h, which was the calculated transport time of the river water to the Sea (approximately 4 km) and a point on the river where alum addition has been considered. After 2 h, the solutions were vacuum filtered through a 0.45-µm pore-sized membrane filter and the filtrates immediately analyzed for pH, acidified to pH < 2 using trace metal grade HNO$_3$, and stored at 4°C prior to analysis for Ca, Mg, Na, K, Si, and PO$_4$. The major cations (Ca, Mg, Na, and K) and dissolved Si were analyzed using inductively coupled plasma optical emission spectroscopy. The samples were tested for ortho-phosphate using a modified ascorbic acid-molybdate blue colorimetric method (Murphy & Riley, 1962) on an Astoria-Pacific Int. Alpkem RFA 300 Autoanalyzer (Clackamas, OR). All samples were diluted as necessary and analyzed using appropriate spikes, blanks, duplicates, and standards.

The floc was collected on membrane filters and washed with a small volume of Nanopure deionized water to remove excess electrolytes. The floc samples were kept moist prior to their use in Experiment 2 (described below) because drying the floc may have affected the crystallinity of the precipitate. Twenty percent of the floc was saved for mineralogical analysis described below.

Experiment 2: desorption of floc-bound P in Salton Sea water

The filtered solids from four of the five replicates from the alum-treated waters were transferred to acid washed (10% HCl) 1-l polyethylene bottles containing 0.5 l of filtered Salton Sea water and a stir bar. A control, containing filtered Salton Sea water and a stir bar, was run in triplicate. The solutions were stirred at 60 rpm and sampled periodically throughout a two-week period. The samples were analyzed for pH and a 10 ml sample of the Sea water was vacuum filtered through a 0.45-µm pore-sized membrane filter and immediately diluted 1:4. The filtrates were acidified to pH < 2 using trace metal grade HNO$_3$ and stored at 4°C prior to analysis for Ca, Mg, Na, K, Si, and PO$_4$ as discussed above. The membrane filters from the sampling, which may have had precipitated material adhering, were returned to the corresponding bottles.

After the 2-week period, the suspensions were vacuum filtered through a 0.45-µm pore-sized membrane filter and the aged floc washed with a small volume of Nanopure deionized water to remove excess electrolytes. Due to of some unusual changes in the chemical composition of the Salton Sea water containing alum floc, we investigated the mineralogy of the aged floc. The mineralogy of floc samples from Experiments 1 and 2 was determined by X-ray diffraction (XRD) on a Siemens D-500 diffractometer using Cu-Kα radiation with a graphite crystal monochromator. X-ray diffraction peaks from the samples were matched with mineral reference standards (JCPDS, 1993) for identification.

Results

Experiment 1: phosphorus removal from New River water

The pH of New River water was measured immediately after the addition of the aluminum sulfate. The 0, 0.1, 0.3, 1.0, 3.0, 5.0, 7.5, 10, and 30-mg l^{-1} Al solutions yielded pH values of 8.0, 8.0, 8.0, 7.8, 7.5, 7.2, 7.0, 6.9, and 6.1, respectively. Aluminum hydrolysis consumes alkalinity and thus, pH decreased with increasing additions of alum.

The changes in soluble phosphorus concentration in the New River with aluminum sulfate additions are shown in Fig. 1. The values for percent soluble P removed with increasing alum additions were 2%, 7%, 22%, 68%, 93%, 98%, 98%, and 99.6%. The graph of P concentration vs. Al added has two distinct

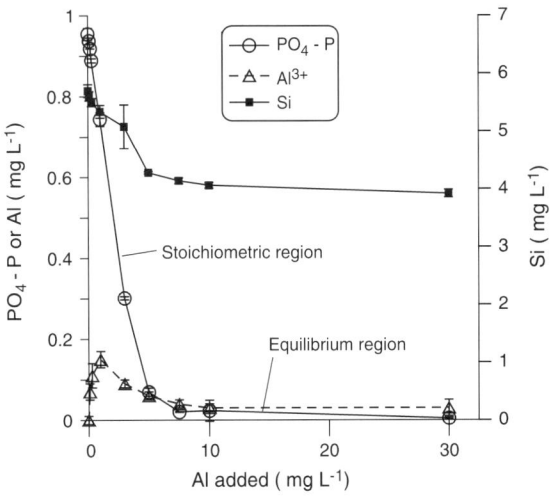

Fig. 1 The effect of alum addition on the soluble reactive P, Al, and Si concentrations in New River water after 2 h post-addition. Error bars are ± one standard deviation ($n = 5$); when not visible, they are smaller than the symbol

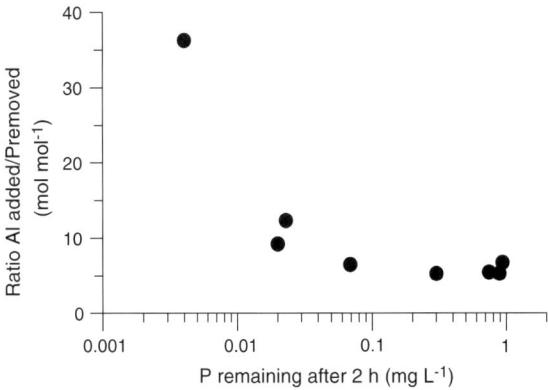

Fig. 2 Molar ratio of Al-added to P-removed versus residual soluble reactive P

regions: the stoichiometric region (high P concentration), where removal is proportional to the addition of the metal salt, and the equilibrium region (low P concentration), where much higher increments of alum dosage are required to remove a given amount of the P (Jenkins & Hermanowicz, 1991). In the stoichiometric region an Al-phosphate precipitate forms, while in the equilibrium region both a mixed solid of Al-phosphate and Al-hydroxide forms. The plot of $Al_{added}/P_{removed}$ versus residual soluble reactive P concentration (Fig. 2) looks like those of previous studies (Jenkins & Hermanowicz, 1991), and indicates that the chemical model of metal phosphate precipitation is generally valid. Achieving P concentrations below 0.02 mg l^{-1} requires Al/P molar ratios above 6.

The concentrations of major cations (Na, Ca, Mg, K) remained constant with any addition of alum (data not shown). However, Si concentration decreased with increasing additions of alum (Fig. 1) suggesting that a hydroxy-aluminosilicate phase may have formed in the stoichiometric region of Al-phosphate precipitation (Exley et al., 2002). Berkowitz et al. (2005) also found reductions in dissolved Si concentrations from surface waters treated with alum; moreover, XRD analysis of the aged floc gave diffraction peaks consistent with aluminosilicate solid phases, although geochemical calculations indicated significant undersaturation of the solutions.

Soluble aluminum in the New River water increased and peaked at the 1.0-mg l^{-1} Al^{3+} addition and decreased at a rate proportional to the Si loss data (Fig. 1). This drop in Al^{3+} at dosages above 1 mg l^{-1} coincides with reductions in pH and decreased solubility of $Al(OH)_3(s)$.

Experiment 2: desorption of floc-bound P in Salton Sea water

The alum floc formed in the New River water was collected and mixed with Salton Sea water to determine the extent of P desorption. The phosphate concentration in the Sea water increased over time, up to 14 days (Fig. 3a). The highest concentration of P occurred in the 3-mg l^{-1} Al dose, with the 1, 5, and 7.5-mg l^{-1} doses also releasing significant amounts of P to the Sea water (Fig. 3a). The percent P released from the floc decreased with increasing Al dose, although the maximum solution concentration occurred in the 3-mg l^{-1}-dose, where 46% of the adsorbed/precipitated P was solubilized (Fig. 3a and Table 1).

The release of P from the floc appears to be related to both the dissolution of the Al floc and to desorption. This is indicated by the general increase in Al released to solution over time, with the exception of the 30-mg l^{-1} dose (Fig. 3b). The highest soluble Al concentrations were observed in the 7.5 and 10-mg l^{-1} doses, reaching a stable Al concentration of 0.5 mg l^{-1} after 14 days. The 30 mg l^{-1}-dose initially produced a higher Al concentration, but declined from 3 days to 14 days,

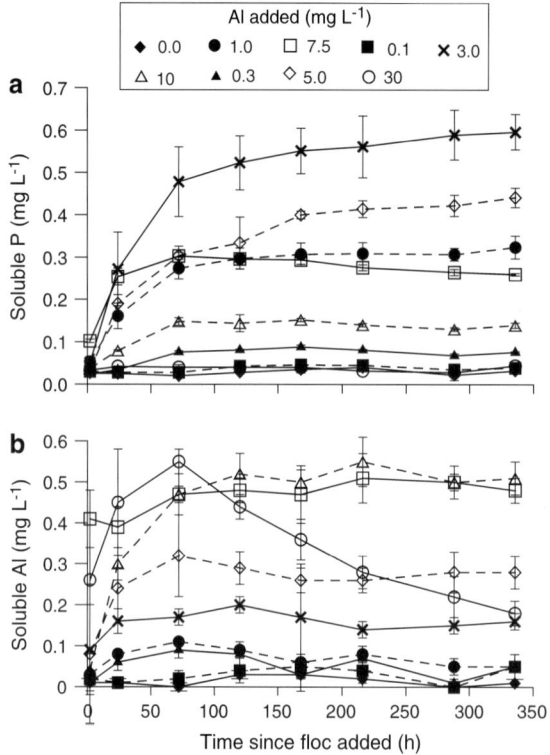

Fig. 3 Changes over time following the addition of alum-floc to Salton Sea water in (**a**) soluble reactive P, and (**b**) dissolved Al. Error bars are ± one standard deviation ($n = 4$ for all samples, except the control where $n = 3$); when not visible, they are smaller than the symbol

(Fig. 4a), suggests a transformation of the Al floc to an aluminosilicate phase in the Sea water (Table 1). This will be discussed further below.

Chemical mechanisms of P desorption

The phosphorus desorption can result from competition with F$^-$, OH$^-$, and soluble organic matter in the Sea water. The concentration of F$^-$ averages 2.1 mg l^{-1} in Salton Sea water (Holdren & Montaño, 2002), and it is well-known that fluoride is an effective competitor for specific adsorption sites on Al-hydroxide (Stumm, 1992).

The Salton Sea water had a pH of ~ 8.6 prior to the floc addition. The pH gradually decreased over time following the addition of the floc, with the largest decline in the Sea water receiving the most aluminum floc, 30 mg l^{-1} (Fig. 4b). This was surprising, although a ligand exchange reaction involving OH$^-$ adsorption and a subsequent release of adsorbed-PO$_4$ could cause a pH drop.

It is unclear how dissolution of the floc and reprecipitation of a new aluminosilicate phase could cause a drop in pH. Three of the major cations, Ca, Mg, and Na, remained unchanged over time (data not shown), however, potassium decreased about 26% from the time the floc was added until the end of the experiment (Fig. 4c). Potassium was not lost from the river water during floc formation, but when the floc was added to the Sea water, the alum floc appears to have reacted with Si and K to form a new phase, possibly a K-Al-Si-OH-SO$_4$ solid. Worley & Das (2000) observed an 8% removal of potassium from the liquid fraction when alum was added to swine manure.

apparently due to precipitation (discussed below). Based on Al concentrations in solution, <4% of the floc dissolved at the higher alum loadings (Table 1). This suggests that desorption, and not floc dissolution, is the dominant source of P to solution. However, the loss of Si from solution at all floc additions

Table 1 The solubilization of phosphate and aluminum, and the loss of Si from solution when Al-floc was added to Salton Sea water. Time of Al-floc contact with Sea water was 336 h

Initial Al^{3+} dose (mg l^{-1})	Mass PO$_4$ on the solid (mg)	% P dissolved into Sea water	Mass Al^{3+} on the solid (mg)	% Al dissolved into Sea water	Mass Si on the solid (mg)	% Si removed from the Sea water
0.1	0.02	100	0.03	83	0.09	69
0.3	0.07	59	0.19	13	0.20	71
1.0	0.21	77	0.85	2.9	0.36	77
3.0	0.65	46	2.9	2.7	0.62	87
5.0	0.89	25	4.9	2.8	1.4	88
7.5	0.93	14	7.5	3.2	1.6	94
10.0	0.93	7.5	10	2.6	1.6	94
30.0	0.95	2.3	30	0.3	1.8	94

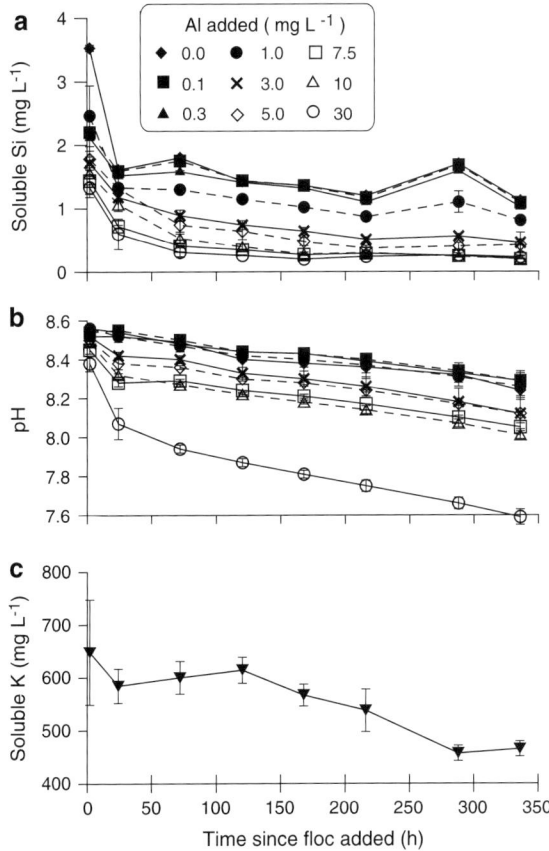

Fig. 4 Changes over time following the addition of New River water Al-floc to Salton Sea water in (**a**) dissolved Si, (**b**) pH, and (**c**) dissolved K^+. The K^+ data are averaged for all alum treatments. Error bars are $\pm$ one standard deviation ($n = 4$ for Si and pH, except the control where $n = 3$; $n = 35$ for K); when not visible, the error bars are smaller than the symbol

The mineralogy of the floc after reaction in Salton Sea water was determined using X-ray diffraction (XRD) analysis. Following the reaction for two weeks in Sea water, a sample of the aged floc (30-mg Al l^{-1} dose) was examined by XRD. The location and height of the XRD peaks suggested that a potassium aluminosilicate phase may be forming. Similar XRD data were reported by Berkowitz et al. (2005) on alum floc formed in lake water containing 6.2 mg l^{-1} dissolved Si. In this study, the Si concentration decreased nearly 75% one hour after alum addition, and after 20 days had decreased 93% (Berkowitz et al., 2005).

In our experiment, over 90% removal of Si from the Sea water was observed in the 7.5, 10, and 30 mg l^{-1} Al^{3+} doses as shown in Table 1. The Si loss data and XRD data support the theory that an aluminosilicate solid is forming both in the New River water and continues in the Salton Sea water, possibly with K^+.

Discussion

Based on these experiments, the optimum treatment to remove more than 90% of the soluble reactive P from the New River water was 5 mg Al l^{-1}. This agrees with the findings of Pilgrim & Brezonik (2005a) who reported ~ 6 mg Al l^{-1} were required to achieve significant phosphorus removal from the inflow to Fish Lake, MN. At the current price of alum ($270/ton) and the average flow in the New River (600 Mm^3 $year^{-1}$), the material cost of alum treatment would be $13 million USD $year^{-1}$.

The addition of freshly formed alum floc to the Salton Sea water resulted in a gradual transformation of the amorphous $Al(OH)_3$ into an aluminosilicate phase. Aluminosilicate formation in surface water has not been widely investigated, but studies have shown silicic acid to have a strong and unique affinity for aluminum (Chappell & Birchall, 1988; Birchall et al., 1989). Water containing dissolved silicon, even at low pH levels, has been shown to eliminate aluminum toxicity to fish fry (Birchall et al., 1989), and increasing dissolved silicon may be as important in preventing toxicity as maintaining a neutral pH (Cooke et al., 1993). Holdren & Montaño (2002) reported the Si concentration of the Salton Sea to be 4.6 mg l^{-1}, which is relatively high for lakes with diatom populations (Berkowitz et al., 2005), suggesting the inputs of dissolved Si and the dissolution of the suspended solids keep the Si elevated. Additional work is needed to characterize the role of dissolved Si and K in the aluminosilicate reaction and the long-term changes that occur to the alum floc as a result of aging, recrystallization, and exposure to high salinity.

The use of alum to reduce the external loading of P to the Sea would result in a significant reduction in algal blooms and would also help alleviate low-oxygen conditions, fish kills, and undesirable smells. However, if the floc from this reaction is allowed to enter the Salton Sea, much of the adsorbed/precipitated P will be solubilized, thus reducing the overall effectiveness of the treatment. On-river treatment to remove P will only be effective if the system is

engineered to trap and remove the floc before it enters the Salton Sea. This would require settling basins and associated maintenance and construction costs.

References

Berkowitz, J., M. A. Anderson & R. C. Graham, 2005. Laboratory investigation of aluminum solubility and solid-phase properties following alum treatment of lake waters. Water Research 39: 3918–3928.

Berkowitz, J., M. A. Anderson & C. Amrhein, 2006. Influence of aging of phosphorus sorption to alum floc in lake water. Water Research 40: 911–916.

Birchall, J. D., C. Exley, J. S. Chappell & M. J. Phillips, 1989. Acute toxicity of aluminum to fish eliminated in silicon-rich acid waters. Nature 338: 146–148.

Boers, P. C. M., T. E. Cappenberg & W. van Raaphorst, 1993. The third international workshop on phosphorus in sediment. Summary and synthesis. Hydrobiologia 253: xi–xviii.

Chappell, J. S. & J. D. Birchall, 1988. Aspects of the interaction of silicic acid with aluminum in dilute solution and its biological significance. Inorganica Chimca Acta – Bioinorganic Chemistry 153: 1–4.

Cooke, G. D., E. B. Welch, S. A. Peterson & P. R. Newroth, 1993. Restoration and Management of Lakes and Reservoirs. Lewis Publishers, New York.

Exley, C., C. Schneider & F. J. Doucet, 2002. The reaction of aluminum with silicic acid in acidic solution: an important mechanism controlling the biological availability of aluminum? Coordination Chemistry Reviews 228: 127–135.

Holdren, G. C. & A. Montaño, 2002. Chemical and physical characteristics of the Salton Sea, California. Hydrobiologia 473: 1–21.

James, W. F., J. W. Barko & W. D. Taylor, 1991. Effects of alum treatment on phosphorus dynamics in a north-temperate reservoir. Hydrobiologia 215: 231–241.

JCPDS (Joint Committee on Powder Diffraction Standards), 1993. Mineral Powder Diffraction file Databook. JCPDS, Swarthmore, PA.

Jenkins, D. & S. W. Hermanowicz, 1991. Principles of chemical phosphate removal. In Sedlak, R. (ed.), Phosphorus and Nitrogen Removal from Municipal Wastewater: Principles and Practice, 2nd ed. Lewis Publishers, New York, Chapter 4: 91–110.

Kennedy, R. H. & G. D. Cooke, 1982. Control of lake phosphorus with aluminum sulfate: dose determination and application techniques. Water Resources Bulletin 18: 389–395.

Murphy, J. & J. P. Riley, 1962. A modified method for the determination of phosphate in natural waters. Analytica Chimica Acta 27: 31–36.

Omoike, A. I. & G. W. vanLoon, 1999. Removal of phosphorus and organic matter removal by alum during wastewater treatment. Water Research 33: 3617–3627.

Pilgrim, K. M. & P. L. Brezonik, 2005a. Treatment of lake inflows with alum for phosphorus removal. Lake and Reservoir Management 21: 1–9.

Pilgrim, K. M. & P. L. Brezonik, 2005b. Evaluation of the potential adverse effects of lake inflow treatment with alum. Lake and Reservoir Management 21: 78–88.

Rodriguez, I. R., C. Amrhein & M. A. Anderson, 2008. Laboratory studies on the coprecipitation of phosphate with calcium carbonate in the Salton Sea, California. Hydrobiologia (this issue).

Rydin, E., B. Huser & E. B. Welch, 2000. Amount of phosphorus inactivated by alum treatments in Washington lakes. Limnology and Oceanography 45: 226–230.

Rydin, E. & E. B. Welch, 1998. Aluminum dose required to inactivate phosphorus in lake sediments. Water Research 32: 2969–2976.

Stumm, W., 1992. Chemistry of the Solid–Water Interface. Wiley, New York: 428 pp.

Swan, B. K., J. M. Watts, K. M. Reifel & S. H. Hurlbert, 2007. Role of the polychaete *Neanthes succinea* in phosphorus regeneration from sediments in the Salton Sea, California. Hydrobiologia 576: 111–125.

Tank, J. L. & W. K. Dodds, 2003. Nutrient limitation of epilithic and epixylic biofilms in ten North American streams. Freshwater Biology 48: 1031–1049.

Watts, J. M., B. K. Swan, M. A. Tiffany & S. H. Hurlbert, 2001. Thermal, mixing, and oxygen regimes of the Salton Sea, California, 1997–1999. Hydrobiologia 466: 159–176.

Welch, E. B. & G. D. Cooke, 1993. Effectiveness of Al, Ca, and Fe salts for control of internal phosphorus loading in shallow and deep lakes. Hydrobiologia 253: 323–335.

Welch, E. B. & G. D. Cooke, 1999. Effectiveness and longevity of phosphorus inactivation with alum. Lake and Reservoir Management 15: 5–27.

Welch, E. B. & G. D. Schrieve, 1994. Alum treatment effectiveness and longevity in shallow lakes. Hydrobiologia 275/276: 423–431.

Worley, J. W. & K. C. Das, 2000. Swine manure solids separation and composting using alum. Applied Engineering in Agriculture 16: 555–561.

 Springer

Hydrobiologia (2008) 604:45–55
DOI 10.1007/s10750-008-9310-7

Laboratory studies on the coprecipitation of phosphate with calcium carbonate in the Salton Sea, California

I. R. Rodriguez · C. Amrhein · M. A. Anderson

Abstract The Salton Sea is a hypereutrophic, saline lake in the desert of southern California. Like many lakes, the primary productivity of the Sea is limited by phosphorus. However, unlike most lakes, the release of P from the sediments is not controlled by the reductive dissolution of Fe(III)-oxide minerals. Most of the iron in the sediments of the Salton Sea is present as Fe(II)-sulfides and silicates. Rather, the sediments are dominated by calcite which is actively precipitating due to alkalinity production via sulfate reduction reactions. We hypothesized that calcite could be an important sink for phosphorus released from the decomposing organic matter. In this work we evaluated the potential for phosphate to coprecipitate with calcite formed in simulated Salton Sea sediment pore water. At calcite precipitation levels and P concentrations typical for the Salton Sea pore water, coprecipitation of P removed 82–100% of the dissolved phosphorus. The amount of P incorporated into the calcite was independent of temperature. The results of this work indicate that the internal loading of P within the Salton Sea is being controlled by calcite precipitation. Management of external P loading should have an immediate impact on reducing algae blooms in the Salton Sea.

Keywords Phosphorus · Calcite precipitation · Adsorption · Internal loading

Guest editor: S. H. Hurlbert
The Salton Sea Centennial Symposium. Proceedings of a Symposium Celebrating a Century of Symbiosis Among Agriculture, Wildlife, and People, 1905–2005, held in San Diego, California, USA, March 2005

I. R. Rodriguez · C. Amrhein (✉) · M. A. Anderson
Department of Environmental Sciences, University of California, Riverside, Riverside, CA 92521, USA
e-mail: amrhein@ucr.edu

Introduction

The Salton Sea is a large, eutrophic, closed-basin lake (70 m below sea level) located in Southern California, 50 km north of the U.S.–Mexico border. It initially formed between 1905 and 1907 when a diversion of the Colorado River failed, filling the low-lying Salton Sink. The Salton Sea is currently 56 km long and about 20-km wide with a maximum depth of 15 m. Its salinity is ~ 46 g l^{-1} (2005) and it receives a high nutrient and salt input from the three major rivers (New, Alamo, and Whitewater Rivers) whose flows are dominated by agricultural runoff and municipal wastewater derived from the Colorado River.

The Sea is eutrophic, characterized by high nutrient concentrations, high chlorophyll a concentrations, low clarity, low dissolved oxygen concentrations, periodic massive fish kills, and noxious odors (Watts et al., 2001; Holdren & Montaño, 2002). Primary productivity in the Sea is phosphorus-limited, and total nitrogen to total phosphorus ratios in the summer months average 200:1, and 25:1 in the winter months (Holdren

& Montaño, 2002). Thus, there is interest in controlling phosphate loading to reduce algal blooms, low dissolved oxygen conditions, and foul odors.

Removing phosphorus from waters flowing into the Sea will only be beneficial if internal loading of phosphorus from the sediments is low. In many lakes, the mineralization of detrital organic matter releases P back to the water column resulting in continuing cycles of algal blooms (Sondegaard et al., 1999). However, in the Salton Sea there is evidence that the mineralization of phosphorus from organic matter and release to overlying water is relatively low (Swan et al., 2007; and unpublished data of the authors). Measurements of soluble reactive phosphorus in the pore water of the sediments had a mean concentration of 0.8 mg l^{-1} PO_4–P with a range of 0.2–5.2 mg l^{-1} (Anderson et al., 2007). These concentrations are only one-quarter of those found in the pore water of other lakes in southern California (Oza, 2003) and one-tenth reported for Mediterranean lagoon sediments (Souchu et al., 1998). The P fluxes from the sediments of the Salton Sea are one-quarter of that reported for nutrient-rich coastal sediments (Selig et al., 2006; Swan et al., 2007). Recent measurements of the P budget for the Sea suggest that 15–20% of the total annual external P inputs are being sequestered or permanently buried within the Sea (unpublished data of the authors). The mechanism of sequestration of phosphorus in the sediments is unknown, however, and is the focus of this current work.

The geochemical model PHRQPITZ indicates that the Salton Sea is supersaturated with respect to calcite and gypsum (Holdren & Montaño, 2002). The calcite content of the sediments averaged 24% and gypsum has been found in sediment samples but the amount has not been quantified (Anderson et al., 2007). Based on mass balance calculations using the inflow data and long-term chemical composition of the Sea (Holdren & Montaño, 2002), we estimate that between 700,000 and 1,000,000 metric tons of calcite are forming annually in the Sea, most likely within the sediments (Wardlaw & Valentine, 2005). Most of this calcite precipitation is being driven by the alkalinity (HCO_3^-) generated during sulfate reduction in the sediments according to the following reaction:

$$SO_4^{2-} + 2CH_2O + H^+ \rightarrow 2HCO_3^- + H_2S(g) \qquad (1)$$

In this reaction, organic matter, CH_2O, is oxidized using sulfate as the terminal electron acceptor and

alkalinity and hydrogen sulfide gas are formed. The production of alkalinity in the sediment pore water greatly accelerates the calcite precipitation reaction. Analysis of pore water from the Salton Sea has shown that alkalinities can be six times higher than the overlying Sea water (unpublished data). In earlier work we estimated sulfate reduction rates based on pore water sulfate concentrations, H_2S gradients in the pore water, and organic matter decomposition potential. These estimates suggest that up to 6,000 metric tons of H_2S per month could be formed in the sediments during the summer months (deKoff et al., 2007). The intense sulfate reduction reaction at the Sea is driven by the high algal growth (estimated at 55,000 metric tons of organic carbon per year) and the high SO_4^{2-} concentration of the Sea. Thus, the coprecipitation of phosphate with calcite is a likely mechanism regulating phosphate release and immobilization within the sediments of the Salton Sea. This coprecipitation reaction could explain the relatively low P concentrations of the sediment pore water and low measured P-flux rates.

The coprecipitation of dissolved phosphate with calcite has been observed in freshwater lakes (Danen-Louwerse et al., 1995). The coprecipitation reaction is caused by the interaction between dissolved phosphate and the calcite surface (adsorption) during crystal growth, followed by incorporation of the surface phosphorus into the bulk structure as crystal growth proceeds (House & Donaldson, 1986). Coprecipitation can result in the removal of as much as 97% of the phosphorus in the epilimnion (House, 1990) while others have estimated that about 35% of total phosphorus removed from the epilimnion of Lake Constance (bordered by Austria, Germany, and Switzerland) could be adsorbed and coprecipitated with calcite (Kleiner, 1988, 1990). At high concentrations, inorganic phosphorus can inhibit the crystal growth of calcite and, depending on the degree of supersaturation, may stop growth completely (House, 1987).

The objective of this study was to evaluate the extent to which phosphate coprecipitation with calcite might be occurring in the sediments of the Salton Sea. The sediment is the zone of active organic matter mineralization and P release in the Salton Sea, and also the zone of maximum calcite precipitation. Thus, phosphate coprecipitation with calcite is a likely mechanism controlling internal loading of P

from the sediments. This is very different compared to most freshwater lakes where phosphate adsorption onto Fe(III)-oxides often controls P release from sediments (Cooke et al., 1993; Horne & Goldman, 1994). In the Salton Sea, sediment iron is largely present as iron sulfide minerals and iron-rich silicate minerals, and the intense reducing conditions have converted most of the reactive Fe(III)-oxides to Fe(II)-sulfides (deKoff et al., 2007). Thus, it is likely that iron minerals play a minor role in P adsorption in Salton Sea sediments, while calcite precipitation is the dominant process.

In this work, we attempt to distinguish between coprecipitation and surface adsorption of phosphate onto the calcite. It is well known that phosphate readily adsorbs onto calcite and the reaction has been described by application of Langmuir and BET equations (Griffin & Jurinak, 1973). Adsorption is considered part of the coprecipitation process as calcite overgrowth buries the P within the calcite crystal (House & Donaldson, 1986; Carreira et al., 2006).

In the following experiments, changes in phosphate concentration were monitored following calcite precipitation from simulated Salton Sea sediment pore water. Sediment pore water was simulated by adding carbonate and phosphate to Salton Sea water to achieve concentrations typical of pore water. The calcite precipitation reaction was studied over a temperature range reported for the Salton Sea. In a separate experiment, phosphate adsorption onto calcite was measured to compare the relative effects of coprecipitation and adsorption.

Material and methods

Eighty liters of Salton Sea surface water were collected on August 6, 2001 from the center of the north basin (N 33° 25′, W 115° 57′). The Sea water was prefiltered through Whatman #5 filters and then through 0.45 μm pore-size membrane filters within 24 h of collection to removal suspended solids and algae that might confound the study. The filtered water was placed in an acid washed (HCl) carboy and refrigerated at 9°C until needed. The filtered Sea water was allowed to equilibrate to room temperature before use.

Sediments were also collected at the same site with a Ponar dredge and placed in 500 ml glass jars

without headspace and kept on ice. Pore water was separated by centrifugation at 3000 RCF for 20 min and analyzed immediately for sulfide by mixing 3 ml of pore water with 3 ml of an antioxidant buffer (Orion SAOB II) and S^{2-} measured with a combination silver/sulfide electrode (Orion Research, Beverly, MA). The pore water was analyzed for major ions as described below.

Effects of temperature on P coprecipitation and adsorption

Simulated sediment pore water was produced by spiking Salton Sea water with phosphate and carbonate. In this first experiment, we have evaluated the effect of temperature and the amount of calcite precipitated on the coprecipitation and/or adsorption of P with the calcite.

The reactions were conducted in 1 l polyethylene bottles that had been acid washed with 10% HCl, and rinsed well with deionized water. One liter of filtered Salton Sea water, a stir bar, and 1 ml of 1000 mg l^{-1} stock P solution (as K_2HPO_4) was added to each bottle (27 total bottles) to give a final PO_4–P concentration of 1.0 mg l^{-1}.

Based on earlier measurements (Anderson et al., 2007), the average pore water phosphate concentration from the Sea was 0.8 mg l^{-1} ($n = 90$), so the 1.0 mg l^{-1} addition used in this experiment was considered typical. All reactions were run in triplicate.

The solutions were maintained at three temperatures (12, 22, and 32°C), which represent the low, mean, and high temperatures measured in the Salton Sea (Holdren & Montaño, 2002). The solutions were allowed to equilibrate in a temperature-controlled water bath overnight. The solutions were then spiked with varying amounts of 1 M Na_2CO_3 stock solution (A.C.S. certified, Fisher Scientific) to increase the alkalinity concentrations in the Sea water by 0, 4, or 7 mM. The added alkalinity was meant to simulate the alkalinity generated during sulfate reduction in the sediments (Eq. 1). These simulated pore water solutions were continuously stirred at 300 rpm and the pH measured regularly throughout 120 h of reaction. After sufficient time for calcite precipitation to have occurred (120 h), the pH was measured and the solutions were vacuum filtered through a 0.45 μm pore-size membrane filter. The filtrates were immediately analyzed for alkalinity

by titration under air to pH 4.4 with standardized 10.0 mM H_2SO_4. An additional volume of filtrate was immediately diluted 1:4 (10 ml filtrate, 30 ml 'Nanopure' water), acidified to pH less than 2 with trace metal grade HNO_3, and stored at 4°C prior to analysis for Ca, Mg, Na, K, Si, and PO_4. The major cations (Ca, Mg, Na, and K) and dissolved Si were analyzed using inductively coupled plasma atomic emission spectroscopy (ICP). Phosphate was measured using the molybdate blue + ascorbic acid colorimetric method on an Astoria-Pacific Int. Alpkem RFA 300 Autoanalyzer (Clackamas, OR). All samples were diluted as necessary and analyzed using QA/QC protocols for the lab that included appropriate spikes, blanks, duplicates, and standards. Spike recoveries and duplicates that agreed to better than 10% were considered acceptable.

The precipitate collected on the membrane filters was washed with a small volume of Nanopure water to remove excess electrolytes. The precipitate was then divided into four parts without drying. The samples were not allowed to dry because of possible changes to the phosphate that had been adsorbed or to the crystallinity of the solids.

A weighed amount of the damp solid was dissolved in a 100 ml volumetric flask with 25% trace metal grade HNO_3 and brought up to volume with Nanopure water. These samples were stored at 4°C prior to analysis for PO_4, Si, Ca, Na, K, and Mg.

Another sample of the damp precipitate (~ 0.1 g) was weighed into tared, metal tins and dried overnight at a temperature of 105°C to determine water content. These oven-dried samples were examined using a Philips XL30 scanning electron microscope (SEM) equipped with an energy dispersive X-ray spectrometer (EDX) for elemental analysis. The mineralogy of the samples was determined by X-ray diffraction on a Siemens D-500 diffractometer using Cu-Kα radiation with a graphite crystal monochromator. Surface area of the dried solids was determined by single-point N_2 adsorption with a Quantasorb Jr. surface area analyzer (Quantachrome, Boynton Beach, FL).

Effects of P concentration on P partitioning into calcite

A second experiment was conducted to determine the effects of varying P concentration on the partitioning

of P into calcite. The experiment was performed similar to that discussed above with a few modifications. The temperature was kept constant at 22°C and only one level of Na_2CO_3 was added to the Sea water (4 mM). Varying amounts of phosphate solution were added to the filtered Sea water to yield concentrations of 0.0, 0.1, 0.25, 0.50, 0.75, and 1.0 mg l^{-1} P. The experiment was run in triplicate and was analyzed as discussed above.

Determination of mechanisms for P removal from solution

A third experiment was done to determine if coprecipitation or adsorption on calcite was the dominant mechanism removing P from solution. As an initial step, 9 g of reagent grade $CaCO_3$ (Mallinckrodt) was added to 20 l of filtered Sea water to seed the growth of calcite and reduce the high level of oversaturation. The water was continuously stirred and pH monitored for 3 days. The water was then allowed to sit quiescent for 2 h to allow for settling and then filtered to remove the seed calcite and any precipitated calcite. The filtrate (1.0 l) was placed in acid washed polyethylene bottles and spiked with 1 mg l^{-1} P. To each bottle, 0.500 g of reagent grade $CaCO_3$ was added, allowed to react for 5 days, and periodically analyzed for pH. The solutions and solid phase were analyzed as in the first two experiments.

Results

The chemical composition of the Salton Sea water used in these experiments was similar to those reported by Holdren & Montaño (2002). The Ca^{2+} and HCO_3^- concentrations were 24 and 4 mM, respectively, and the SO_4^{2-} concentration was 110 mM, which is four times higher than ocean water.

The sediment pore water composition collected from the Salton Sea north basin was found to be $Ca^{2+} = 16$ mM, $HCO_3^- = 25$ mm, and $SO_4^{2-} = 64$ mm. The lower Ca^{2+} and SO_4^{2-}, and elevated HCO_3^- concentrations indicate that active sulfate reduction, coupled with calcite precipitation, was occurring in the sediments. This is further verified by the hydrogen sulfide concentration which was 4 mM in the sediment pore water collected at the site.

Effects of temperature on P coprecipitation and adsorption

The coprecipitation of phosphate with calcite was studied at three different temperatures that spanned the temperature range found in the Salton Sea. Figure 1 shows the change in pH of the Salton Sea water spiked with 1 mg l^{-1} P and varying amounts of carbonate. The filtered Sea water had an initial pH of 8.73 ± 0.01 (standard deviation of three replicates) and when spiked with 1 mg l^{-1} P, the pH decreased to 8.65. Adding Na$_2$CO$_3$ increased the initial pH of the water (Eq. 2), but due to calcite precipitation the pH decreased over time as bicarbonate was consumed (Eq. 3):

$$Na_2CO_3 + H_2O \rightarrow 2Na^+ + HCO_3^- + OH^- \qquad (2)$$

$$Ca^{2+} + HCO_3^- \rightarrow CaCO_3 \, (s) + H^+ \qquad (3)$$

By the end of the experiment, pH values were lowest in the 7 mM Na$_2$CO$_3$ solutions, and increased with decreasing additions of Na$_2$CO$_3$. Even though the carbonate-spiked solutions had higher initial alkalinities, after 120 h, the pH (Fig. 1) and alkalinities (Fig. 2a) were lowest in the most heavily spiked solutions. These data indicate that precipitation occurred to a greater extent in the carbonate-spiked solutions due to the removal of precipitation-inhibiting ions, like phosphate and dissolved organic carbon (House, 1987).

Temperature affected the amount of calcite formed as indicated by changes in pH, alkalinity, and calcium (Figs. 1, 2a–c). The reactions at 12°C typically had the highest pH, alkalinity, and dissolved Ca concentrations, while the reactions at 32°C had the lowest, indicating a larger amount of precipitate had formed at the higher temperature. A loss of calcium from solution was seen in all treatments including the water with no added alkalinity (Fig. 2b), although the amount of precipitate collected on the filters from the solutions without added carbonate was insufficient for characterization. In the 4 mM carbonate-spiked water, 22, 28, and 29% of initial calcium was removed at 12, 22, and 32°C, respectively, and 35, 36, and 43% of initial calcium was removed at the 7 mM addition. The mass of calcium carbonate formed as a function of Na$_2$CO$_3$ spike and temperature is shown in Fig. 2c. These data are supported by the work of House (1999) who also showed that

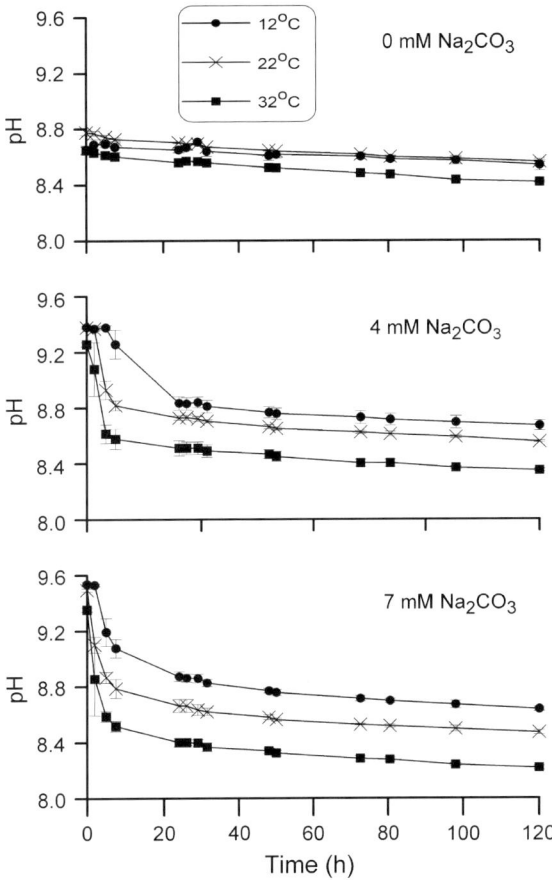

Fig. 1 The effects of temperature and alkalinity additions on the pH of Salton Sea water. Error bars are ± one standard deviation ($n = 3$); when not visible, they are smaller than the symbol

higher temperatures increased the amount of calcite formed. Unlike most solids, the solubility of calcium carbonate decreases with increasing temperature (Stumm & Morgan, 1996, p. 152).

The effect of Na$_2$CO$_3$ addition and resulting calcite precipitation on phosphate concentration in the Salton Sea water is shown in Fig. 2d. Phosphate removal was correlated with calcite formation and there was little effect of temperature other than its effect on calcite precipitation. At the 4 mM carbonate addition, 82% of initial phosphate was removed, while at the 7 mM addition, 94% of the initial phosphate precipitated with the calcite.

Based on the analysis of the solid phase, the phosphorus in the calcite averaged 1.6 ± 0.1 mg P per g CaCO$_3$ at the 4 mM Na$_2$CO$_3$ addition, and 1.1 ± 0.0 mg P per g CaCO$_3$ at the 7 mM addition.

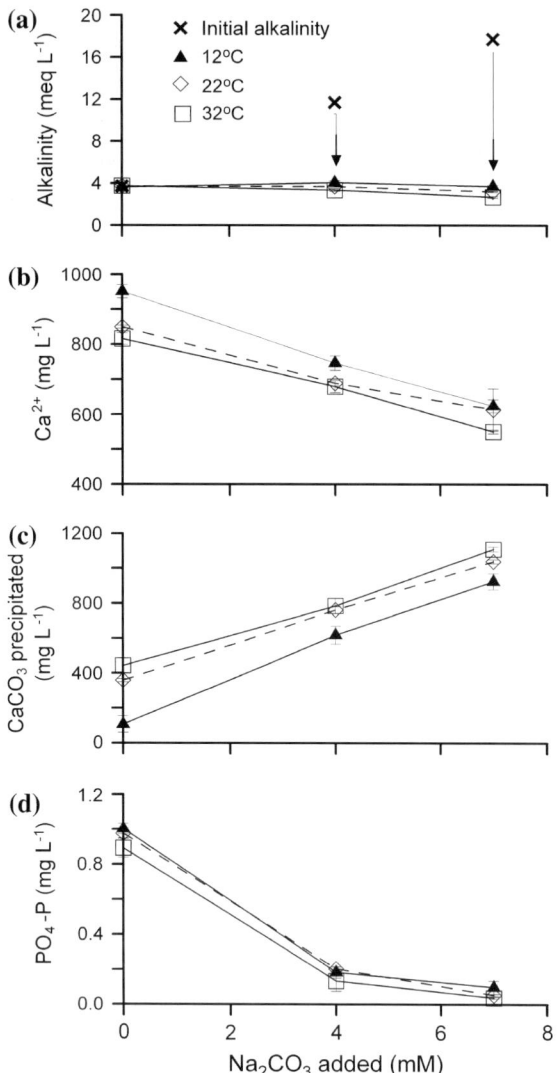

Fig. 2 Effects of temperature and alkalinity additions (as Na_2CO_3) on Salton Sea water composition after 120 h of reaction: (**a**) alkalinity, (**b**) Ca, (**c**) mass of calcium carbonate precipitated, and (**d**) soluble reactive phosphorus. All solutions were open to the atmosphere. Error bars are ± one standard deviation ($n = 3$); when not visible, they are smaller than the symbol

There was no effect of temperature on the amount of P in the solid, confirming the solution data shown in Fig. 2d. The phosphorus content was lower in the 7 mM Na_2CO_3 additions compared to the 4 mM-spiked solutions simply due to a dilution effect. That is, ∼0.3 g more calcite formed in the 7 mM-spiked solutions when compared with the 4 mM solution (∼0.7 g) (Fig. 2c), and the coprecipitated P became depleted as calcite precipitation progressed.

The P/Ca ratios based on changes in the solution composition were compared to the P/Ca ratios based on solids analysis (Fig. 3). There was a good agreement between these two independent measurements indicating conservation of mass in our analyses for all three experiments.

Adsorption studies conducted by Griffin & Jurinak (1973) demonstrated that the amount of phosphate adsorbed by calcite was found to increase with increasing temperature, and higher temperatures increased the adsorption reaction rate (Griffin & Jurinak, 1973). In our study the amount of P incorporated with the calcite was independent of temperature. This suggests that coprecipitation, not adsorption, may be the dominant mechanism of removal.

The solution concentrations of Si, Mg, Na, and K were also measured after 120 days reaction. Dissolved Si, Na, and K remained constant in all solutions but there was a decrease in Mg concentration with increasing additions of Na_2CO_3 and with increasing temperature (data not shown). It is well known that Mg can substitute for Ca in the calcite structure (Drever, 1997) and based on the change in Ca and Mg concentrations, the mole fraction of Mg in the precipitated calcite ranged from 0.02 to 0.13.

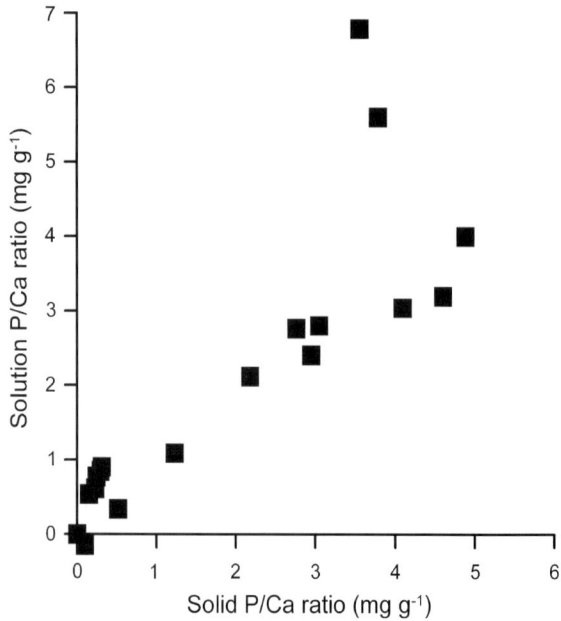

Fig. 3 Relationship between the P/Ca ratio of the solid as determined by two different methods: analysis of the solid (*x*-axis) and changes in the Sea water composition (*y*-axis)

Direct analysis of the solid phase gave a Mg/Ca mole fraction content from 0.0 to 0.12, confirming the solution analysis (Table 1).

Mucci & Morse (1983) studied the coprecipitation of Mg^{2+} into calcite formed in ocean water and derived an equation to estimate the mole fraction of Mg in the calcite based on the Mg/Ca molar ratio in solution. Using these relationships and the Mg/Ca molar ratio of the Salton Sea water (2.44), the calcite formed in the Sea should have a Mg/Ca mole ratio of 0.054. This would give an average formula for the calcite mineral as: $Ca_{0.949}Mg_{0.051}CO_3$. However, this Mg content would be the theoretical composition of the calcite that first formed in the Sea water. In our experiment, as precipitation progressed the Mg/Ca of the solution increased, thereby increasing the Mg/Ca ratio of the solid.

X-ray diffraction analysis can also be used to determine the extent of Mg substitution into calcite. Magnesium incorporation into calcite causes a collapse of the crystal lattice, which can be measured using X-ray diffraction. The relationship between the Mg mole fraction in the calcite and the lattice spacing is linear up to a mole fraction of about 0.20 (Doner & Lynn, 1989). X-ray diffraction analysis of actual Salton Sea sediment indicates that the natural Salton Sea calcite has a Mg/Ca molar ratio between 0.031 and 0.040. The difference between laboratory precipitated calcite and sediment calcite is likely due to differences in the rate of precipitation.

The specific surface area was measured on the solids formed at different temperatures (Table 1) and compared to a reagent-grade $CaCO_3$ reference material. The reference material measured $0.59 \ m^2 \ g^{-1}$ and has been previously measured as $0.6 \ m^2 \ g^{-1}$

(Amrhein et al., 1985). The surface area of the solids increased with increasing temperature. This suggests that the higher temperature produces a faster rate of precipitation and a smaller particle size. Scanning electron micrographs (SEM) of the precipitated calcite were taken to further characterize the solid. The SEM images of the precipitates from three different temperatures are shown in Fig. 4. The precipitates were of high surface area and the crystal morphology did not change with a change in temperature. The morphology of the reference material was the classic calcite rhombohedral shape (Fig. 4d).

Effects of P concentration on P partitioning into calcite

In the second experiment, we determined the effects of varying phosphate concentrations on the incorporation of P into calcite. Figure 5 shows that pH, alkalinity, and Ca concentration increased with increasing amounts of added-P, which indicates that phosphate appears to have inhibited calcite formation. The final P concentrations in the first three samples (0, 0.10, and 0.25 mg/l initial P) were at or below detection limit after calcite precipitation. At the higher initial P concentrations (0.5, 0.75, and 1.0 mg/l), 87–91% of added P was removed from solution, indicating highly effective removal of soluble P via calcite precipitation. The fact that calcium concentrations increased with increasing added-P indicates that Ca-phosphate solid phases (such as $Ca_4H(PO_4)_3 \cdot 2.5H_2O$, $Ca_3(PO_4)_2$, $CaHPO_4$, or $Ca_5(PO_4)_3OH$) were not forming, and the added-P

Table 1 Analysis of the solids precipitated at varying temperatures. Initial $P = 1.0 \ mg \ l^{-1}$

Temp. (°C)	Added alkalinity (mM)	Mg in the solid (mg g^{-1})	Mg/Ca mole fraction	Surface area (m² g^{-1})	P/Ca (mg g^{-1}) ± s.d.[b]	Adsorbed P (mg P m^{-2} calcite)
12	4	0	0.00	n.a.[a]	4.88 ± 0.84	0.48
	7	10.5	0.04	4.04	2.76 ± 0.11	0.27
22	4	4.8	0.02	n.a.	4.09 ± 0.17	0.27
	7	24.8	0.10	6.12	3.05 ± 0.20	0.20
32	4	22.2	0.09	n.a.	4.59 ± 0.16	0.24
	7	30.0	0.12	7.80	2.95 ± 0.12	0.15

[a] n.a. = not analyzed; assumed equal to the 7 mM samples

[b] s.d. = standard deviation

Fig. 4 SEM pictures
of precipitated calcite at
varying temperatures:
(**a**) 12°C; (**b**) 22°C;
(**c**) 32°C; (**d**) and reagent
grade calcite

most likely retarded the calcite precipitation (House, 1987).

Phosphorus in the solid increased with increasing P in solution (Table 2) and the P/Ca ratios for Experiments 1 and 2 ranged from 2.8 to 4.9 mg g^{-1} when the initial P was 1.0 mg l^{-1} (Tables 1 and 2). These levels are in very good agreement with those found by Anderson (2004) for a high pH, alkalinity-rich lake water treated with Ca^{2+} (initial dissolved P = 0.7 mg l^{-1} and P/Ca ratio in the solid = 3.4 mg g^{-1}).

Anderson et al. (2007) reported the inorganic phosphorus and calcite contents of 90 Salton Sea sediment samples collected on a grid pattern around the Sea. Based on these analyses, the average P/Ca ratio of the natural Salton Sea calcite is 4.8 ± 3.1 mg g^{-1} (standard deviation of 90 samples). This calcite formed in the presence of an average porewater P concentration of 0.8 mg/l (Anderson et al., 2007), which is comparable to our experiments. The calcite P/Ca ratios of our laboratory experiments agree quite well with the range and mean found in the Salton Sea. This further confirms that coprecipitation of P with calcite is an important mechanism controlling P immobilization in the Salton Sea sediments.

Phosphorus to calcium ratios in solutions precipitating calcite have been reported by other researchers. Kleiner (1988) reported a P/Ca incorporation ratio of

3.4 mg g^{-1}, with initial P concentrations of 0.04 and 0.06 mg l^{-1}. Hieltjes & Lijklema (1979) found a P/Ca ratio of 2–3 mg g^{-1} with an initial P of 0.13–0.17 mg l^{-1}. House & Donaldson (1986) reported ratios from 4 to 0.4 when the initial P concentration changed from 0.05 to 0.001 mg l^{-1}, while Dittrich et al. (1997) reported an incorporation ratio of 1–5 mg g^{-1}. The P/Ca values reported in this current work are within these ranges for phosphorus incorporation into calcite, although our initial P concentrations are somewhat higher (Tables 1 and 2).

Determination of mechanisms for P removal from solution

We conducted a third experiment to further differentiate between adsorption of P on calcite and the coprecipitation of P with the calcite. Rather than forming calcite directly in the Sea, calcite crystals were added to Sea water that had been pre-equilibrated to reduce the degree of calcite oversaturation. Very little of the added P was removed by the calcite with 32%, 28%, 13%, 10%, and 6% adsorbed in the 0.10, 0.25, 0.50, 0.75, and 1.0 mg l^{-1} P additions, respectively (Fig. 6a). These results are consistent with those reported by Anderson (2004), wherein

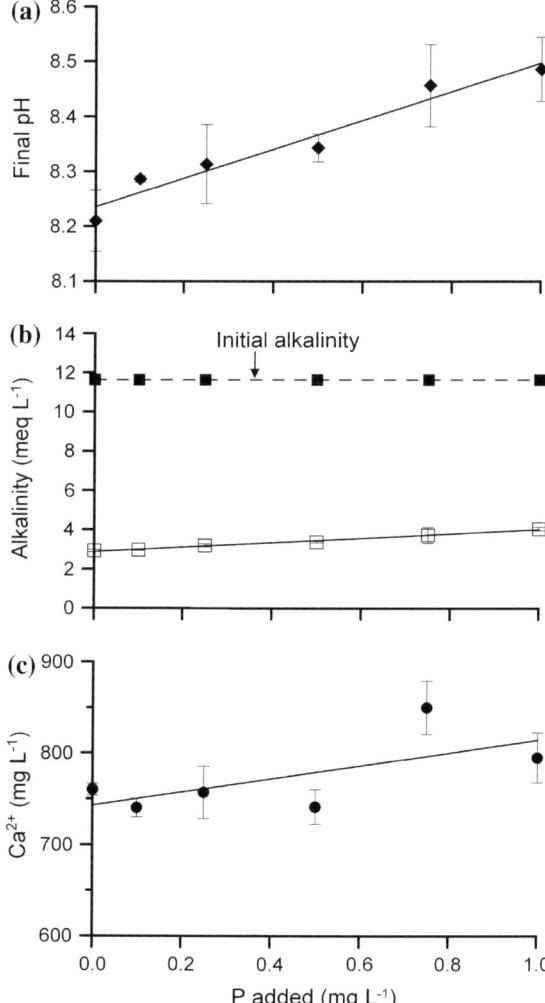

Fig. 5 Effects of phosphate additions to simulated Salton Sea sediment pore water on (**a**) pH, (**b**) alkalinity, and (**c**) Ca^{2+} in Experiment 2. All solutions were spiked with 4.0 mM Na_2CO_3 to initiate calcite precipitation. Error bars are ± one standard deviation ($n = 3$); when not visible, they are smaller than the symbol

substantially greater removal of phosphate was achieved when $CaCO_3$ was formed in place through addition of agricultural gypsum or $Ca(OH)_2$ to Lake Elsinore water relative to that adsorbed onto preformed $CaCO_3$.

We measured a small decrease in Ca concentration of the Sea water indicating an average of 60 mg l^{-1} $CaCO_3$ had precipitated during the 5-day equilibration period. The amount precipitated was ~20% of the amount formed in the non-Na_2CO_3 spiked solutions at 22°C (Fig. 2c). However, the surface area of the added calcite increased from 0.6 to 1.2 ± 0.3 m^2 g^{-1} due to this fresh precipitate. Comparing the P removal on a surface area basis (Fig. 6b) indicates that adsorption was a minor component of the phosphate loss during calcite formation in Salton Sea water.

Discussion

These laboratory experiments demonstrate the potential for phosphate-P removal through coprecipitation with $CaCO_3$ in Salton Sea water. Adsorption reactions appeared to be a minor component of the P loss from solution. There was an excellent agreement between the P/Ca ratios of the laboratory-precipitated calcite and the natural sediment calcite, which supports our hypothesis that phosphorus coprecipitation with calcite is occurring in the sediments of the Salton Sea. Sulfate concentrations in the Salton Sea are more than four times higher than ocean water and this leads to active sulfate reduction within the sediments. Sulfate reduction generates bicarbonate alkalinity which stimulates calcite precipitation. The sediments are also the area of maximum P release from organic matter mineralization. Organic matter

Table 2 The coprecipitation of P with calcium carbonate at varying P concentrations with 4 mM Na_2CO_3 spike, and the adsorption of P onto calcite. All reactions at 22°C

P added mg l^{-1}	Coprecipitation experiment		Adsorption experiment	
	P/Ca ratio (mg g^{-1}) ± s.d.[a]	mg P/m^2 calcite	P/Ca ratio (mg g^{-1}) ± s.d.[a]	mg P/m^2 calcite
0.00	0.10 ± 0.07	0.01	0.00 ± 0.00	0.00
0.10	0.52 ± 0.03	0.03	0.15 ± 0.02	0.06
0.25	1.23 ± 0.23	0.08	0.31 ± 0.01	0.12
0.50	2.18 ± 0.16	0.14	0.30 ± 0.02	0.12
0.75	3.55 ± 0.27	0.23	0.24 ± 0.03	0.10
1.00	3.78 ± 0.58	0.25	0.23 ± 0.01	0.09

[a] s.d. = standard deviation

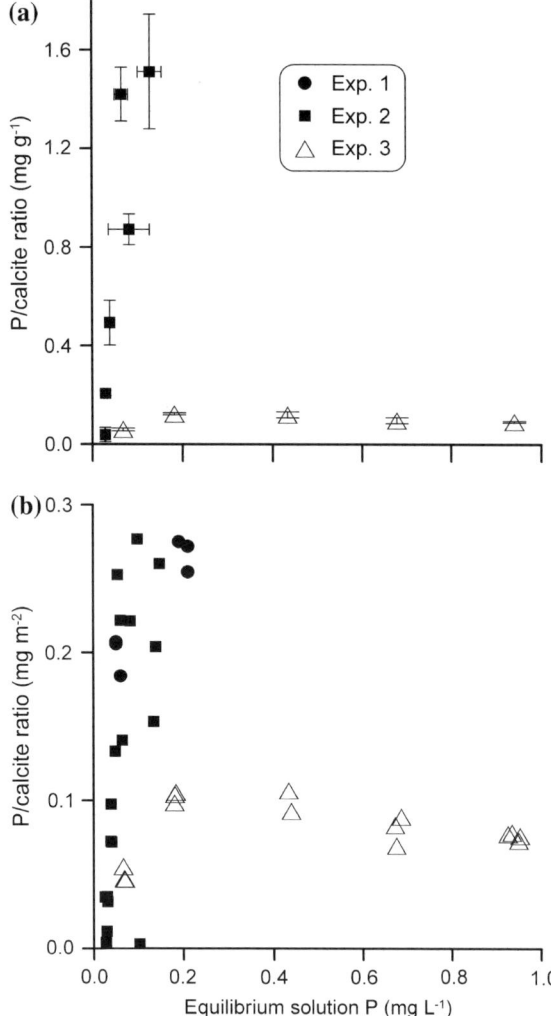

external P loading by reducing P in the New and Alamo Rivers should have an immediate impact on reducing algae blooms in the Salton Sea.

Fig. 6 Calcite associated P per unit mass basis (**a**) and per unit surface area basis (**b**), as a function of equilibrium P concentration. In (**a**) the error bars are ± one standard deviation ($n = 3$); when not visible, they are smaller than the symbol. Error bars are not plotted in (**b**). All data were collected at 22°C

decomposition and alkalinity production via sulfate reduction leads to the highest possible P incorporation into calcite. This results in permanent removal of P because calcite within the sediments is unlikely to dissolve without dramatic changes in the chemistry of the Sea (e.g., substantial pH reductions). Thus, the internal-loading of P is substantially reduced in the Salton Sea due to this calcite coprecipitation process. This is different from many lakes where binding by ferric iron phases is thought to regulate P cycling. Moreover, these findings suggest that management of

References

Amrhein, C., J. J. Jurinak & W. M. Moore, 1985. Kinetics of calcite dissolution as affected by carbon dioxide partial pressure. Soil Science Society of America Journal 49: 1393–1398.
Anderson, M. A., 2004. Impacts of metal salt addition on water chemistry of Lake Elsinore, California: 2. Calcium salts. Lake and Reservoir Management 20: 270–279.
Anderson, M. A., L. Whiteaker, E. Wakefield & C. Amrhein, 2007. Properties and distribution of sediment in the Salton Sea, California. Hydrobiologia (this issue).
Carreira, J. A., B. Vinegla & K. Lajtha, 2006. Secondary CaCO₃ and precipitation of P–Ca compounds control the retention of soil P in arid ecosystems. Journal of Arid Environments 64: 460–473.
Cooke, G. D., E. B. Welch, S. A. Peterson & P. R. Newroth, 1993. Restoration and Management of Lakes and Reservoirs. Lewis Publishers, New York.
Danen-Louwerse, H. J., L. Lijklema & M. Coenratts, 1995. Coprecipitation of phosphate with calcium carbonate in Lake Veluwe. Water Research 29: 1781–1785.
deKoff, J., M. A. Anderson & C. Amrhein, 2007. Geochemistry of iron in the Salton Sea, California. Hydrobiologia (this issue).
Dittrich, M., T. Dittrich, I. Sieber & R. Koschel, 1997. A balance analysis of phosphorus elimination by artificial calcite precipitation in a stratified hardwater lake. Water Research 31: 237–248.
Doner, H. E. & W. C. Lynn, 1989. Carbonate, halide, sulfate, and sulfide minerals, Chap. 6. In: Dixon, J. B. & S. B. Weed (eds), Minerals in the Soil Environment. Soil Science Society America, Madison, WI: 279–330.
Drever, J. I., 1997. The Geochemistry of Natural Waters: Surface and Groundwater Environments. Prentice Hall Publisher, New Jersey, 436 pp.
Griffin, R. A. & J. J. Jurinak, 1973. The interaction of phosphate with calcite. Soil Science Society of America Proceedings 37: 847–850.
Hieltjes, A. H. M. & L. Lijklema, 1979. Nalevering van fosfaat door sedimenten (III): interactie van fosaat in sedimenten en poriënwater. (In Dutch) H₂O 12: 599–602.
Holdren, G. C. & A. Montaño, 2002. Chemical and physical characteristics of the Salton Sea, California. Hydrobiologia 473: 1–21.
Horne, A. J. & C. R. Goldman, 1994. Limnology, 2nd Edn. McGraw-Hill, New York, 576 pp.
House, W. A., 1987. Inhibition of calcite crystal growth by inorganic phosphate. Journal of Colloid and Interface Science 119: 505–511.
House, W. A., 1990. The prediction of phosphate coprecipitation with calcite in freshwaters. Water Research 24: 1017–1023.

House, W. A., 1999. The physico-chemical conditions for the precipitation of phosphate with calcium. Environmental Technology 20: 727–733.

House, W. A. & L. Donaldson, 1986. Adsorption and coprecipitation of phosphate on calcite. Journal of Colloid and Interface Science 112: 309–324.

Kleiner, J., 1988. Coprecipitation of phosphate with calcite in lake water: a laboratory experiment modeling phosphorus removal with calcite in Lake Constance. Water Research 22: 1259–1265.

Kleiner, J., 1990. Calcite precipitation—regulating mechanisms in hardwater lakes. International Association of Theoretical and Applied Limnology 24: 136–139.

Mucci, A. & J. W. Morse, 1983. The incorporation of Mg^{2+} and Sr^{2+} into calcite overgrowths: influences of growth rate and solution composition. Geochemica et Cosmochimica Acta 47: 217–233.

Oza, H. I., 2003. Nutrient levels and phytoplankton abundance in Canyon Lake and Lake Elsinore, CA. M.S. Thesis. University of California, Riverside, CA.

Selig, U., H. Baudler, M. Krech & G. Nausch, 2006. Nutrient accumulation and nutrient retention in coastal waters— 30 years investigation in the Darss-Zingst Bodden chain. Acta Hydrochimica et Hydrobiologica 34: 9–19.

Stumm, W. & J. Morgan, 1996. Aquatic Chemistry: Chemical Equilibria and Rates in Natural Waters, 3rd Edn. John Wiley & Sons, Inc., New York.

Sondergaard, M., J. P. Jensen & E. Jeppesen, 1999. Internal phosphorus loading in shallow Danish lakes. Hydrobiologia 409: 145–152.

Souchu, P., A. Gasc, Y. Collos, A. Vaquer, H. Tournier, B. Bibent & J. M. Deslous-Pauli, 1998. Biogeochemical aspects of bottom anoxia in a Mediterranean lagoon (Thau, France). Marine Ecology—Progress Series 164: 125–146.

Swan, B. K., J. M. Watts, K. M. Reifel & S. H. Hurlbert, 2007. Role of the polychaete *Neanthes succinea* in phosphorus regeneration from sediments in the Salton Sea, California. Hydrobiologia 576: 111–125.

Wardlaw, G. D. & D. L. Valentine, 2005. Evidence for salt diffusion from sediments contributing to increasing salinity in the Salton Sea, California. Hydrobiologia 533: 77–85.

Watts, J. M., B. K. Swan, M. A. Tiffany & S. H. Hurlbert, 2001. Thermal, mixing, and oxygen regimes of the Salton Sea, California, 1997–1999. Hydrobiologia 466: 159–176.

Hydrobiologia (2008) 604:57–75
DOI 10.1007/s10750-008-9311-6

A linked hydrodynamic and water quality model for the Salton Sea

Eu Gene Chung · S. Geoffrey Schladow ·
Joaquim Perez-Losada · Dale M. Robertson

Abstract A linked hydrodynamic and water quality model was developed and applied to the Salton Sea. The hydrodynamic component is based on the one-dimensional numerical model, DLM. The water quality model is based on a new conceptual model for nutrient cycling in the Sea, and simulates temperature, total suspended sediment concentration, nutrient concentrations, including PO_4^{-3}, NO_3^{-1} and NH_4^{+1}, DO concentration and chlorophyll a concentration as functions of depth and time. Existing water temperature data from 1997 were used to verify that the model could accurately represent the onset and breakup of thermal stratification. 1999 is the only year with a near-complete dataset for water quality variables for the Salton Sea. The linked hydrodynamic and water quality model was run for 1999, and by adjustment of rate coefficients and other water quality parameters, a good match with the data was obtained. In this article, the model is fully described and the model results for reductions in external phosphorus load on chlorophyll a distribution are presented.

Keywords Restoration · Nutrients ·
Sediment resuspension · Internal loading

Guest editor: S. H. Hurlbert
The Salton Sea Centennial Symposium. Proceedings of a Symposium Celebrating a Century of Symbiosis Among Agriculture, Wildlife and People, 1905–2005, held in San Diego, California, USA, March 2005

E. G. Chung · S. G. Schladow
Department of Civil and Environmental Engineering, University of California, Davis, CA 95616, USA

S. G. Schladow (✉)
Tahoe Environmental Research Center, University of California, Davis, CA 95616, USA
e-mail: gschladow@ucdavis.edu

J. Perez-Losada
Departament of Physics, University of Girona, Girona 17071, Spain

D. M. Robertson
U.S. Geological Survey, Middleton, WI 53562, USA

Introduction

The Salton Sea is a highly saline, terminal lake, located in the southeastern desert of California. The Sea serves as a repository for agricultural drain water from the Coachella and Imperial Valleys of California and the Mexicali Valley of Mexico. It constitutes the largest body of water in the state, encompassing 963 km², with a maximum depth of 15.5 m, a maximum length and width of 56 and 26 km respectively, a maximum water elevation of −69 m mean sea level (MSL) and a salinity of approximately 48,000 mg l^{-1} (State of California Resources Agency, 2006). Its total volume is about 9.25×10^9 m³ (Cook et al., 2002). In its present state, the Salton Sea is a highly eutrophic lake,

 Springer

Hydrobiologia (2008) 604:57–75

characterized by high nutrient concentrations, high algal biomass, high fish productivity, low clarity, frequent very low dissolved oxygen (DO) concentrations, massive fish kills, and noxious odors (Holdren & Montaño, 2002). Nitrogen is a typical limiting nutrient of Great Basin terminal lakes including Pyramid Lake (Galat et al., 1981), Walker Lake (Cooper & Koch, 1984), Big Soda Lake (Cloern et al., 1983), Mono Lake (Herbst, 1998), and Great Salt Lake (Stephens & Gillespie, 1976). However, Holdren & Montaño (2002) reported orthophosphate to be frequently detected below the limit of 0.005 mg l^{-1} in the Sea, and concluded that algal growth is limited by phosphorus.

The Sea is presently listed under Section 303(d) of the Clean Water Act as impaired, with the cause of the impairment being cited as high nutrient and suspended sediment concentrations. In 2003, a Quantification Settlement Agreement (QSA) was signed by the Coachella Valley Water District, Imperial Irrigation District, and Metropolitan Water District of Southern California. The QSA addresses water allocation issues between the holders of water rights to Colorado River water and enables California to stay within its 5.427 billion cubic meter annual apportionment of Colorado River water. It also establishes a water transfer from agricultural water users to urban water users that will reduce the physical size of the Salton Sea. Concomitantly, the California State Legislature passed legislation to facilitate environmental restoration of the Salton Sea, agricultural lands surrounding the Salton Sea, and the tributaries and drains within the Imperial and Coachella valleys that deliver water to the Salton Sea (State of California Resources Agency, 2006).

A detailed study of the water quality of the Salton Sea and its tributaries was conducted in 1999 by the U.S. Department of Interior's Bureau of Reclamation (USBR; Holdren & Montaño, 2002). Based on the observed nitrogen-to-phosphorus ratios that ranged from 25:1 to 400:1 and dissolved nutrient concentrations, it can be concluded that phosphorus is the algal growth-limiting nutrient (Holdren & Montaño, 2002; Robertson et al., 2008). Furthermore, the majority of the external phosphorus to the system ($\sim$ 1,450,000 kg yr^{-1}) comes from the three main inflows, the New River ($\sim$ 51%), the Alamo River ($\sim$ 38%), and the Whitewater River ($\sim$ 4%) (Robertson et al., 2008). The remainder of the external load was from

agricultural drains that flow directly to the Sea ($\sim$ 6%) and atmospheric deposition ($\sim$ 1%).

As expected in highly eutrophic conditions, surface DO concentrations were frequently above saturation values while bottom DO concentrations were often zero during times of thermal stratification. Under these conditions in most other lakes, orthophosphate-P concentrations in the hypolimnion increase dramatically, a phenomenon referred to as "internal loading" (Wetzel, 2001). Such an increase was never observed in the Salton Sea. In fact, concentrations near the bottom were usually less than near the surface during periods of anoxia. The low bottom phosphorus concentrations were hypothesized by Holdren & Montaño (2002) to be caused by the chemical sequestration of phosphorus due to precipitation of hydroxyapatite ($Ca_5(PO_4)_3OH$) and fluorapatite ($Ca_5(PO_4)_3F$) based on results from geochemical programs, PHRQPITZ and PHREEQC. Therefore, there was no indication of internal loading of phosphorus from the deep sediments, even during extended periods of anoxia, based on the data collected in 1999.

Phosphorus may, however, be released from the sediments during resuspension events. Internal phosphorus loading has been related to sediment resuspension in other shallow eutrophic lakes and reservoirs by resuspension of particulate phosphorus and by the entrainment of pore waters with phosphate into the water column (Osgood, 1988; Romero et al., 2002; Somlyody & van Straten, 1986). Sediment resuspension in shallow lakes is induced mainly by wave activity due to wind increasing the bottom shear stress above a critical value (Aalderink et al., 1985; Luettich et al., 1990; Somlyody & van Straten, 1986). The resuspended sediments can increase turbidity, and reintroduce nutrients directly to the euphotic zone (Hamilton & Mitchell, 1997; Imboden, 1974; Kristensen et al., 1992; Nagid et al., 2001; Reddy et al., 1996; Sondergaard et al., 1992).

M. Anderson (University of California, Riverside, personal communication) evaluated the extent of sediment resuspension in the Salton Sea by modeling the susceptibility of different parts of the lake to wind. Using an approach proposed by Hakanson & Jansson (1983), he found that large areas of the Sea were susceptible to sediment resuspension. As part of the same research, Anderson found that under laboratory conditions, agitated sediment could release high concentrations of phosphorus.

The importance of an episodic event such as sediment resuspension having a large role in phosphorus dynamics of the Sea is also suggested by the data of Holdren & Montaño (2002), shown in Fig. 1. In this figure, it is evident that the total phosphorus concentration in the Sea varies by over an order of magnitude between successive samplings (2–4 weeks). In Fig. 2, the daily flux of total phosphorus from all the gauged stream is shown, along with the calculated daily flux required to account for the observed changes in the Sea total phosphorus concentrations (assuming a constant flux between sampling events). This clearly shows that the observed phosphorus concentrations can only be explained by large increases and decreases consistent with resuspension and settling. The low sampling

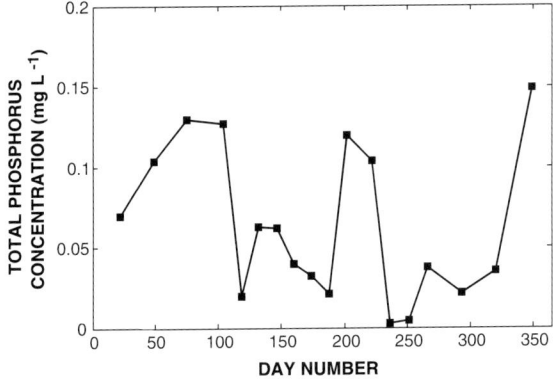

Fig. 1 Whole-lake, volume-averaged total phosphorus concentration in the Salton Sea during 1999, based on water quality data from Holdren & Montaño (2002)

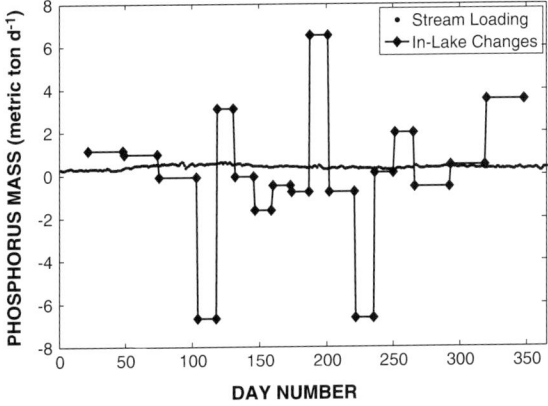

Fig. 2 Mass flux of phosphorus from stream inflows to the Salton Sea (based on data from Robertson et al. (2008) compared to the inferred phosphorus flux required to account for measured changes in volume averaged Sea concentrations, based on data from Holdren & Montaño (2002)

frequency (2–4 weeks) has most likely distorted the apparent duration of these events, and in reality the fluxes are higher than suggested and the duration of the events shorter.

In order to assist both the development of a total maximum daily load (TMDL; a detailed plan to reduce P loading to a specified value; U.S. EPA, 2000) for phosphorus input to the Salton Sea (as required under the Clean Water Act), and to assess future conditions of the Sea under reduced water inflows and changed physical size, a linked hydrodynamic and water quality model was developed. For the long-term effect of changes in external phosphorus loading to the Sea, in particular, Robertson & Schladow (2008) used several empirical eutrophication models to demonstrate how long-term phosphorus load reductions should affect the average annual water quality of this terminal saline lake, and how much of a reduction is needed for the Salton Sea to become a moderately eutrophic lake. Results of the empirical models indicate the potential long-term changes in the water quality of the Salton Sea associated with phosphorus load reductions, and with which there may be much variability internal loading due to hydrodynamic processes in the Sea.

The hydrodynamic behavior of the Salton Sea, such as a double gyre circular pattern and polymixis, is mainly driven by the wind blowing over the long fetch of the lake (Cook et al., 2002). The wind forcing also affects the water temperature, DO, and nutrient cycling of the Salton Sea (Carpelan, 1958; Watts et al., 2001). In other shallow water systems, water quality has been shown to be affected by sediment resusepension (Aalderink et al., 1985; Hamilton & Mitchell, 1997; Luettich et al., 1990; Somlyody, 1986). The possibility that resuspension of sediment and phosphorus may play a major role in the ecological status of the Salton Sea adds uncertainty to the effects of future restoration efforts.

The unique biogeochemistry of the Salton Sea necessitated that the water quality model, based on a new conceptual model of nutrient dynamics, together with a sediment resuspension model, be incorporated into the water quality version of the dynamic lake model, (DLM-WQ; Hamilton & Schladow, 1997; Schladow & Hamilton, 1997). This conceptual model takes into account both the sequestering of dissolved phosphorus through the formation of hydroxyapatite, and the role of sediment resuspension as the major source of phosphorus to the water column. In what

follows, we present the basis of the Salton Sea conceptual nutrient-cycling model, a description of the modified coupled hydrodynamic and water quality model incorporated with a new algorithm for sediment resuspension, the results of the model calibration and validation using the limited field dataset from the Salton Sea, and the implications of the model results for reducing eutrophication in the Sea.

Methods

Conceptual model of water quality in the Salton Sea

Conceptual models for nutrients, DO, and suspended solids were developed for the Salton Sea, based on the extremely limited dataset were available. These conceptual models were incorporated into the numerical model, DLM-WQ that is described in the following section.

Phosphorus

A schematic of the conceptual model for phosphorus that was developed specifically for the Salton Sea is shown in Fig. 3a. State variables include orthophosphate-P (PO_4^{-3}), particulate phosphorus (PP), and internal phosphorus (IP) in the biomass. IP is assumed to be exclusively composed of phytoplankton and expressed as the portion of phosphorus in the phytoplankton. Inflows are assumed to add PO_4^{-3} to the Sea. The PO_4^{-3} pool is also increased through sediment release, such as sediment flux, and through mechanical release associated with the entrainment of PO_4^{-3} from the pore water into the water column by sediment resuspension. PO_4^{-3} is either taken up biologically or chemically sequestered by precipitation and/or adsorption to form PP. Constants are used to describe the rates at which these processes occur. IP can either settle (via phytoplankton settling), be permanently lost to the sediments, such as permanent deposition as fish bone, or be added to the PP pool via respiration and mortality. PP is assumed to be only the phosphorus contained in non-biotic suspended particles, such as the phosphorus contained in resuspended sediments. PP can either be lost via settling or increased by sediment resuspension. The PO_4^{-3} addition by the inflows is part of the model input data.

Sediment and mechanical release are assumed to occur only above the thermocline, which is estimated by the hydrodynamic component of DLM-WQ.

Nitrogen

A schematic of the conceptual model for nitrogen is shown in Fig. 3b. State variables include nitrate (NO_3^{-1}), ammonium (NH_4^{+1}), particulate nitrogen (PN), and internal nitrogen (IN) in the biomass. The IN is assumed to be exclusively composed of phytoplankton. IN can either settle (via phytoplankton settling) or be added to the PN pool via respiration and mortality as IP. PN is also assumed to be the only nitrogen in non-biotic suspended particles. NO_3^{-1} is assumed to be added with inflow and nitrification of NH_4^{+1}, and consumed by the uptake by IN and by denitrification. Nitrogen fixation is not considered in this model, because the fixation typically contributes nitrogen to a lake when there is an extremely low availability of nitrogen, which is not the case of the Salton Sea (Tõnno & Nõges, 2003). Nitrogen reactions are assumed to be the first-order kinetics, and constants are used to describe the rates at which these processes occur. The nitrification reaction depends on NH_4^{+1} concentrations and sufficient oxygen levels (greater than 1–2 mg l^{-1}). The nitrification inhibition factor, f_{NTR}, is expressed by a function of the concentration of DO (Chapra, 1997). The ammonia (NH_3) preference factor for the phytoplankton uptake, $\beta_{NH}3$, is used in this model and it is assumed that there is no preference for either form of nitrogen but based on the relative proportions of the concentrations of NH_3 and NO_3^{-1} in the water column (Bowie et al., 1985).

NH_4^{+1} is assumed to be added with inflow, sediment releases, mechanical releases, and with ammonification of PN, and removed by nitrification, volatilization of ammonia and the uptake by IN. Ammonification is the decomposition of organic nitrogen back to NH_4^{+1} by heterotrophic microbes (Schlesinger, 1991). Volatilization is considered only for the surface layer and can occur only in the form of ammonia (Jellison et al., 1993).

Dissolved oxygen (DO)

A schematic of the conceptual model for DO is shown in Fig. 3c. This model considers oxygen reaeration,

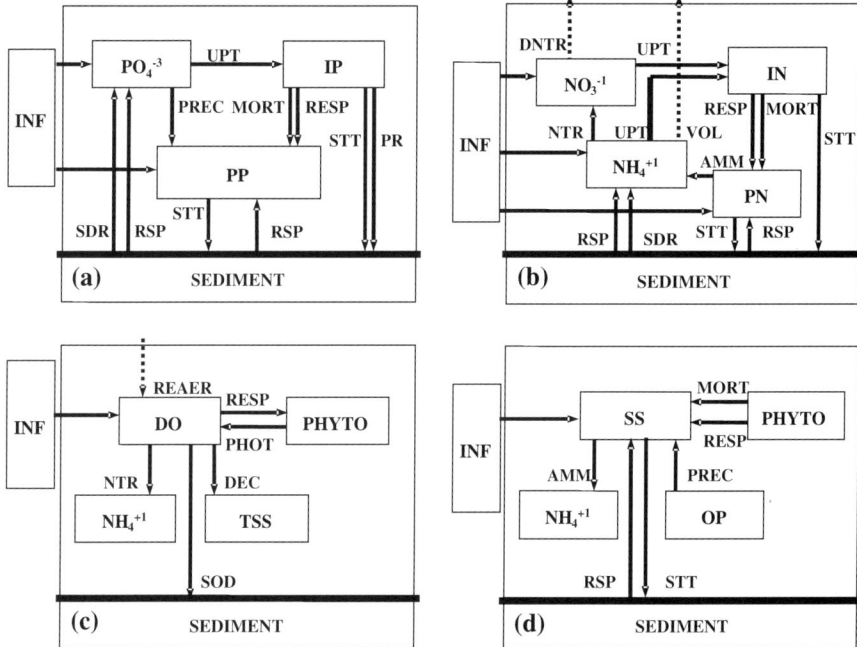

Fig. 3 Conceptual models for the Salton Sea: (**a**) phosphorus; (**b**) nitrogen; (**c**) DO; (**d**) suspended solids. Solid lines indicate processes in the water column and between water and sediment, and dashed lines processes between air and water surface. INF, Inflow; UPT, uptake; RESP, respiration, MORT, mortality; PR, permanent removal; SDR, sediment release;

STT, settling; RSP, sediment resuspension; PREC, precipitation + adsorption; NTR, nitrogen; DNTR, denitrification; VOL, volatilization; AMM, ammonification; REAER, reaeration; SOD, sediment oxygen demand; NTR, nitrogen; PHOT, photosynthesis; DEC, decomposition

biochemical oxygen demand (BOD) for organic decomposition, nitrogenous biochemical oxygen demand (NBOD) for nitrification, and the sediment oxygen demand (SOD) for benthic sediment and organisms. DO is assumed to be provided with inflow, photosynthesis, and reaeration by atmospheric O_2 gas (gas transfer) across the air–water interface. Reaeration is controlled by a liquid film because of its high Henry's Law constant. The saturation concentration of DO is affected by temperature and salinity, and the equations in Table 3 describe these relationships. The oxygen-transfer coefficient for reaeration for lakes was estimated as a function of wind speed (Chapra, 1997).

DO is consumed by the respiration of phytoplankton, nitrification, SOD, and decomposition of suspended solids (SS) (e.g., carbonaceous matter). SOD is produced by the oxidation of organic matter in bottom sediment and occurs only in the bottom layer of the model. SOD is affected by the organic content of the sediments and the DO concentration of the overlying waters. In this model, SOD is assumed to have an areal SOD rate at 20°C multiplied by a temperature adjustment factor (Chapra, 1997).

Suspended solids

A schematic of the conceptual model for suspended solids (SS) that was developed specifically for the Salton Sea is shown in Fig. 3d. Changes in SS are assumed to be related to physical processes, such as sediment resuspension and settling, and to biochemical processes, such as precipitation of PO_4^{-3} with hydroxyapatite ($Ca_5(PO_4)_3OH$) and fluorapatite ($Ca_5(PO_4)_3F$). SS is also assumed to be provided by mortality of phytoplankton, but consumed by ammonification of SS.

Description of hydrodynamic and water quality model

The hydrodynamic and water quality model is the combination of a one-dimensional hydrodynamic model, a sediment resuspension model, and a water quality model that incorporates the conceptual nutrient models described above. The hydrodynamic model simulates the vertical distribution of temperature and salinity in lakes and reservoirs, and is based on DLM.

(Heald et al., 2005; Losada, 2001; McCord & Schladow, 1998; McCord et al., 2000). Here, the water quality version of the model, DLM-WQ (Hamilton & Schladow, 1997; Schladow & Hamilton, 1997), which couples the transport and mixing processes to a set of biological and chemical processes that describe the growth of phytoplankton, the cycling of nutrients and the fate of particulate material was used.

DLM-WQ is a process-based, deterministic lake/reservoir simulation model based on a Lagrangian layer scheme in which the lake is modeled by a series of horizontal layers of uniform property, but variable thickness. The hydrodynamic model describes the individual mixing and transport processes associated with surface heat, mass, and momentum exchange, mixed layer dynamics, inflows, outflows, and hypolimnetic mixing. The surface heat, mass, and momentum exchange constitutes the main driving forces to the model. A parameterized turbulent kinetic energy model is used to represent mixing in the surface mixed layer. A turbulent kinetic energy budget, based on meteorological forcing across the air–water interface, is calculated at each internal timestep (3 h) of the model. Turbulent kinetic energy production is parameterized through four distinct processes—wind stirring, convective overturn, interfacial shear production, and Kelvin–Helmholtz billowing. Inflows to the Salton Sea are invariably less saline (and less dense) than the Sea itself. Under these circumstances, DLM-WQ adds the inflow water to the surface of the water body. Mixing in the hypolimnion is modeled as a diffusive-like process, with the actual events being parameterized by eddy diffusivity. The formulation in DLM-WQ follows the premise that the diffusivity is directly proportional to the dissipation of turbulent kinetic energy, and inversely proportional on the strength of stratification.

A new algorithm for sediment resuspension was incorporated in the Salton Sea version of DLM-WQ. As very little is known about the sediments or resuspension fluxes in the Salton Sea, a simple modeling approach following Somlyody (1986) was adopted. Here, the resuspension flux (E) which is calculated as the same timestep, is described as:

$$E = k_1 W^{m1}$$

where W is the wind speed, and k_1 and $m1$ are calibration constants. A critical velocity, W_c is

defined, below which resuspension is assumed not to occur. The resuspension flux is assumed to be the depth-averaged suspended solids, which means each layer has the same resuspension flux. The dissolved nutrients in the sediment porewater, such as PO_4^{-3} and NH_4^{+1}, are also mechanically released when the sediments are resuspended. Once resuspended, particles are allowed to resettle at a prescribed velocity (Table 2).

Biological and chemical components modeled in DLM-WQ are considered to be analogous to the physical components (e.g., temperature and salinity) in the mixing processes that are simulated by DLM-WQ. When layers merge, the new concentration is a volumetric average of the component layers. If water is removed from a layer by an outflow, the concentration of the water quality components in the layer remains unchanged and only the layer volume is adjusted. Mixing of chemical and biological components is, therefore, done in conjunction with modeling of temperature, salinity, and density, and uses the same timestep.

The phytoplankton in the biological component within DLM-WQ is represented by chlorophyll a, the standard biomass measurement (OECD, 1982). Phytoplankton increases with growth, and decreases by respiration, mortality, and settling. Grazing by zooplankton and filter feeding by barnacles is not explicitly considered in the model due to an absence of data, but the impact of these loss terms would be reflected through mortality. Phytoplankton growth is modeled as a function of the growth rate, temperature, light, and potential limiting nutrients (phosphorus and nitrogen). The light response function, $f(I)$, assumes that phytoplankton responds to various light conditions instantaneously and with no light history effect. The nutrient functions, $f(P)$ and $f(N)$, are based on the Michaelis–Menten kinetics and assume that the external concentrations of available nutrients determine the growth rate of phytoplankton (Bowie et al., 1985). A fixed stoichiometric approach is adopted in the biological component of the model (Chapra, 1997; Hamilton & Schladow, 1997; Jørgensen & Bendoricchio, 2001; Lehman, 1975). Phytoplankton is assumed to settle a fixed amount each day, although hydrodynamic mixing processes are capable of mixing the water column and keeping the phytoplankton in suspension.

Model application

The hydrodynamic and water quality model, DLM-WQ was applied to the Salton Sea with processes that were developed or modified for the unique conditions of the Salton Sea. A list of symbols of the water quality model in DLM-WQ and values of the coefficients (literature and calibration values) are shown in Tables 1 and 2,

Table 1 List of symbols of the water quality model in DLM-WQ

Symbols	Description
Hydrodynamic symbols	
A_{ns}	Surface area (m^2)
AS_i	Area of sediments in contact with layer i (m^2)
H	Depth of thermocline or 10 m if themocline >10 m
H_i	Depth of layer i (m)
V_i	Volume of water in layer i (m^3)
V_{ns}	Volume of water surface layer (m^3)
Phytoplankton symbols	
$Chla_i$	Concentration of chlorophyll a in layer i (µg/l)
$f(I_i)$	Function of minimum value of light
$f(P_i)$	Function of minimum value of phosphorus
$f(N_i)$	Function of minimum value of nitrogen
I	Ambient light intensity (irradiance level, not PAR)
Chemical symbols	
DO_i	Dissolved oxygen concentration in layer i (mg/l)
DO_{ns}	Dissolved oxygen concentration in surface layer (mg/l)
f_{NTR}	Nitrification inhibition factor
IN_i	Internal nitrogen concentration in layer i (µg/l), ($= a_N \times Chla$)
IP_i	Internal phosphorus concentration in layer i (µg/l), ($= a_P \times Chla$)
$K_{1,NH4}$	Mass-transfer velocity of NH_3 in the liquid laminar layer (m/day)
$K_{1,O}$	Mass-transfer velocity in the liquid laminar layer (m/day)
$NH3_{WS}$	Saturation ammonia concentration (mg/m^3)
$NH3_{ns}$	Dissolved ammonia concentration in surface layer (mg/l)
NH_{4i}	Concentrations of ammonia in layer i (mg/l)
NO_{3i}	Concentrations of nitrate in layer i (mg/l)
O_S	Saturation oxygen concentration (mg/m^3)
O_{SF}	Saturation concentration of dissolved oxygen in fresh water at 1 atm (mg/l)
O_{SS}	Saturation concentration of dissolved oxygen in saltwater at 1 atm (mg/l)
OP_i	Orthophosphate concentration in layer i (µg/l)
PN_i	Particulate nitrogen concentration in layer i (µg/l)
PP_i	Particulate phosphorus concentration in layer i (µg/l)
S	Salinity (g/l = ppt, ‰)
S'_B	Areal SOD rate
T_a	Absolute temperature (K)
U_w	Wind speed measured 10 m above the water surface (m/s)
β_{NH3}	Ammonia preference factor
Sediment resuspension	
SS_i	Suspended solid concentration in layer i (µg/l)
W	Wind speed (m/s) if $W > W_C$

Table 2 The coefficients, constants, and forcing parameters of DLM-WQ (Bowie et al., 1985; Losada, 2001)

Symbols	Description	Lit. range min/max	Calibrated value	Units
Phytoplankton				
a_P	Constant ratio of phosphorus to chlorophyll a	0.5–1.0	0.5	–
a_N	Constant ratio of nitrogen to chlorophyll a	7.0–15.0	15.0	–
a_C	Constant ratio of carbon to chlorophyll a	10.0–112.0	61.0	–
G_{MAX}	Maximum growth rate of phytoplankton	0.2–8.0	4.0	1/day
I_{SAT}	Saturation light intensity	48.4–193.7	86.0	Watts/m^2
k_{OP}	Half saturation constant of phosphorus	0.5–80.0	15.0	µg/l
$k_{NO3+NH4}$	Half saturation constant of nitrogen	1.5–400.0	75.0	µg/l
k_{RESP}	Respiration rate of phytoplankton	0.02–0.80	0.2	1/day
k_{MORT}	Mortality of phytoplankton	0.003–0.17	0.03	1/day
θ_{Chla}^{T-20}	Non-dimensional temperature multipliers of phytoplankton	1.0–1.14	1.068	–
W_{Chla}	Settling velocity of chlorophyll a	0.5–3.0	0.5	m/day
Chemical constants				
k_{AMM}	Ammonification rate constant	0.001–0.4	0.001	1/day
k_{DEC}	Decomposition rate constant	0.004–1.5	0.0025	1/day
k_{DNT}	Denitrification rate constant of NO$_3$	0.0–1.0	0.25	1/day
k_{PR}	Coefficient of permanent removal of P	–	1.09e-5	1/day
k_{NIT}	Nitrification rate constant of NH$_4$	0.001–3.0	0.2	1/day
f_{NTR}	Nitrification inhibition coefficient	–	0.6	l/mg
r_{OC}	Ratio of mass of oxygen consumed per mass of carbon assimilated	–	2.67	mgO/mgC
r_{ON}	Amount of oxygen consumed per unit mass of nitrogen oxidized in the total process of nitrification	–	4.57	gO/gN
Sediment				
$S_{B,20}$	Areal SOD rate at 20°C	0.02–15.0	0.2	mg/l
θ_{RXN}^{T-20}	Non-dimensional temperature multipliers of reactions	1.0–1.14	1.08	–
θ_{SOD}^{T-20}	Non-dimensional temperature multipliers of SOD	1.04–1.13	1.065	–
NH4$_S$	NH$_4$ concentration in the pore water of sediment	–	10.0	µg/l
OP$_S$	OP concentration in the pore water of sediment	–	2.0	µg/l
ϕ	Porosity	–	0.8	–
θ_{SD}^{T-20}	Non-dimensional temperature multipliers of sediment	1.0–1.14	1.08	–
θ_{SS}^{T-20}	Non-dimensional temperature multipliers of suspended solids	1.0–1.14	1.08	–
r_{CS}	Constant ratio particulate C of SS	–	4.17e-3	–
r_{NS}	Constant ratio particulate N of SS	–	1.10e-3	–
r_{PS}	Constant ratio particulate P of SS	–	2.50e-3	–
S_N	Sediment release rate of NH$_4$	13.1–95.4	7.5	µg/m^2/d
S_P	Sediment release rate of OP	0.166–7.78	0.08	µg/m^2/d
W_{SS}	Settling velocity of SS	0.5–3.3	3.25	m/day
k_1	Sediment resuspension constant	–	600	µg/l
m_1	Sediment resuspension constant	–	1.2	m/s
W_C	Critical wind speed	–	2.0	m/s

Table 3 Differential equations for the state variables in the water quality component of DLM-WQ

Chlorophyll a

$$\frac{\partial \text{Chla}_i}{\partial t} = \left[G_{\text{MAX}} \theta_{\text{Chla}}^{\text{T}-20} \text{Chla}_i \text{Min}\{f(I_i), f(P_i) f(N_i)\} \right] - \left[k_{\text{RESP}} \theta_{\text{Chla}}^{\text{T}-20} \text{Chla}_i + k_{\text{MORT}} \theta_{\text{Chla}}^{\text{T}-20} \text{Chla}_i + \frac{W_{\text{Chla}} \text{Chla}_i}{H_i} \right]$$

$$f(I_i) = \frac{I}{I_{\text{sat}}} \text{EXP}\left(1 - \frac{I}{I_{\text{sat}}}\right); \quad f(P_i) = \frac{\text{SRP}}{k_{\text{SRP}} + \text{SRP}} = \frac{(\text{PO}_4^{-3})}{k_{\text{op}} + (\text{PO}_4^{-3})}; \quad f(N_i) = \frac{(\text{NO}_3 + \text{NH}_4)}{k_{\text{NO}_3 + \text{NH}_4} + (\text{NO}_3 + \text{NH}_4)}$$

Phosphorus

$$\frac{\partial (\text{PO}_4^{-3})_i}{\partial t} = \left[S_{\text{p}} \theta_{\text{SD}}^{\text{T}-20} \frac{\text{AS}_i}{V_i} + \text{OP}_S \frac{\phi}{1-\phi} \frac{k_1 W^{m1}}{H} \right] - \left[G_{\text{max}} \theta_{\text{Chla}}^{\text{T}-20} a_{\text{P}} \text{Chla}_i \cdot \text{Min}\{f(I_i), f(P_i), f(N_i)\} + k_{\text{PREC}} \theta_{\text{SS}}^{\text{T}-20} (\text{PO}_4^{-3})_i \right]$$

$$\frac{\partial \text{PP}_i}{\partial t} = \left[k_{\text{RESP}} \theta_{\text{Chla}}^{\text{T}-20} a_{\text{P}} \text{Chla}_i + k_{\text{MORT}} \theta_{\text{Chla}}^{\text{T}-20} a_{\text{P}} \text{Chla}_i + r_{\text{PS}} \frac{k_1 W^{m1}}{H_i} + k_{\text{PREC}} \theta_{\text{SS}}^{\text{T}-20} (\text{PO}_4^{-3})_i \right] - \left[\frac{W_{\text{SS}} \text{PP}_i}{H_i} \right]$$

$$\frac{\partial \text{IP}_i}{\partial t} = \left[G_{\text{MAX}} \theta_{\text{Chla}}^{\text{T}-20} a_{\text{P}} \text{Chla}_i \cdot \text{Min}\{f(I_i), f(P_i), f(N_i)\} \right]$$
$$- \left[k_{\text{RESP}} \theta_{\text{Chla}}^{\text{T}-20} a_{\text{P}} \text{Chla}_i + k_{\text{MORT}} \theta_{\text{Chla}}^{\text{T}-20} a_{\text{P}} \text{Chla}_i + \frac{W_{\text{Chla}} a_{\text{P}} \text{Chla}_i}{H_i} + k_{\text{PR}} \theta_{\text{Chla}}^{\text{T}-20} a_{\text{P}} \text{Chla}_i \right]$$

Nitrogen

$$\frac{\partial (\text{NO}_3^{-1})_i}{\partial t} = \left[k_{\text{NIT}} \theta_{\text{RXN}}^{\text{T}-20} (\text{NH}_4^{+1})_i f_{\text{NIT}} \right]$$
$$- \left[(1 - \beta_{\text{NH3}}) G_{\text{max}} \theta_{\text{Chla}}^{\text{T}-20} a_{\text{N}} \text{Chla}_i \cdot \text{Min}\{f(I_i), f(P_i), f(N_i)\} + k_{\text{DNT}} \theta_{\text{RXN}}^{\text{T}-20} (\text{NO}_3^{-1})_i \right]$$

$$\beta_{\text{NH3}} = \frac{\text{NH}_3}{\text{NH}_3 + (\text{NO}_3^{-1})}; \quad f_{\text{NTR}} = 1 - \exp(-k_{\text{NITR}} \times \text{DO})$$

$$\frac{\partial (\text{NH}_4^{+1})_i}{\partial t} = \left[S_N \theta_{\text{SD}}^{\text{T}-20} \frac{\text{AS}_i}{V_i} + \text{NH4}_S \frac{\phi}{1-\phi} \frac{k_1 W^{m1}}{H} + k_{\text{AMM}} \theta_{\text{RXN}}^{\text{T}-20} \text{PN}_i \right] - \left[K_{1,\text{NH3}} \frac{A_{\text{ns}}}{V_{\text{ns}}} (\text{NH3}_{\text{ns}} - \text{NH3}_{\text{WS}}) \right]$$
$$- \left[\beta_{\text{NH3}} G_{\text{max}} \theta_{\text{Chla}}^{\text{T}-20} a_{\text{N}} \text{Chla}_i \cdot \text{Min}\{f(I_i), f(P_i), f(N_i)\} + k_{\text{NIT}} \theta_{\text{RXN}}^{\text{T}-20} (\text{NH}_4^{+1})_i f_{\text{NIT}} \right]$$

$$K_{\text{L,NH4}} = 483.19 V^{0.8} T^{-1.4}$$

$$\frac{\partial \text{IN}_i}{\partial t} = \left[G_{\text{max}} \theta_{\text{Chla}}^{\text{T}-20} a_{\text{N}} \text{Chla}_i \cdot \text{Min}\{f(I_i), f(P_i), f(N_i)\} \right] - \left[k_{\text{RESP}} \theta_{\text{Chla}}^{\text{T}-20} a_{\text{N}} \text{Chla}_i + k_{\text{MORT}} \theta_{\text{Chla}}^{\text{T}-20} a_{\text{N}} \text{Chla}_i + \frac{W_{\text{Chla}} a_{\text{N}} \text{Chla}_i}{H_i} \right]$$

$$\frac{\partial \text{PN}_i}{\partial t} = \left[k_{\text{RESP}} \theta_{\text{Chla}}^{\text{T}-20} a_{\text{N}} \text{Chla}_i + k_{\text{MORT}} \theta_{\text{Chla}}^{\text{T}-20} a_{\text{N}} \text{Chla}_i + r_{\text{NS}} \frac{k_1 W^{m1}}{H_i} \right] - \left[k_{\text{AMM}} \theta_{\text{RXN}}^{\text{T}-20} \text{PN}_i + \frac{W_{\text{SS}} \text{PN}_i}{H_i} \right]$$

DO

$$\frac{\partial \text{DO}_i}{\partial t} = \left[K_{1,\text{O}} \frac{A_{\text{ns}}}{V_{\text{ns}}} (O_{\text{S}} - \text{DO}_{\text{ns}}) + G_{\text{max}} \theta_{\text{Chla}}^{\text{T}-20} r_{\text{OC}} a_{\text{C}} \text{Chla}_i \cdot \text{Min}\{f(I_i), f(P_i), f(N_i)\} \right]$$
$$- \left[k_{\text{RESP}} \theta_{\text{Chla}}^{\text{T}-20} r_{\text{OC}} a_{\text{C}} \text{Chla}_i + k_{\text{NIT}} \theta_{\text{RXN}}^{\text{T}-20} r_{\text{ON}} (\text{NH}_4^{+1})_i f_{\text{NIT}} + S'_{\text{B},20} \theta_{\text{SOD}}^{\text{T}-20} \frac{\text{AS}_i}{V_i} + k_{\text{DEC}} \theta_{\text{RXN}}^{\text{T}-20} r_{\text{OC}} r_{\text{CS}} \text{SS}_i \right]$$

$$K_l = 0.864 U_{\text{w}}; \quad \ln O_{\text{S}} = \ln O_{\text{SF}} - S\left(1.7674e-2 - \frac{1.0754e1}{T_{\text{a}}} + \frac{2.1407e3}{T_{\text{a}}^2}\right)$$

$$\ln O_{\text{SF}} = -139.34411 + \frac{1.575701e5}{T_{\text{a}}} - \frac{6.642308e7}{T_{\text{a}}^2} + \frac{1.243800e10}{T_{\text{a}}^3} - \frac{8.621949e11}{T_{\text{a}}^4}$$

Suspended solids

$$\frac{\partial \text{SS}_i}{\partial t} = \left[\frac{k_1 W^{m1}}{H_i} + \frac{k_{\text{PREC}} \theta_{\text{SS}}^{\text{T}-20} (\text{PO}_4^{-3})_i}{r_{\text{PS}}}\bigg|_i + k_{\text{MORT}} \theta_{\text{Chla}}^{\text{T}-20} a_{\text{C}} \text{Chla}_i + k_{\text{RESP}} \theta_{\text{Chla}}^{\text{T}-20} a_{\text{C}} \text{Chla}_i \right] - \left[\frac{W_{\text{SS}} \text{SS}_i}{H_i} + \frac{k_{\text{AMM}} \theta_{\text{RXN}}^{\text{T}-20} \text{PN}_i}{r_{\text{NS}}} \right]$$

respectively. The governing differential equations for the biological and chemical state variables of DLM-WQ are summarized in Table 3.

Data inputs

DLM-WQ requires four kinds of input data which include descriptive physical data for the lake, hydrodynamic forcing data, water quality parameters, and initial conditions for all the modeled variables. The

descriptive physical data for the Salton Sea are surface areas and cumulative volumes as a function of depth, as well as longitudinal slope and the channel cross section of the inflowing rivers (in order to calculate inflow entrainment). The slope and cross section of the inflowing rivers were estimated from topographical maps. A Hydrographic GPS Survey of the Salton Sea was conducted by the USBR in 1995. Depth contours for the Salton Sea were based on these survey data (Fig. 4).

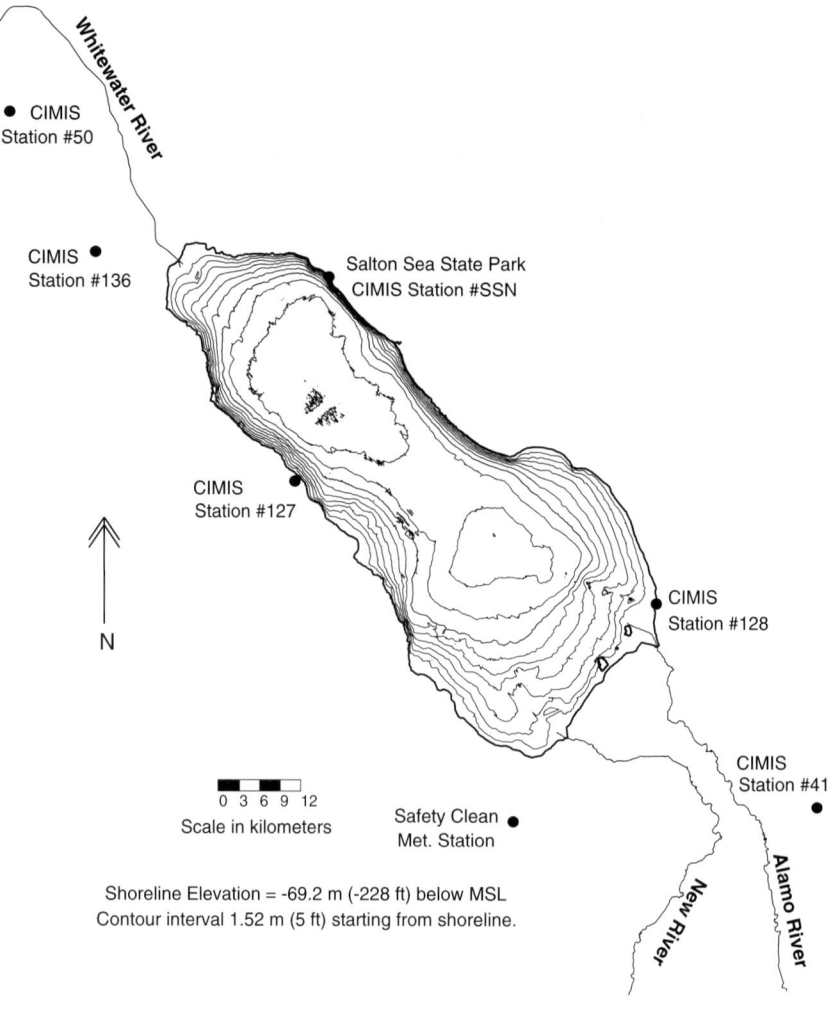

Fig. 4 Elevation contours below mean seal level for the Salton Sea and location of meteorological stations around the Salton Sea (from Cook, 2000)

Table 4 CIMIS stations around the Salton Sea

Station name	No.	Nearby city	Latitude (N)	Longitude (W)	Instrument measurement height (m)	Calibration frequency
Calipatria	41	Calipatria	33°02′37″	115°24′56″	2	Twice/yr
Salton Sea West	127	Salton City	33°19′38″	115°57′00″	2	Twice/yr
Salton Sea East	128	Niland	33°13′12″	115°34′48″	2	Twice/yr
Oasis	136	Oasis	33°31′32″	116°09′15″	2	Twice/yr
Mecca	141	Mecca	33°32′17″	115°59′30″	2	Twice/yr
Salton Sea State Park	154 (SSN)	Desert Beach	33°30′16″	115°54′52″	2	Twice/yr

Meteorological data required by the model are daily averages of solar short-wave radiation, vapor pressure, air temperature, relative humidity, wind speed, precipitation, and long-wave radiation. Since daily values are used in the model, all of the fluxes are equally distributed across the full 24 h, with the exception of short-wave radiation, which applied only over the daylight hours and applied with a sine function. The California Irrigation Management Information System (CIMIS) collects meteorological data from several stations around the Salton Sea. The locations of CIMIS meteorological stations around the Salton Sea are given in Table 4, and shown on Fig. 4.

Daily average streamflows in the three major tributaries have been continually monitored since 1965 by the U.S. Geological Survey (Fig. 1), while in-stream water-quality data were collected only for short periods of time by various agencies. These data were used to estimate daily and annual flows and nutrient loading to the Sea. The annual loads are summarized in Robertson et al. (2008).

For the modeling, the meteorological data from CIMIS station #127 were used. The recorded CIMIS data for wind speed, air temperature, and relative humidity were adjusted to 10 m from the 2-m measurement height using standard methods (Cook, 2000).

Model calibration and validation

Initial conditions for all the water quality state variables are needed to fully calibrate and validate the model. The state variables include temperature, salinity, DO, nutrients (nitrogen and phosphorus), chlorophyll *a*, and suspended solids. Only 2 years with incomplete data were available for the Salton, thereby precluding the possibility of performing a complete calibration and a validation with two independent sets of data. During 1997, Cook (2000) collected temperature data for several periods using thermistors at several depths within the Sea. In 1999, the USBR collected water quality profiles at 2- to 4-week intervals at several locations in the Sea (Holdren & Montano, 2002). During that same year, San Diego State University (SDSU) also collected chlorophyll *a* data (Hurlbert et al., San Diego State University, unpubl. raw data).

As thermal stratification and the meteorology largely control the physical mixing and many of the biogeochemical processes in the Sea, 1997 data were used to perform a calibration of the temperature model, and 1999 data were used to validate the temperature model. The 1999 data were used to calibrate all the other water quality variables by adjusting rate coefficients and parameters.

Simulated and measured (Cook, 2000) water temperatures for 1997 are shown in Fig. 5. The dataset extends from May 20, (day 140) to July 27, (day 208), and includes the time when the Sea was stratified. The dataset was from thermistors at four depths: 2.7, 4.1, 7.7, and 14 m from the surface. The surface thermistor was not operating during this period.

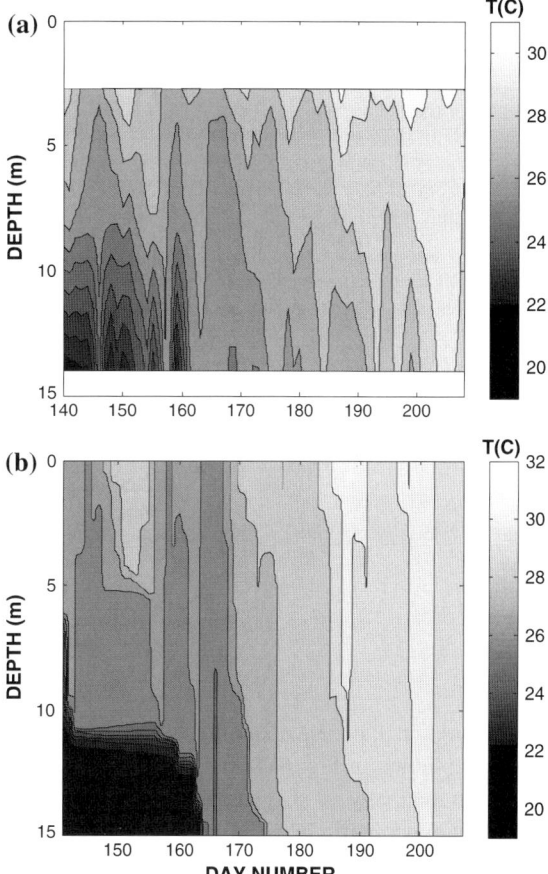

Fig. 5 Depth–time contours of temperature of the Salton Sea from 20 May 1997 to 27 July 1997: (**a**) Measured temperature data from a station near the center of the Salton Sea. No data were available at the surface and bottom since temperature loggers were placed near the top (10%), upper-quarter (25%), mid (50%), bottom (90%) points of the water column (Cook, 2000); (**b**) Modeled temperature distribution using DLM-WQ

The overall trends between simulated and measured data agree well (Fig. 5). The maximum of the difference between daily averaged temperatures is 1.9°C. Both datasets show a deep thermocline initially, which breaks down between days 160 and 165. The weaker stratification that persists after day 165, as well as the near-isothermy at the end of the period is resolved reasonable well by the model. The coarse spacing of instruments (up to 6 m) with no surface thermistor and the interpolation required to produce contours, yields the appearance of more gradual changes than the simulated results, which has a vertical resolution of 0.1–0.2 m. Direct comparisons between the actual thermistor traces and the

Hydrobiologia (2008) 604:57–75

Fig. 6 Temperature traces of the Salton Sea at four elevations. Dashed line shows measurement near the center of the Salton Sea with a depth of 15 m and solid line shows DLM-WQ simulation from 20 May 1997 to 27 July 1997

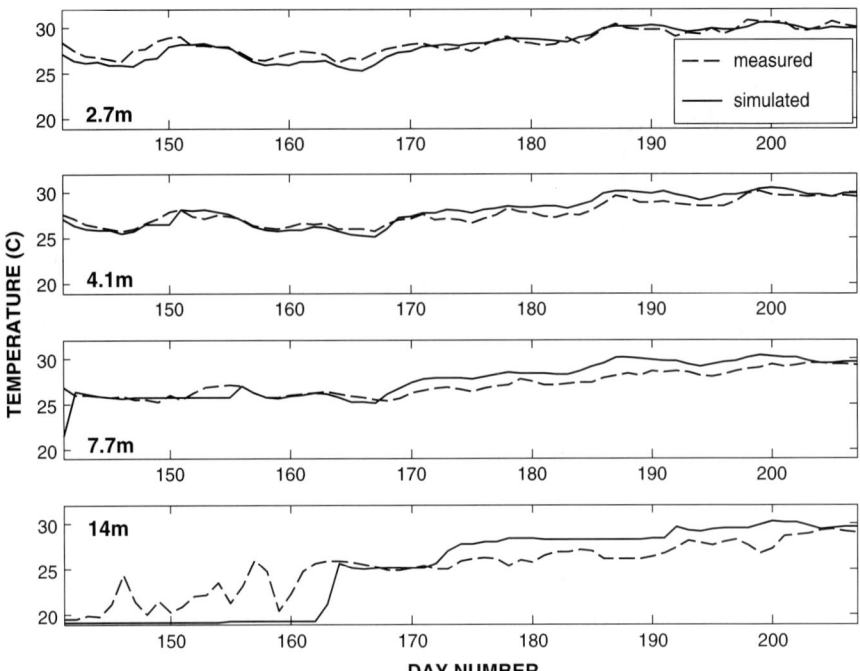

temperatures at the corresponding elevations in the model are shown in Fig. 6. Simulated and measured data are in relatively good agreement. The differences of mean temperatures between simulated and measured at 2.7 m and 4.1 m from the surface are only 1.3°C and 0.63°C, respectively.

The water temperature component of DLM-WQ was validated using the 1999 data. The remainder of the water quality parameters was calibrated using these data, but a true validation will have to await the availability of a suitable dataset. Measured temperature data (sampled at 2- to 4-week intervals) are compared with the DLM-WQ simulations run for the period January 22, 1999 (day 22) to December 18, 1999 (day 352) in Fig. 7. These dates mark the first and last days of the USBR data sampling for 1999. Theses data are in good agreement, particularly considering the crude measurement intervals. Both datasets show unstratified initial and final conditions and stratification between days 160 and 250.

The calibrated model output for chlorophyll a, DO, PO_4^{-3}, PP, NO_3^{-1}, NH_4^{+1}, PN, and SS are compared with the measured data in Figs. 8 and 9. Model results (shown by lines) are compared with the measured data (shown by symbols) for surface (Fig. 8) and for bottom (Fig. 9), respectively. This difference may account for some of the variation. The

seasonal trends and some of the short-term variations are captured fairly well by the model for chlorophyll a (Figs. 8 and 9a). Part of the differences in modeled and measured data may be due to the phytoplankton in DLM-WQ being represented as one functional group, not by the true variation in functional groups that are known to exist in the Salton Sea because only total chlorophyll a data were available for the Sea.

The general trends in DO concentration, with the exception of the high concentration of DO in the winter (around day 100), are captured fairly well by the model (Figs. 8 and 9b). The algal bloom around day 75 was not simulated very well in the chlorophyll a simulations. That may account for the concentrations of DO at that time being low. Simulated and measured DO near the bottom both exhibit the occurrence of summer anoxia.

Simulated and measured PO_4^{-3} (Figs. 8 and 9c) demonstrates the same general trends, with the exception of the low PO_4^{-3} in the spring. This difference may be related to the absence of the algal bloom in the spring in the DLM-WQ simulation. The model simulated the very low concentration of PO_4^{-3} in the summer and high PO_4^{-3} in the winter fairly well. Care must also be taken in interpreting the measured results, because the lines joining the symbols may not represent real trends because of

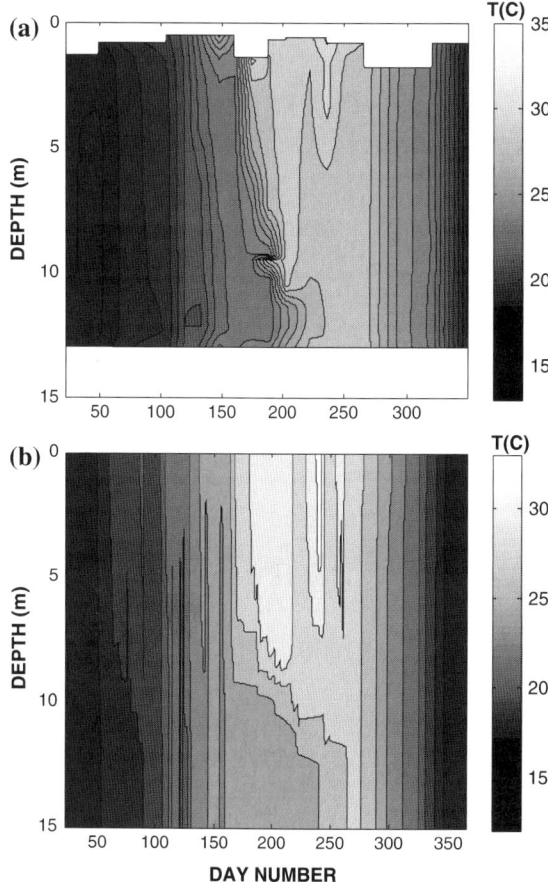

Fig. 7 DLM-WQ calibration, using the coefficient values from Table 2: (**a**) Measured temperature data in 1999. No data were available at the surface and bottom; (**b**) model simulation of temperature in 1999

interpolation. The modeled PO_4^{-3} data show rapid changes in concentration (from 0 to 50 μg l^{-1}) caused by sediment resuspension, which is the same general range as the measurements. The 2- to 4-week sampling interval precludes resolving such events.

Simulated and measured PP data (Figs. 8 and 9d) both demonstrate similar late winter and early spring trends and summer peaks. However, the simulated results do not agree with the measured data between days 240 and 289. These differences may be related to difference in total suspended solids (TSS) because the ratio of phosphorus to TSS is based on an estimation of PP/TSS from Holdren & Montaño (2002), not from direct measurements.

Simulated and measured NO_3^{-1} data (Figs. 8 and 9e) both demonstrate similar trends throughout the

year, with the exception of the high measured bottom values in the summer. These high values are directly related to the simulated DO and NH_4^{+1} concentrations because nitrification from NH_4^{+1} to NO_3^{-1} does not occur under anoxic conditions and NH_4^{+1} is accumulated near the bottom of the lake. Simulated NH_4^{+1} in Figs. 8 and 9f exhibit a peak in the summer that was not found in the measured data. The measured PN also had an abnormally high value in summer that was not simulated (Figs. 8 and 9g). These differences in the measured and simulated data suggest that further refinement of the nitrogen model may be required. However, as noted above, the Salton Sea is strongly phosphorus limited, therefore, these modification should not strongly affect the changes in productivity simulated by the model.

Simulated and measured SS data demonstrate similar trends, with the exception of the high concentrations measured in spring and winter (Figs. 8 and 9h). As mentioned previously, if sediment resuspension is a dominant process, then rapidly varying values are to be expected.

Results

Figure 10 shows modeled time and depth distributions of chlorophyll *a*, DO, PO_4^{-3}, PP, NO_3^{-1}, NH_4^{+1}, PN, and SS, respectively. These results show that variations in water quality are closely related to the hydrodynamic behavior of the Sea. For example, when the Sea is stratified between days 160 and 260, simulated bottom concentrations of chlorophyll *a* reach a minimum concentration, while DO and NO_3^{-1} are depleted, PO_4^{-3} is sequestered, and NH_4^{+1} increases. On the other hand, during periods with strong winds, mixing events tend to keep the Sea non-stratified and sediment resuspension maintains high concentrations of PO_4^{-3}, PP, and SS. These results indicate that many of the gradual changes in water quality inferred from the measured data are actually a series of changes associated with short-term mixing events.

Discussion

While there are presently insufficient independent data to confirm the short-term mixing patterns, the

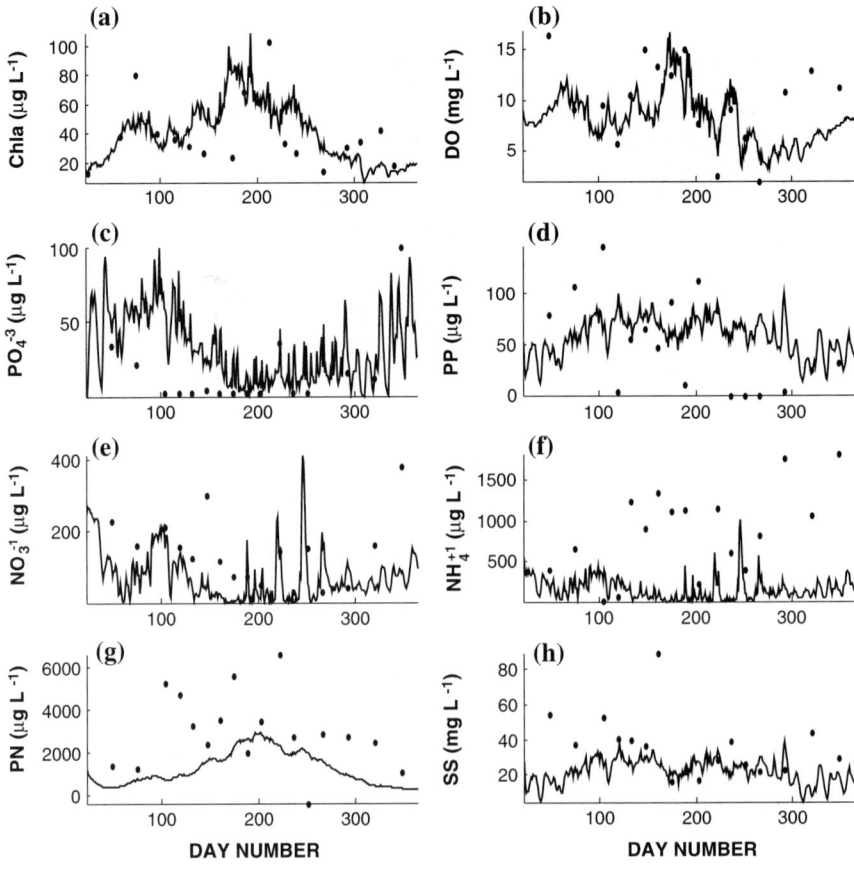

Fig. 8 Comparison between measured data (symbols) and simulated data (lines) of the Salton Sea in 1999 at the surface: (a) Chlorophyll a; (b) DO; (c) PO_4^{-3}; (d) PP; (e) NO_3^{-1}; (f) NH_4^{+1}; (g) PN; (h) SS

model does simulate plausible results and can be used to guide future measurement strategies. Therefore, one can use the results of the model to provide an indication of variations in properties that are not measured, to provide a better understand the causes of these short-term variations, and to obtain an indication of possible changes in the Salton Sea in response to changes in nutrient loading from the watershed.

In order to explore the relative importance of external phosphorus loading from tributaries and internal phosphorus loading from resuspension on the chlorophyll a concentrations in the Sea, the calibrated model was used to simulate the effects of an external tributary phosphorus load reduction by 90% (all other inputs to the model remained constant). Simulated chlorophyll a concentrations for both surface and bottom of the Sea are shown for the base case (the calibration result) and the decreased phosphorus loading scenario in Fig. 11. The small difference in the results of these two simulations indicates that the

short-term effect of the decrease in the external phosphorus load on chlorophyll a concentrations should be relatively small. The small effect on chlorophyll a concentrations caused by the large reduction in external P loading (90%) was a result of the importance of sediment resuspension in the model and its contribution to the nutrient budget. If the external nutrient supply was reduced, however, in time the amount of phosphorus in the near-shore sediment would also decrease, and that natural burial processes would eventually reduce the phosphorus supply from the sediments. The rate at which this might occur is unknown, particularly if there is a reduction in stream sediment loads along with the reduction in stream phosphorus loads.

The calibrated model can also be used to determine which factor(s) presently limit phytoplankton growth. Model results indicate that nutrient limitation only occurs in the upper 1–3 m of the Salton Sea. The remainder of the water column is light limited (Fig. 12). These simulations also indicate that there

Fig. 9 Comparison between measured data (symbols) and simulated data (lines) of the Salton Sea in 1999 at the depth 12 m except chlorophyll a: (**a**) Chlorophyll a at 6 m depth; (**b**) DO; (**c**) PO_4^{-3}; (**d**) PP; (**e**) NO_3^{-1}; (**f**) NH_4^{+1}; (**g**) PN; (**h**) SS

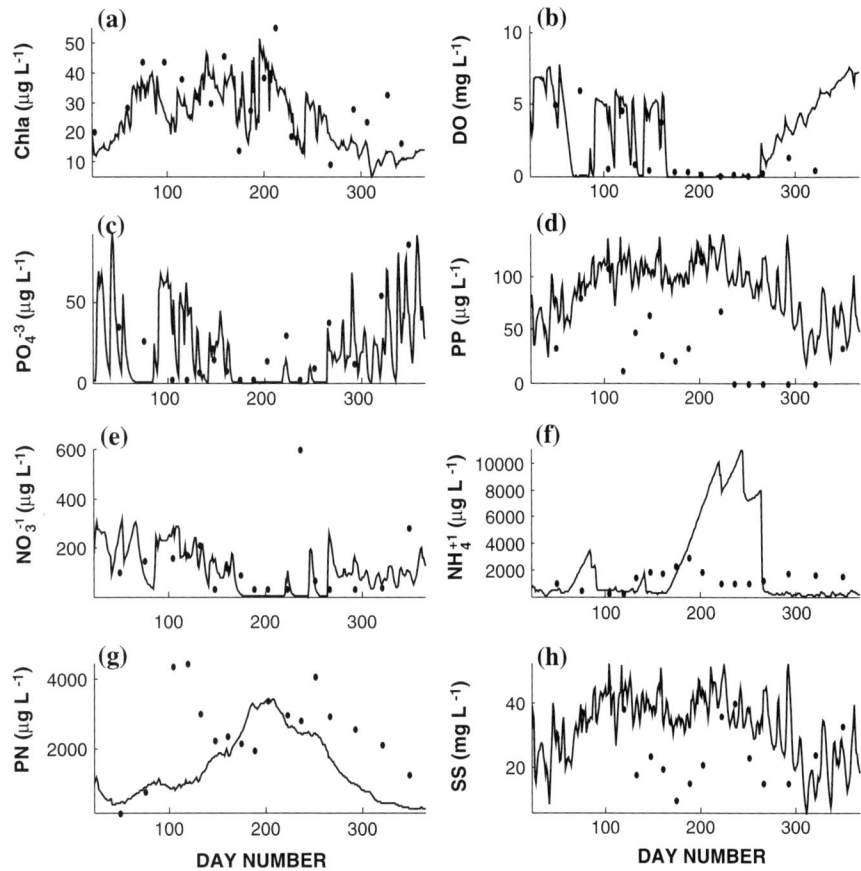

should be little short-term effect of the 90% reduction in the external loading of phosphorus on phytoplankton growth in the Sea. This is the result of the dominating importance of resuspension events and the minor effects of nutrients only limiting phytoplankton growth near the surface of the Salton Sea.

The long-term effect of changes in external phosphorus loading to the Salton Sea needs to be investigated for its ecological restoration. The long-term effect of changes in external phosphorus loading to the Salton Sea can be determined through the use of empirical eutrophication models that relate phosphorus loading with measured water quality, such as near-surface total phosphorus, chlorophyll a, and Secchi depths (Walker, 1986). However, results of the empirical models indicate the potential long-term changes in the water quality of the Salton Sea associated with external phosphorus load reductions as well as with internal load reduction, especially, related to hydrodynamic processes, such as sediment

resuspension (Robertson & Schladow, 2008). Therefore, effects of resuspension in this projected future condition may need to be further investigated for ecological restoration by using the hydrodynamic and water quality model developed in this study for the Salton Sea.

Conclusions

New conceptual nutrient models were developed specifically for the Salton Sea to take account of the Sea's unique biogeochemistry and the importance of sediment resuspension to nutrient cycling. These conceptual models represent balances between what is considered to be the best present understanding of nutrient dynamics in the Salton Sea, and the extremely limited dataset available for model calibration and validation. Although most of the major problems in the Salton Sea are associated with its water quality, no

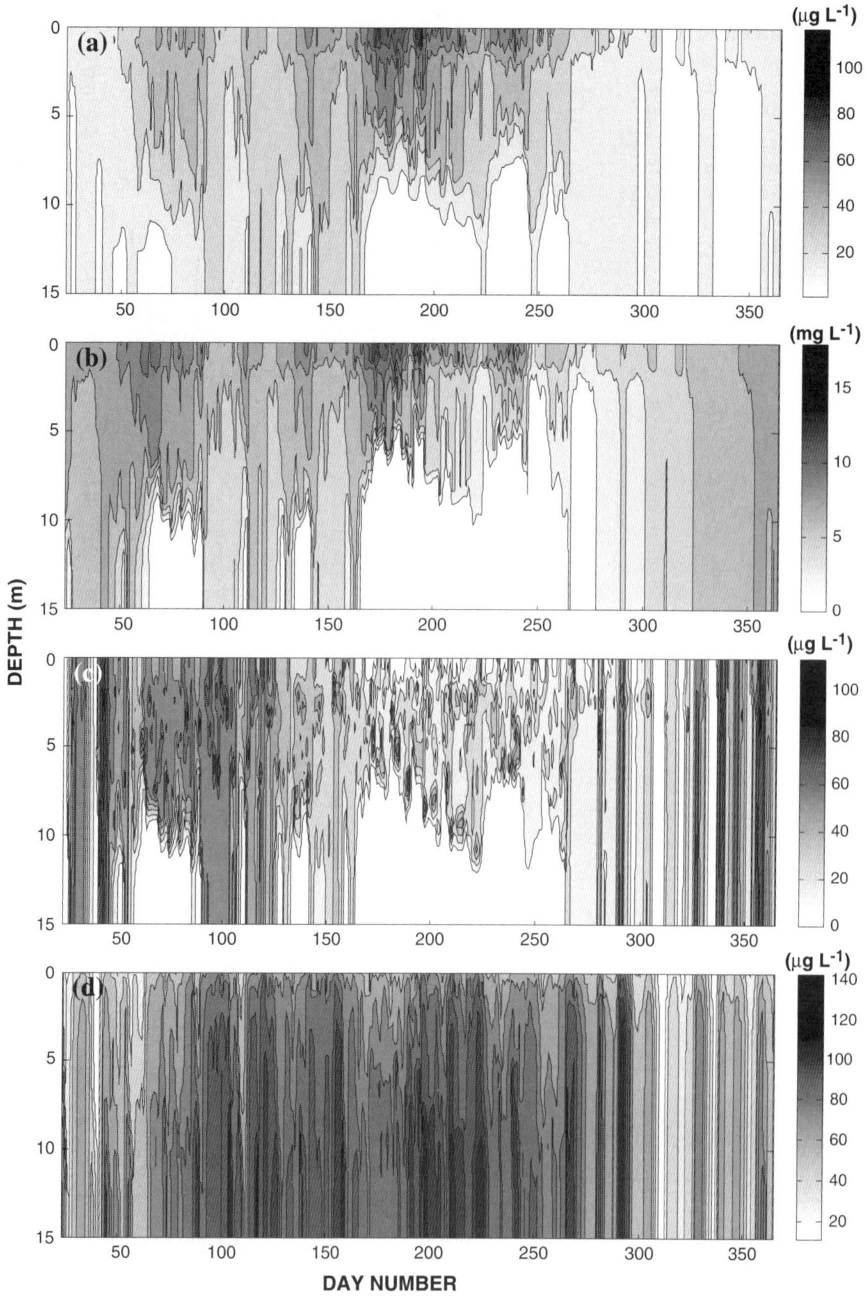

Fig. 10 Modeled depth–time contours of DLM-WQ in 1999 for the Salton Sea: (**a**) chlorophyll a(μg l^{-1}); (**b**) DO (mg l^{-1}); (**c**) PO$_4^{-3}$(μg l^{-1}); (**d**) PP (μg l^{-1}); (**e**) NO$_3^{-1}$(μg l^{-1}); (**f**) NH$_4^{+1}$(μg l^{-1}); (**g**) PN (μg l^{-1}); (**h**) SS (mg l^{-1})

long-term monitoring program exists and only a few detailed water quality studies have been conducted. Therefore, extensive and long-term efforts should be made in the future measurement strategies, for example at least monthly water quality monitoring program which measures water quality variables at every 1–2 m depth for at least three stations (north, center, and south basins) in the Salton Sea.

The conceptual models were embedded in the hydrodynamic and water quality model, DLM-WQ, together with a sediment resuspension model. Simulations with the model suggest that sediment resuspension of nutrients in both particulate and dissolved form (from sediment porewater), rather than external loading, is presently the most dominant factor in the Sea's nutrient cycling. Results of the

Fig. 10 continued

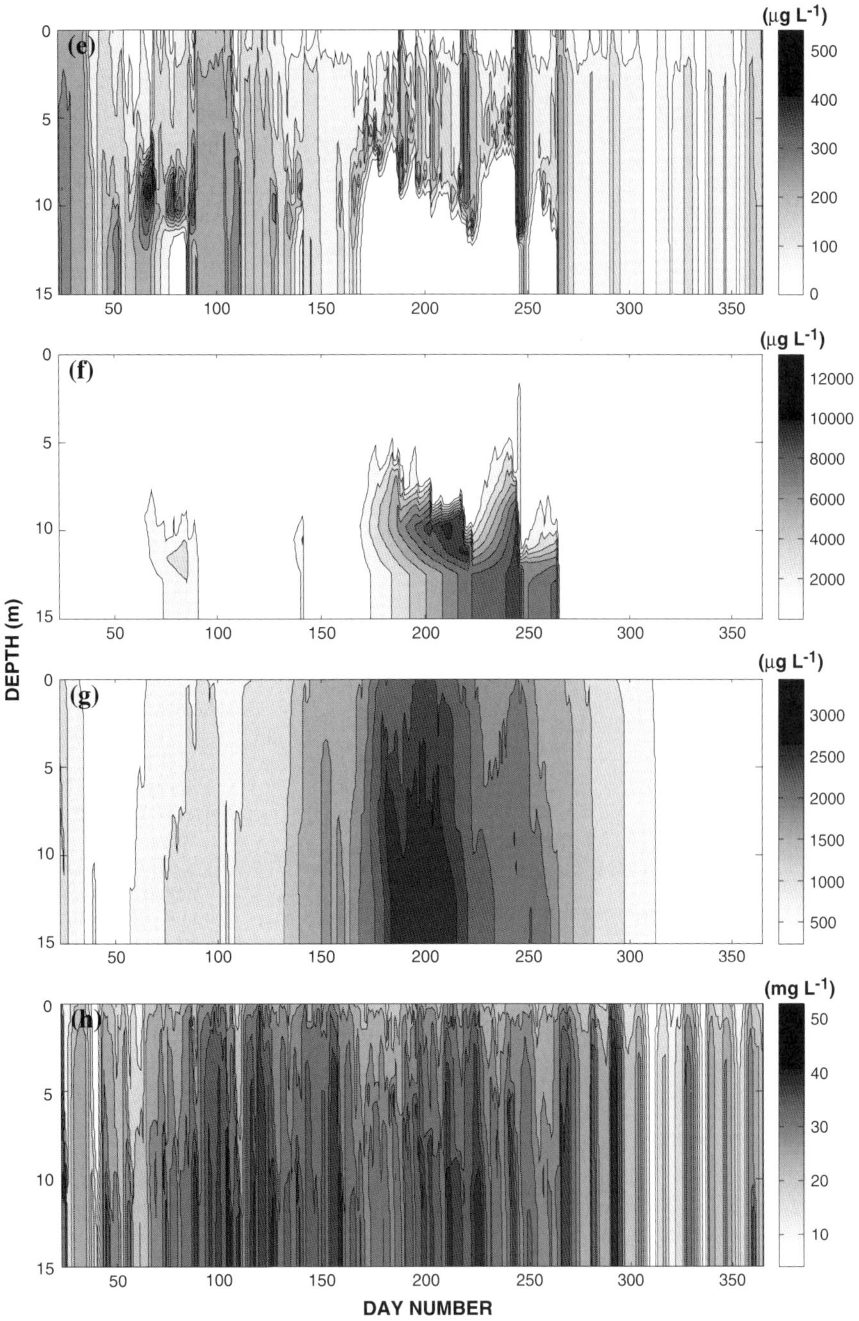

model also indicate that most of the phytoplankton growth was not only limited by nutrients, but also by light below about 1–3 m in the water column. Both these results suggest that efforts to impose TMDLs for phosphorus and turbidity in the inflowing streams may not have the desired short-term outcome. The turbidity reduction may in fact increase chlorophyll *a* concentrations by overcoming the light limitation. While reductions in stream phosphorus may not have an immediate impact on productivity because of the dominance of sediment resuspension, eventually natural sediment burial processes will lead to a reduction in system phosphorus. Insufficient information is known about the nature of the Salton Sea's

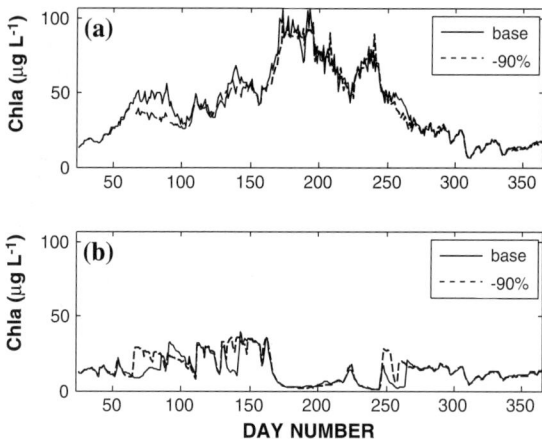

Fig. 11 Comparison of chlorophyll *a* concentrations between for base case (solid lines) and 90% reduction of tributary total phosphorus concentration (dashed lines): (**a**) Surface; (**b**) bottom

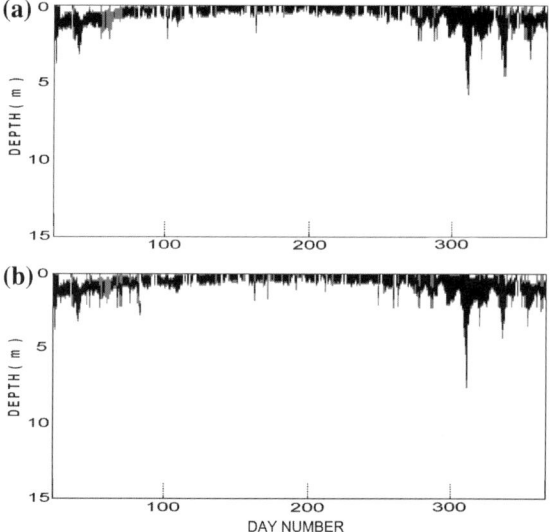

Fig. 12 Limiting factor maps of phytoplankton growth. White indicates light limitation on algal growth, black indicates phosphorus limitation and gray indicates nitrogen limitation: (**a**) Base case; (**b**) 90% reduction of phosphorus loading

sediments to conjecture the rate at which this may occur. Even though empirical eutrophication models can be used to demonstrate how long-term phosphorus load reductions should affect the average annual water quality of the Salton Sea, the effects of sediment resuspension in this projected future condition may need to be further investigated for ecological restoration by using the linked

hydrodynamic and water quality model developed in this study.

The model results also confirmed previous descriptions of the Sea being polymictic. This is one of the reasons why sediment resuspension is an important factor in the nutrient cycling. This propensity to mix in its present configuration should serve as a warning of the potential for unintended consequences in the future, when the Sea's configuration may be altered by engineering changes. A shallower Sea would be likely to mix more, thereby increasing sediment resuspension.

Acknowledgments We wish to thank Chris Holdren (USBR) and Stuart Hurlbert (SDSU) for providing data. The Colorado River Basin Regional Water Quality Control Board (CRBRWQCB) provided funding for this study.

References

Aalderink, R. H., L. Lijklema, J. Breukelman, W. Vanraaphorst & A. G. Brinkman, 1985. Quantification of wind induced resuspension in a shallow lake. Water Science and Technology 17: 903–914.

Bowie, G. L., W. B. Mills, D. B. Porcella, C. L. Campbell, J. R. Pagenkopf, G. L. Rupp, K. M. Johnson, P. W. H. Chan & S. A. Gherini, 1985. Rates, Constants, and Kinetics Formulations in Surface Water Quality Modeling, 2nd Edn. U.S. EPA, EPA 600-3-85-040.

Carpelan, L. H., 1958. The Salton Sea. Physical and chemical characteristics. Limnology and Oceanography 3: 373–386.

Chapra, S. C., 1997. Surface Water-Quality Modeling. McGraw-Hill Companies, Inc., Singapore.

Cloern, J. E., B. E. Cole & R. S. Oremland, 1983. Seasonal changes in the chemistry and biology of a meromictic lake (Big Soda Lake, Nevada, U.S.A.). Hydrobiologica 105: 195–206.

Cook, C. B., 2000. Internal dynamics of terminal basin lake: a numerical model for management of the Salton Sea. PhD Dissertation, University of California, Davis, USA.

Cook, C. B., G. T. Orlob & D. W. Huston, 2002. Simulation of wind-driven circulation in the Salton Sea: implications for indigenous ecosystems. Hydrobiologia 473: 59–75.

Cooper, J. J. & D. L. Koch, 1984. Limnology of a desertic terminal lake, Walker Lake, Nevada, U.S.A. Hydrobiologia 118: 275–292.

Galat, D. L., E. L. Lider, S. Vigg & S. R. Robertson, 1981. Limnology of a large, deep, North American terminal lake, Pyramid Lake, Nevada, U.S.A. Hydrobiologia 82: 281–317.

Hakanson, L. & M. Jansson, 1983. Principles of Lake Sedimentology. Springer-Verlag, New York.

Hamilton, D. P. & S. F. Mitchell, 1997. Wave-induced shear stresses, plant nutrients and chlorophyll in seven shallow lakes. Freshwater Biology 38: 159–168.

Hamilton, D. P. & S. G. Schladow, 1997. Prediction of water quality in lakes and reservoirs. 1. Model description. Ecological Modelling 96: 91–110.

Heald, P. C., S. G. Schladow, J. E. Reuter & B. Allen, 2005. Modeling MTBE and BTEX in lakes and reservoirs used for recreational boating. Environmental Science and Technology 39: 1111–1118.

Herbst, D. B., 1998, Potential salinity limitations on nitrogen fixation in sediments from Mono Lake, California. International Journal of Salt Lake Research 7: 261–274.

Holdren, G. C. & A. Montaño, 2002. Chemical and physical characteristics of the Salton Sea, California. Hydrobiologia 473: 1–21.

Imboden, D. M., 1974. Phosphorus model of lake eutrophication. Limnology and Oceanography 19: 297–304.

Jellison, R., L. G. Miller, J. M. Melack & G. L. Dana, 1993. Meromixis in hypersaline Mono Lake, California. 2. Nitrogen fluxes. Limnology and Oceanography 38: 1020–1039.

Jorgensen, S. E. &. G. Bendoricchio, 2001. Fundamentals of Ecological Modelling, Elsevier Science Ltd., Amsterdam, 526 p.

Kristensen, P., M. Sondergaard & E. Jeppesen, 1992. Resuspension in a shallow eutrophic lake. Hydrobiologia 228: 101–109.

Lehman, J. T., D. B. Botkin & G. E. Likens, 1975. The assumptions and rationales of a computer model of phytoplankton population dynamics. Limnology and Oceanography 20: 343–364.

Losada, J. P., 2001. A deterministic model for lake clarity; application to management of Lake Tahoe (California–Nevada), USA. PhD dissertation, University of Girona, Spain.

Luettich, R. A., D. R. F. Harleman & L. Somlyody, 1990. Dynamic behavior of suspended sediment concentrations in a shallow lake perturbed by episodic wind events. Limnology and Oceanography 35: 1050–1067.

McCord, S. A. & S. G. Schladow, 1998. Numerical simulations of degassing scenarios for CO_2-rich Lake Nyos, Cameroon. Journal of Geophysical Research—Solid Earth, 103(B6): 12355–12364.

McCord, S. A., S. G. Schladow & T. G. Miller, 2000. Modeling artificial aeration kinetics in ice covered lakes. Journal of Environmental Engineering-ASCE 126: 21–31.

Nagid, E. J., D. E. Canfield & M. V. Hoyer, 2001. Wind-induced increases in trophic state characteristics of a large (27 km(2)), shallow (1.5 m mean depth) Florida lake. Hydrobiologia 455: 97–110.

OECD, 1982. Eutrophication of Waters. Monitoring, Assessment and Control. OECD, Paris.

Osgood, R. A., 1988. Lake mixis and internal phosphorous dynamics. Archiv Für Hydrobiologie 113: 629–638.

Reddy, K. R., M. M. Fisher & D. Ivanoff, 1996. Resuspension and diffusive flux of nitrogen and phosphorus in a hypereutrophic lake. Journal of Environmental Quality 25: 363–371.

Robertson, D. M. & S. G. Schladow, 2008. Response in the water quality of the Salton Sea, California, to changes in phosphorus loading: an empirical modeling approach. Hydrobiologia (this issue).

Robertson, D. M., S. G. Schladow & G. C. Holdren, 2008. Long-term changes in the phosphorus loading to and trophic state of the Salton Sea, California. Hydrobiologia (this issue).

Romero, J. R., I. Kagalou, J. Imberger, D. Hela, M. Kotti, A. Bartzokas, T. Albanis, N. Evmirides, S. Karkabounas, J. Papagiannis & A. Bithava, 2002. Seasonal water quality of shallow and eutrophic Lake Pamvotis, Greece: implications for restoration. Hydrobiologia 474: 91–105.

Schladow, S. G. & D. P. Hamilton, 1997. Prediction of water quality in lakes and reservoirs. 2. Model calibration, sensitivity analysis and application. Ecological Modelling 96: 111–123.

Schlesinger, W. H., 1991. Biogeochemistry, an Analysis of Global Change. Academic Press, Inc., London, 443 p.

Somlyody, L., 1986. Wind induced sediment resuspension in shallow lakes. In Bhra, T. F. E. C. (ed.), Water Quality Modelling in the Inland Natural Environment. The Fluid Engineering Centre, 28-298.

Somlyody, L. & G. van Straten, 1986. Modeling and Managing Shallow Lake Eutrophication with Application to Lake Balaton. Springer-Verlag, New York.

Sondergaard, M., P. Kristensen & E. Jeppesen, 1992. Phosphorus release from resuspended sediment in the shallow and wind-exposed Lake Arreso, Denmark. Hydrobiologia 228: 91–99.

State of California Resources Agency, 2006. Salton sea ecosystem restoration program draft programmatic environmental impact report. State Clearinghouse # 2004021120.

Stephens, D. W. & D. M. Gillespie, 1976. Phytoplankton production in the Great Salt Lake, Utah, and a laboratory study of algal response to enrichment. Limnology and Oceanography 21: 74–87.

Tõnno, I. & T. Nõges, 2003. Nitrogen fixation in a large shallow lake: rates and initiation conditions. Hydrobiologia 490: 23–30.

U.S. EPA, 2000. Nutrient Criteria Technical Guidance Manual: Lakes and Reservoirs. Report no. EPA-822-B00-001, Washington, DC, variously paginated.

Walker, W. W., 1986. Empirical Methods for Predicting Eutrophication in Impoundments; Report 3, Phase III: Applications Manual. Technical Report E-81-9, U.S. Army Engineer Waterways Experiment Station, Vicksburg, MS.

Watts, J. M., B. K. Swan, M. A. Tiffany & S. H. Hurlbert, 2001. Thermal, mixing, and oxygen regimes of the Salton Sea, California, 1997–1999. Hydrobiologia 466: 159–176.

Wetzel, R. G., 2001. Limnology. Lake and River Ecosystems, 3rd Edn. Academic Press, San Diego, 1006.

Hydrobiologia (2008) 604:77–84
DOI 10.1007/s10750-008-9309-0

Barnacle growth rate on artificial substrate in the Salton Sea, California

J. B. Geraci · C. Amrhein · C. C. Goodson

Abstract The Salton Sea is one of the few saline, inland lakes in the world with a population of barnacles, *Balanus amphitrite*. It is also one of California's most impaired water bodies due to excessive nutrient loading which leads to phytoplankton blooms and low dissolved oxygen. Currently, *B. amphitrite* growth is limited due to lack of hard substrate in and around the Sea. We have hypothesized that artificial substrate could support the growth of *B. amphitrite* and their filter-feeding would lead to improved water quality. Periodic harvesting of the barnacles would result in the permanent removal of nitrogen and phosphorus from the Sea. A 44-day in-situ experiment was carried out in the Salton Sea to assess the rate of barnacle growth and phosphorus and nitrogen sequestration on burlap sheets suspended vertically from a floating line. Burlap panels were collected weekly and the barnacles analyzed for Ca, total-P, inorganic-P, total-N, total-C, $CaCO_3$, and organic matter content. After 44 days of growth, the barnacle mats weighed 7.4 kg m^{-2} on a dry weight basis, with 80% of the mass as shell material. The nutrient sequestration was 9.4 g P m^{-2} and 100 g N m^{-2}. Approximately half of the P was inorganic and appears to be coprecipitated with the calcium carbonate shell material. Results indicate that harvesting barnacles grown on artificial substrate in the Salton Sea would not be an effective method for removing N or P from the lake because of the relative proportions of shell material and organic material.

Keywords *Balanus amphitrite* · Growth · Phosphorus · Nitrogen · Calcium carbonate

Abbreviations

mat	Barnacle encrusted burlap sheets
$CaCO_3$	Calcium carbonate

Guest editor: S. H. Hurlbert
The Salton Sea Centennial Symposium. Proceedings of a Symposium Celebrating a Century of Symbiosis Among Agriculture, Wildlife and People, 1905–2005, held in San Diego, California, USA, March 2005

J. B. Geraci · C. Amrhein (✉) · C. C. Goodson
Department of Environmental Sciences, University of California, Riverside, CA 92521, USA
e-mail: amrhein@ucr.edu

Introduction

The Salton Sea is a closed-basin, saline lake in the desert of southeastern California. Due to the combination of abundant sunlight, warm temperatures, and excessive nutrients from agricultural wastewater and municipal effluent, the Sea is hypereutrophic. The salinity is approximately 1.3-times that of ocean water, with total dissolved solids of 47 g l^{-1} in spring 2004. The dominant fish in the Sea in recent decades has been a hybrid tilapia (*Oreochromis mossambicus* Peters × *O. urolepis hornorum* (Trewavas)), but several species

of marine fish are also present, including orangemouth corvina (*Cynoscion xanthulus* Jordan and Gilbert), bairdiella (*Bairdiella icistia* Jordan and Gilbert) and sargo (*Anisotremus davidsoni* Steindachner). Abundances of all these species have dropped substantially since 2000 (Riedel et al., 2002; Caskey et al., 2007; Crayon, personal communication).

The Salton Sea is one of the few inland saline lakes in the world that hosts a barnacle population, *Balanus amphitrite* Darwin, or "striped" barnacle (Detwiler et al., 2002). It is believed that the barnacles were unintentionally introduced in the 1940s by military seaplanes (Carpelan, 1961), and have since adapted to the unique conditions of the Salton Sea. *Balanus amphitrite* production in the Salton Sea is limited to solid surfaces such as docks, buoys, debris, and scattered rock formations, all of which are relatively scarce in the Sea (Detwiler et al., 2002; Tiffany et al., 2002). The Sea floor consists mainly of fine, organic-rich sediments, and the hard surfaces are colonized by either barnacles or filamentous green algae (Detwiler et al., 2002; Tiffany et al., 2002). This lack of adequate substrate leaves cyprids swimming freely in the water column with nowhere to settle. Eventually, fish either consume these cyprids or they simply die and add to the decomposing organic matter in the sediments. In both cases, nutrients are recycled into the Sea.

The dearth of suitable substrate is readily noticed when buoy-lines are left in the Sea for a few days. Barnacles immediately colonize these lines, especially from January to May, and their growth rate is

surprisingly rapid due to the warm water and abundance of phytoplankton (Tiffany et al., 2002). We have hypothesized that artificial substrate could be used to culture barnacles, which would lead to an improvement in water clarity, and periodic harvesting could reduce the nutrient load in the Sea.

In this work, we evaluated the effectiveness of burlap fabric for the settlement and support of *Balanus amphitrite* in the Salton Sea. We measured the growth rate and nutrient removal rates on this suspended substrate over a 44-day period during the spring spawning season of 2004.

Materials and methods

The site chosen for the experiment was a large alcove on the northwest side of the Salton Sea (Fig. 1). The site receives full sunlight, and surface currents move into the project area from the northeast and southeast, depending on weather conditions. The water depth at the site ranged from 1.1 to 1.4 m. The experiment was conducted during two periods from February to June 2004.

A 30-m length and 6-mm diameter polypropylene rope was attached to 15 floats spaced 1 m apart. The floats were constructed from 2-l plastic soda bottles and polystyrene foam blocks. Each end of the rope was attached to a large warning buoy anchored to the Sea floor with concrete weights. The length of the rope was positioned at a 45° angle to the prevailing current. Fourteen panels of untreated burlap fabric

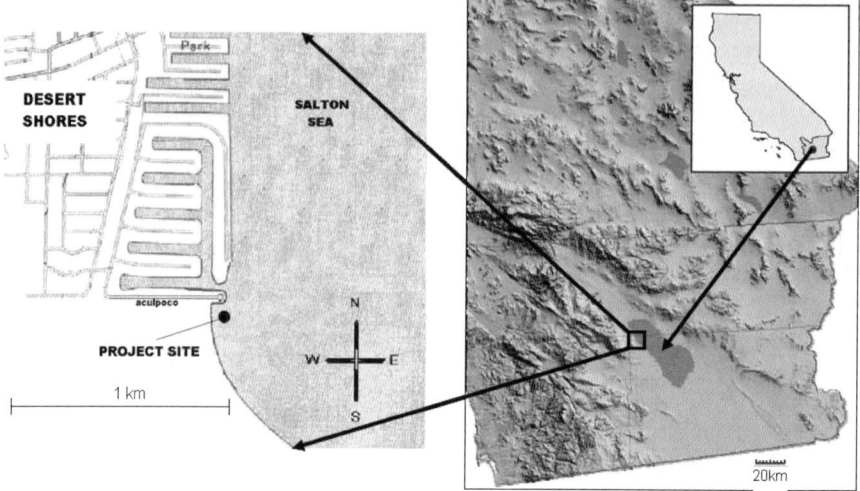

Fig. 1 Location of the Salton Sea and the project site near Desert Shores, California. Map courtesy of Diana Stralberg, PRBO Conservation Science

bacterial reactions (Rodriquez et al., 2008). Thus, phosphorus coprecipitated into the shell material appears to be a mechanism by which P is permanently sequestered from the water column.

The amount of organic-P was calculated as the difference between total-P and inorganic-P. Figure 5 shows that organic-P was proportional to organic matter content, following the nitrogen data. The total-N to organic-P ratio averaged 18 g g^{-1}, which is in the range for fishmeal (Barrias & Oliva-Teles, 2000).

Discussion

It is well known that zebra mussels (*Dreissena polymorpha*) can dramatically improve water clarity (Reed-Anderson et al., 2000; Budd et al., 2001), and markedly impact the phosphorus budget of aquatic systems (James et al., 2001; Canale & Chapra, 2002). This suggests that barnacles, which also feed on plankton, could be used as "living filters" to remove P and N from the Salton Sea.

The calcification of the barnacles and slowing of P and N accumulation that occurred after 30 days suggests that harvesting of the barnacles on the artificial substrate should occur monthly during the active growing period (March–June) to maximize nutrient removal. After 30 days the wet weight of barnacle-encrusted fabric was 10 kg m^{-2}. Due to this large mass, it is likely that any artificial substrate would have to be in shallow water and anchored to the Seabed. The flotation required to maintain such large masses of barnacle mats could be difficult to install and maintain, and so a solid, non-floatation support may be the best alternative.

Phosphorus is the nutrient limiting algae growth in the Salton Sea so P removal and sequestration in shell material could have the greatest impact on improving water quality (Holdren and Montaño, 2002). The relative effectiveness of using barnacles to remove P from the Salton Sea can be calculated using the average P content of the mats (9.4 g P m^{-2} burlap) and the annual P loading (1,340,000 kg P year^{-1}, Holdren and Montaño, 2002). Based on these two numbers, it would take 142 million square meters of burlap or barnacle substrate to collect the annual P load. This is an area equal to ~15% of the Sea's surface area, making it unlikely that this much substrate could be constructed.

One kilometer-long line supporting 1,000 burlap panels could be expected to collect only 20 kg P, if harvested three times over the spring growing season. Compared to the annual loading of over a million kilograms P, deploying artificial substrate does not appear an effective method to clean up the Salton Sea.

Recently, there have been plans developed to control the salinity of the Salton Sea by building a dike or dam to isolate a portion of the lake that could be maintained at ocean water salinity. The length of the dike/dam complex could range from 14 to 100 km depending upon the design, which would introduce substantial new hard-surfaces for barnacle colonization. A large mid-Sea dam could be expected to provide up to 100,000 m^2 of hard surface. In this case the barnacles would not be harvested but might be expected to die each summer, followed by replacement in the spring. The organic P would be recycled to the water and the P in the shell material (45% of the total) would be permanently deposited in the sediments, removing approximately 1,000 kg P, or ~0.1% of the annual P load.

Conclusions

Barnacle growth rate in the Salton Sea was rapid, accumulating on the artificial substrate (burlap) at 0.23 kg m^{-2} day^{-1}. After 44 days, each burlap panel (0.8 × 0.9 m) weighed over 14 kg and the buoy system was unable to support the mass. The mass was dominated by CaCO$_3$, making up 80% of the total dry weight.

The nitrogen and phosphorus in the barnacle material averaged 15 g N kg^{-1} and 1.3 g P kg^{-1}, which is relatively low compared to organic matter. Harvesting barnacles as a means of removing N and P from the Salton Sea would be ineffective due to the low proportion of organic matter and high proportion of calcium carbonate. Construction of a dam or dike system in the Salton Sea for salinity management would provide additional hard surface support for barnacles, but in the total nutrient budget, barnacle growth is likely to be insignificant.

Acknowledgments We want to thank Dr. Michael Anderson and Katheryne Jo Dyal at the University of California, Riverside, for preliminary studies on nutrient removal using barnacles reared in laboratory aquaria. We want to thank the

California Regional Water Quality Control Board, Colorado River Basin, Region 7 for their support of the senior author during the study.

References

Anderson, M. A., L. Whiteaker, E. Wakefield & C. Amrhein, 2008. Properties and distribution of sediment in the Salton Sea, California. Hydrobiologia. doi:10.1007/s 10750-008-9308-1

Aspila, K. I., H. Agemian & A. S. Y. Chau, 1976. A semi-automated method for the determination of inorganic, organic and total phosphate in sediments. Analyst 101: 187–197.

Barrias, C. & A. Oliva-Teles, 2000. The use of locally produced fish meal and other dietary manipulations in practical diets for rainbow trout *Oncorhynchus mykiss* (Walbaum). Aquaculture Research 31: 213–218.

Budd, J. W., T. D. Drummer, T. F. Nalepa & G. L. Fahnenstiel, 2001. Remote sensing of biotic effects: zebra mussels (*Dreissena polymorpha*) influence on water clarity in Saginaw Bay, Lake Huron. Limnology and Oceanography 46: 213–223.

Canale, R. P. & S. C. Chapra, 2002. Modeling zebra mussel impacts on water quality of Seneca River, New York. Journal of Environmental Engineering. 128: 1158–1168.

Carpelan, L. H., 1961. Physical and chemical characteristics. In Walker, B. W. (ed.), The ecology of the Salton Sea, California, in Relation to the Sportfishery. California Department of Fish & Game Fisheries Bulletin 113: 17–32.

Caskey, L. L., R. R. Riedel, B. Costa-Pierce, J. Butler & S. H. Hurlbert, 2007. Population dynamics, growth, and distribution of tilapia (*Oreochromis mossambicus*) in the Salton Sea, 1999–2002, with notes on orangemouth corvina (*Cynoscion xanthulus*) and bairdiella (*Bairdiella icistia*). Hydrobiologia 576: 185–203.

Detwiler, P. M., M. F. Coe & D. M. Dexter, 2002. The benthic invertebrates of the Salton Sea: distribution and seasonal dynamics. Hydrobiologia 473: 139–160.

Hart, C. M., M. R. González, E. P. Simpson & S. H. Hurlbert, 1998. Salinity and fish effects on Salton Sea microecosystems: zooplancton and newton. Hydrobiologia 381: 129–152.

Holdren, G. C. & A. Montaño, 2002. Chemical and Physical Characteristics of the Salton Sea, California. Hydrobiologia 473: 1–21.

James, W. F., J. W. Barko & H. L. Eakin, 2001. Phosphorus recycling by zebra mussels in relation to density and food resource availability. Hydrobiologia 455: 55–60.

Linsley, R. H. & L. H. Carpelan, 1961. Invertebrate Fauna. In Walker, B. W. (ed.), The ecology of the Salton Sea, California, in Relation to the Sportfishery. California Department of Fish & Game Fisheries Bulletin 113: 43–47.

Loeppert, R. H. & D. L. Suarez, 1996. Carbonate and gypsum. In Sparks, D. L. (ed.), Methods of Soil Analysis. Part 3. Chemical Methods. Soil Science Society of America, Madison, WI: 437–474.

Nelson, D. W. & L. E. Sommers, 1996. Total carbon, organic carbon, and organic matter. In Sparks, D. L. (ed.), Methods of Soil Analysis. Part 3. Chemical Methods. Soil Science Society of America, Madison, WI: 961–1010.

Reed-Anderson, T., S. R. Carpenter, D. K. Padilla & R. C. Lathrop, 2000. Predicted impact of zebra mussel (*Dreissena polymorpha*) invasion on water clarity in Lake Mendota. Canadian Journal of Fisheries and Aquatic Sciences 57: 1617–1626.

Riedel, R., L. Caskey & B. A. Costa-Pierce, 2002. Fish biology and fisheries ecology of the Salton Sea, California. Hydrobiologia 473: 229–244.

Rodriguez, I. R., C. Amrhein & M. A. Anderson, 2008. Laboratory studies on the coprecipitation of phosphate with calcium carbonate in the Salton Sea, California. Hydrobiologia. doi:10.1007/s 10750-008-9310-7

Simpson, E. P. & S. H. Hurlbert, 1998. Salinity effects on the growth, mortality and shell strength of *Balanus amphitrite* from the Salton Sea, California. Hydrobiologia 381: 179–190.

Tam, B. P., K. S. Mai, S. X. Zheng, Q. C. Zhou, L. H. Liu & Y. Yu, 2005. Replacement of fish meal by meat and bone meal in practical diets for the white shrimp *Litopenaeus vannamai* (Boone). Aquaculture Research 36: 439–444.

Tiffany, M. A., B. K. Swan, J. M. Watts & S. H. Hurlbert, 2002. Metazooplankton dynamics in the Salton Sea, California, 1997–1998. Hydrobiologia 473: 103–120.

Watts, J. M., B. K. Swan, M. A. Tiffany & S. H. Hurlbert, 2001. Thermal, mixing, and oxygen regimes of the Salton Sea, California, 1997–1999. Hydrobiolgia 466: 159–176.

Hydrobiologia (2008) 604:85–95
DOI 10.1007/s10750-008-9315-2

SALTON SEA

Relating fish kills to upwellings and wind patterns in the Salton Sea

**B. Marti-Cardona · T. E. Steissberg ·
S. G. Schladow · S. J. Hook**

Abstract In recent years, the extreme eutrophication of the Salton Sea has been associated with massive fish kills and associated bird kills. Analysis of the magnitude and direction of high wind events indicates that major fish kills are preceded by strong and persistent wind events, with a 24-h accumulated wind magnitude above a critical threshold of approximately 90 m/s. Twelve of the 14 cases of reported fish kills analyzed were found to be preceded by such wind conditions. The winds could potentially produce upwellings of hypolimnetic water at the upwind end of the Sea, resulting in the entire water column being low in dissolved oxygen and high in concentrations of hydrogen sulfide and ammonium. Remotely sensed thermal infrared data from the MODIS instrument on the Terra satellite was available for 5 of the 14 fish kills analyzed. Evaluation of satellite-derived surface temperature maps for these 5 fish kills shows that upwellings did take place after the wind events, affecting a large fraction of the Sea's area. The location of the upwelling and the fish kills coincided in all cases, confirming the relationship among wind patterns, upwellings, and fish kills in the Salton Sea. The importance of physically mediated processes, such as upwellings, need to be considered in evaluating future remediation strategies for the Salton Sea.

Keywords Infrared · Remote sensing · MODIS

Guest editor: S. H. Hurlbert
The Salton Sea Centennial Symposium. Proceedings of a Symposium Celebrating a Century of Symbiosis Among Agriculture, Wildlife and People, 1905–2005, held in San Diego, California, USA, March 2005

B. Marti-Cardona · T. E. Steissberg · S. G. Schladow
Department of Civil and Environmental Engineering, University of California at Davis, Davis, CA 95616, USA

S. G. Schladow (✉)
Tahoe Environmental Research Center, University of California at Davis, Davis, CA 95616, USA
e-mail: gschladow@ucdavis.edu

S. J. Hook
Jet Propulsion Laboratory, California Institute of Technology, Pasadena, CA 91109, USA

Introduction

The Salton Sea is a terminal, saline lake located in the Sonoran Desert of southeastern California (Fig. 1). It serves the role of an agricultural repository for irrigation return water and is also an important component of the regional and global ecology. The Sea is a key stop on the Pacific flyway for many species of migratory birds, and provides important habitat for many important species (see for example Patten et al., 2003; Shuford et al., 2002). Its importance to bird migration has increased with the loss of over 91% of the original wetlands in California during the twentieth century (Mitch & Gosselin, 2000).

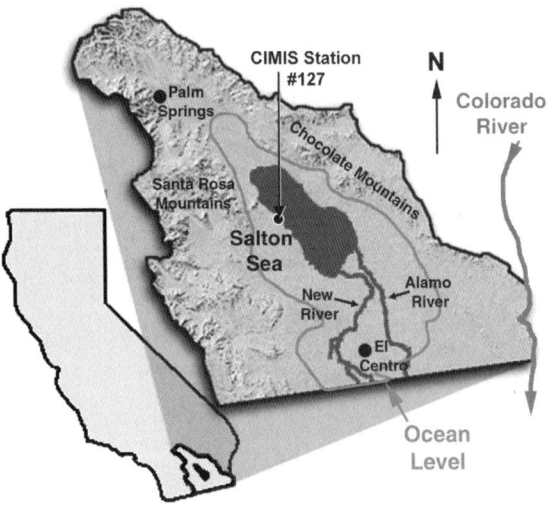

Fig. 1 Location of the Salton Sea Basin

In recent years, the extreme eutrophication of the Sea has been associated with massive fish kills and associated bird kills. Measurements have shown the Sea to have high concentrations of phosphorus and nitrogen, persistent periods of anoxic conditions below the thermocline, and a very low Secchi depth, all conditions symptomatic of a hypertrophic status (Holdren & Montaño, 2002). Anoxic conditions alone could lead to fish mortality. When anoxia occurs in the presence of high nitrogen concentrations and high sulfate concentrations, there is the added potential for the formation of ammonium (M. Anderson & C. Amrhein, personal communication) and hydrogen sulfide (Watts et al., 2001), respectively, both of which are potentially lethal to fish. Other potential causes of fish die-offs include toxic algae (Reifel et al., 2002; Tiffany et al., 2001) and parasite infections (Kuperman et al., 2001).

The purpose of this contribution is not to identify the precise lethal agent associated with fish kills at the Salton Sea, but rather to understand the conditions that lead to the onset of a massive fish kill. The fish kills are by their nature episodic, and do not appear to be an ongoing, daily process. They occur over a short period of time, possibly hours, and tend to be highly localized geographically (although they have been known to occur at different locations around the Sea). The isolation and low population around the Salton Sea make their observation difficult, and little is known of their precise dynamics.

Due in large part to the episodic nature of the fish kills at the Salton Sea, it is hypothesized that they are actually triggered by physical factors, not by chemical or biological factors. In particular, we believe they arise as a result of wind-induced upwelling events. Upwelling in a density (temperature) stratified water body results from a large surface wind stress being balanced by a horizontal pressure gradient, causing denser water to rise at the upwind lake boundary (Monismith, 1985, 1986; Stevens & Imberger, 1996; Farrow & Stevens, 2003). The process is shown schematically in Fig. 2. When an upwelling occurs, the entire water column at the upwind end of the water body has the chemical characteristics of the hypolimnion for the duration of the event. In the case of the Salton Sea, a very large, weakly stratified system, the time scale of an upwelling is on the order of 10–30 h. Upwelling is often an important part of ecosystem functioning, since it transports nutrients to the euphotic zone (MacIntyre, 1993, 1998; MacIntyre & Jellison, 2001), which facilitates phytoplankton growth. However, when the lower waters are anoxic and contain high concentrations of known fish toxins, a situation arises when fish may be rapidly engulfed in water that cannot sustain them for an appreciable length of time.

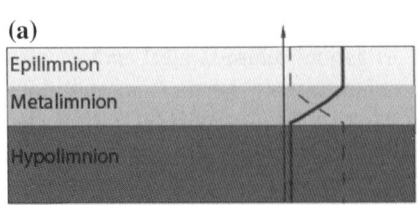

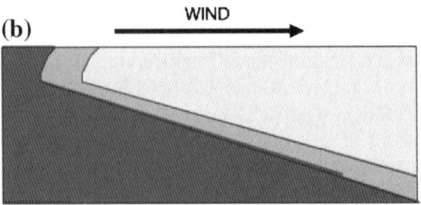

Fig. 2 (**a**) The arrangement of the epilimnion, metalimnion, and hypolimnion in a thermally stratified lake. The solid line represents the vertical distribution of temperature or dissolved oxygen. The dashed line represents the distribution of reduced substances such as hydrogen sulfide or ammonium. (**b**) Under constant winds, the metalimnetic water upwells at the downwind end, distributing reduced substances and low dissolved oxygen water throughout the water column

In order to test our hypothesis, we have examined the connection between documented fish kills at the Salton Sea, the measured wind patterns (in both magnitude and direction) at the Salton Sea, and the evidence of upwelling events from remotely sensed imagery of the surface temperature. If this hypothesis is correct, there should be clear evidence of fish kills occurring on the windward side of the Sea following periods of strong, sustained winds, and there should be evidence of cooler, upwelled water being present at the surface of the Sea during such events.

Materials and methods

Site description

The Salton Sea is an inland, saline lake located in the Sonoran Desert of southeastern California. The Sea's surface is approximately 56 km long and 20 km wide and has the largest surface area of any inland water body in the state of California. In recent years the maximum depth of the Sea is 16 m and the average depth 9 m. The prevailing wind direction is from the southwest.

The water flowing into the Sea is primarily agricultural, municipal, and industrial discharge from the Imperial and Coachella Valleys, and the city of Mexicali, which includes approximately 4,000,000 tons of dissolved salts every year. Since the only outflow for the Sea's water is evaporation, salt concentration has steadily increased over time. Today the salinity of the Salton Sea is approximately 45 g/l,

30% greater than that of the ocean, posing a threat to the sustainability of its ecosystem (Cook et al., 2002).

Records of fish kills in the Salton Sea

The U.S. Fish and Wildlife Service (USFWS) has compiled data on the frequency and magnitude of fish kills in the Salton Sea from January 2000 to December 2002. These data include the dates when the events were reported, estimates of the number, length, and total weight of the fish killed at each event, and the approximate location where the carcasses were found. The area around the Salton Sea is vast and sparsely populated during the hot summer months. It is highly likely that a fish-kill event would not be reported for several days. As there is almost no boating activity on the Sea, fish-kill reports were always based on observations from shore. Figures 3 and 4 depict the estimated number of fish dead by species and aggregated according to the USFWS records (Anderson, T., personal communication, Sonny Bono Wildlife Refuge, US Fish and Wildlife Service). Figure 3 shows that massive die-offs typically occur during the spring and summer months, when the lake is thermally stratified. The dominant fish species involved in these events were tilapia (*Oreochromis mossambicus*) and croaker (*Bairdiella icistia*). On some occasions the fish kills were exclusively a single species, while at other times there were multiple species reported. The 14 particular events that were analyzed in this research are highlighted. These are typically the larger events (greater than 100,000 fish dead).

Fig. 3 Estimates of magnitudes and dates of Salton Sea fish kills between January 1, 2000 and December 31, 2002

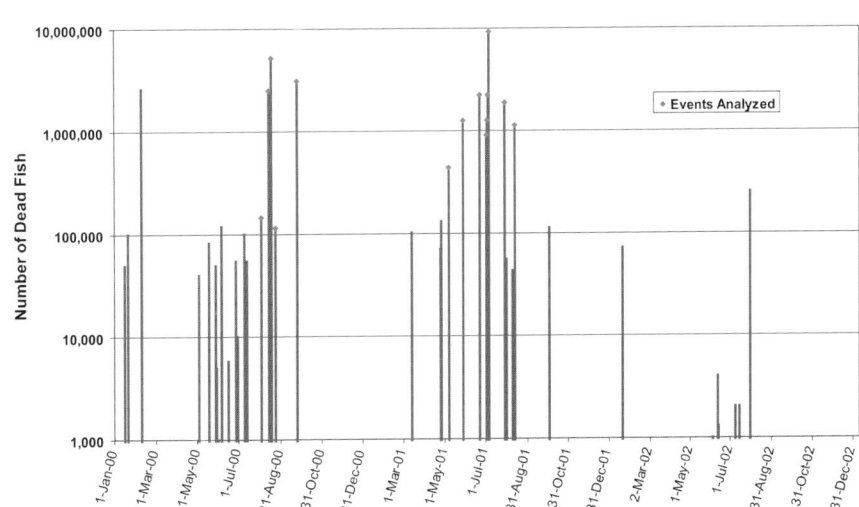

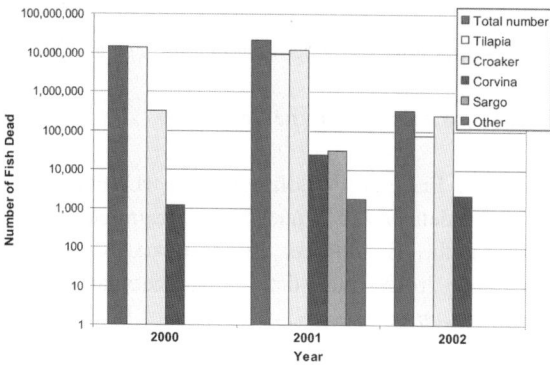

Fig. 4 Estimates of magnitude of Salton Sea fish kills by species between January 1, 2000 and December 31, 2002

Wind data analysis

To study the relationship between fish kills and wind patterns in the Salton Sea, hourly records of wind speed and direction in the Salton Sea were obtained from the California Irrigation Management Information System (CIMIS). The detailed analysis of the wind patterns around the dates of the fish kills used data from CIMIS Station 127, midway down the west coast of the Sea. This station is the most centrally located with respect to the Sea and therefore the most representative of winds in the region. Figure 5a represents the wind direction and speed at every hour for 5 different 24-h periods. Wind direction at Station 127 tends to rotate clockwise through the four quadrants of the wind rose on a daily basis.

Analysis of similar plots prior to the massive fish die-offs reveals that 1–5 days before the kills the winds are more concentrated in a particular quadrant of the wind rose and show higher speeds for a period of about 20 h. Figure 5b is an example for the 5 days prior to a fish kill in August 2001.

To better quantify the persistent direction of the winds prior to a fish-kill event, the 24-h accumulated wind magnitude has been calculated and plotted for the days prior to the fish die-offs. This quantity is defined as the vector sum of the 1-h wind magnitudes in a 24-h period. When the wind direction rotates in a 24-h period the wind vectors tend to cancel, yielding a small sum or accumulated wind magnitude. When winds blow strongly and persistently within narrow angles of the wind rose, the summation yields a large accumulated wind magnitude.

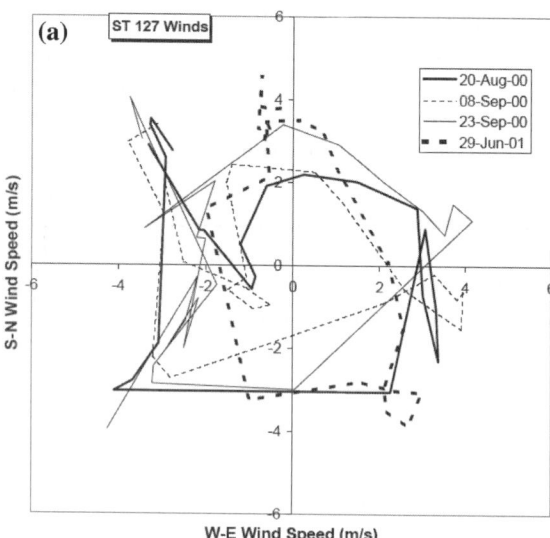

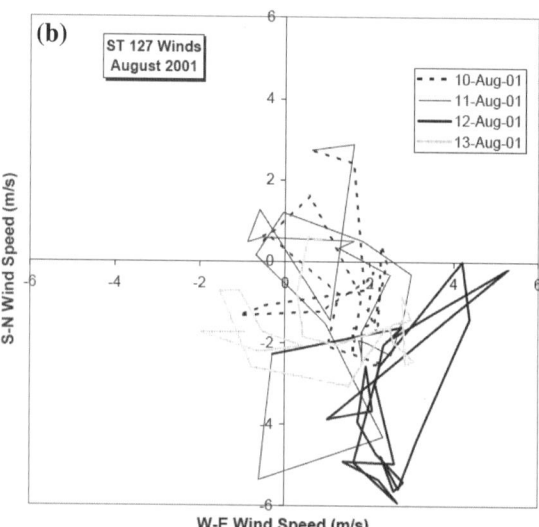

Fig. 5 (**a**) Examples of hourly wind data at Station 127 during normal conditions, with the clockwise swirl of the wind throughout the day. Note: the positions of the points represent the direction where the wind comes from, and their distance to the origin represents the wind speed. (**b**) Hourly wind data at Station 127 before the fish kill reported on August 13, 2001

Remote sensing detection of upwellings

Under stratified, quiescent conditions, the surface of the Salton Sea would be expected to have near-homogenous temperatures over its entire area. When a wind-driven upwelling occurs, as shown schematically in Fig. 2, a horizontal temperature gradient should be evident across the surface of the lake, with the cooler water located in the region of the

upwelling (Watts et al., 2001). The surface temperature can be measured using the thermal infrared (TIR) emission of the water surface. The TIR part of the electromagnetic spectrum (8–12 μm) is a function of the temperature and emissivity of the surface. Remote sensing instruments can measure the TIR radiance emitted from a surface from which the surface temperature can be derived. This temperature corresponds to the upper 10–1,000 μm of the lake water and is sometimes referred to as skin temperature. Spaceborne instruments measuring the TIR radiance emitted from the Earth's surface include the Advanced Spaceborne Thermal Emission and Reflection Radiometer (ASTER) (Yamaguchi et al., 1998) and the Moderate Resolution Imaging Spectroradiometer (MODIS) (Salomonson et al., 1989).

To investigate the occurrence of surface temperature gradients in the Salton Sea before fish-kill episodes, satellite imagery for the days prior to the major die-offs was analyzed. The remote sensing data consisted of thermal infrared imagery from the MODIS instrument on the Terra satellite. The MODIS instrument is also installed onboard the Aqua satellite, but this satellite was not launched until May 2002. MODIS data for the period covering from 6 days before to 1 day after the fish-kill events were used in this study. These data were only available for 5 of the 14 die-offs analyzed, namely: September 26, 2000 (3,090,000 fish dead), May 8, 2001 (442,000 fish dead), May 29, 2001 (1,269,200 fish dead), July 30, 2001 (1,851,000 fish dead), and August 13, 2001 (1,134,000 fish dead).

The MODIS sensor onboard the Terra satellite platform images the Earth both in the morning and in the evening. The sensor includes five thermal infrared bands, and three of these are used for measuring the energy emitted by the surface (band 29: 8.400–8.700 μm, band 31: 10.780–11.280 μm, and band 32: 11.770–12.270 μm). Each band has 1 km spatial resolution at nadir, which is sufficient to enable temperature mapping of a large water body such as the Salton Sea. Further details on the MODIS instrument are given in Salomonson et al. (1989).

The MODIS products used in this analysis were MODIS Calibrated Radiances, 5-Min Level 1B Swath, 1 km (MOD021KM) and Geolocation Fields, 5-Min_L1A Swath, and 1 km (MOD03). The MODIS standard Land Surface Temperature (LST-MOD11) and Sea Surface Temperature (SST-MOD28)

products were examined but not used, as their multi-band algorithms left residual noise in the temperature maps, due to uncorrelated noise in the individual bands (Bowen et al., 2002). This yielded images that were less clear than in the single-band brightness temperature images. Furthermore, the LST-MOD11 algorithm misidentified cool lake water pixels as clouds, resulting in the erroneous elimination of a large fraction of the water pixels of all the nighttime images and a smaller fraction of the water pixels in most of the daytime images.

Filtering of the individual bands prior to employing a multi-band correction algorithm yields low-noise temperature maps (Brown & Minnett, 1999). Principal Component Analysis (PCA) (Preisendorfer, 1988) was used to filter the data by rotating MODIS bands 31 and 32 into principal component space, discarding the second component (uncorrelated noise), and rotating the first component back into normal space to obtain the filtered results for bands 31 and 32. A custom skin temperature split-window algorithm, derived for Lake Tahoe using in situ thermistor data and MODIS Level 1B TIR images, was employed to atmospherically correct the MODIS thermal infrared images of the Salton Sea (Hook et al., 2003, 2007). This method produced clear, low-noise images that properly represented the thermal gradients. The resulting CST maps produced with PCA-filtered data were less noisy and contained fewer artifacts than CST maps obtained by low-pass filtering with either a Finite Impulse Response (FIR) filter or 3 × 3 pixel averaging prior to atmospheric correction.

The 1-km spatial resolution CST images were interpolated to a 90-m grid using bilinear interpolation for the final temperature maps to allow comparison with 90-m ASTER data. A land mask was also added to the images from a composite of two ASTER TIR images.

Mixed land-water pixels in the perimeter of the Sea create regions of unrealistically high temperatures near the shoreline of the daytime images and unrealistically low temperatures near the shoreline in the nighttime images. However, removal of these pixels would remove the necessary endpoints for interpolation near the shoreline, eliminating significant portions of the water surface in the interpolated images. Therefore, these pixels were retained, but to limit this effect, pixels having temperatures greater

than 45°C were set to 45°C, while pixels less than 5°C were set to 5°C prior to interpolation.

Results

The 24-h accumulated wind magnitude was calculated for the days prior to the 14 fish kills that have been analyzed. Figures 6a–c is an example of the results. The accumulated wind results show that most fish kills are reported within days of the peaks in this magnitude (recognizing that there may be several days between the occurrence of a fish kill and it being reported). For 9 of the 14 massive fish kills, they followed an accumulated wind magnitude in excess of 90 m/s. For 3 of the fish kills, they occurred after an accumulated wind magnitude of between 55 and 90 m/s. The remaining 2 fish kills took place on two consecutive days, in July 2001, and did not follow a period of strong or persistent winds.

MODIS thermal infrared images of the Salton Sea were available for 5 of the 14 fish kills analyzed. These were September 26, 2000, May 8, 2001, May 29, 2001, July 30, 2001, and August 13, 2001. Surface temperature maps of the Salton Sea were derived from the satellite thermal infrared data for the period covering from 6 days before to 1 day after these five events. Windy periods tend to coincide with overcast conditions, so the images corresponding to the peak accumulated wind dates, when the upwellings were initiated, are impaired by clouds in 3 of the 5 mapped events. However, the temperature maps capture the occurrence of upwellings, which persist for a few days after the event, either through an explicit temperature gradient in the direction of the wind or by showing a sudden, general cooling of the water surface after a wind peak, suggestive of vertical mixing processes.

The surface temperature maps for three fish-kill events are included and described below. The maps on the left and right columns were derived from the morning and the evening images, respectively. The temperature anomaly is plotted in each case, defined as the difference in temperature between each Sea pixel and the median of all Sea pixels in the image. The median temperature for the image is shown at the side of each image. Plotting the temperature anomaly enables the use of the same temperature scale for all figures. A negative temperature anomaly indicates a cool region. Black regions on areas of water indicate

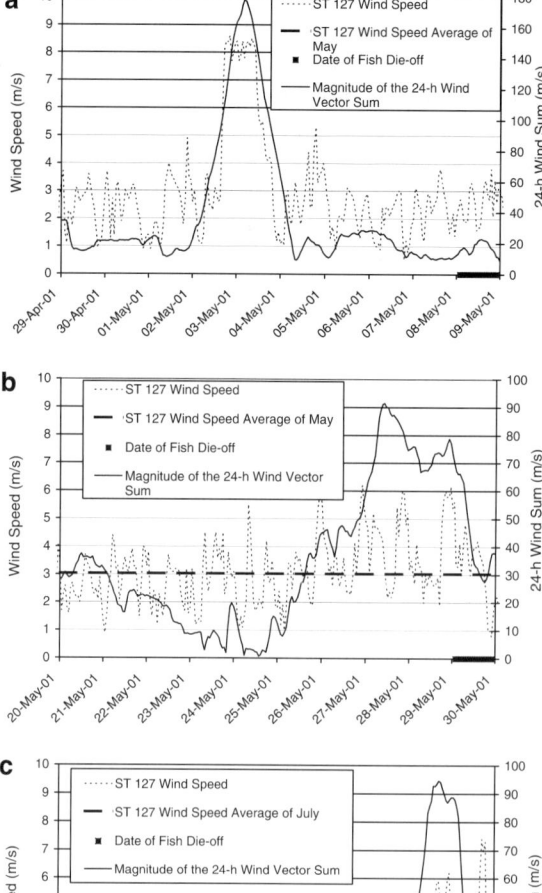

Fig. 6 (**a**) Accumulated wind magnitude before the fish kill reported on May 8, 2001. (**b**) Accumulated wind magnitude before the fish kill reported on May 29, 2001. (**c**) Accumulated wind magnitude before the fish kill reported on July 30, 2001

clouds, which have been masked to black, while white and black areas around the shore are where the temperature gradients were off-scale. The arrows between both plots indicate the wind direction, with the circular arrows indicating days when the wind followed the clockwise rotation described previously. The red circle on the image indicates the region in which the fish kill was reported to have occurred. It is shown on the map at the time of maximum apparent upwelling intensity.

Fish kill on May 8, 2001

Strong winds blowing from the East and Northeast on May 2, 2001 and from the Northeast the next day caused a 24-h wind summation peak of 180 m/s on May 3, 2001 (Fig. 6a). A total of 440,000 dead fish were reported on May 8 at the northeast of the Sea, between Desert Shores and the State Park Headquarters. No images are impaired by clouds during this period, and a temperature gradient consistent with the wind direction can be observed in the temperature maps of May 2 and May 3 (Fig. 7). The vertical mixing resulting from the upwelling is revealed by the general cooling of the surface water, which in the evening of May 3 is about 7°C colder than on the previous day. The period between the windstorm and the die-off is characterized by low-speed, rotating winds.

Fish kill on May 29, 2001

A windstorm took place between May 26, and May 29, 2001, reaching an accumulated wind maximum of 90 m/s on May 27 (Fig. 6b). A total of 1,300,000 fish were reported dead on May 29 at the north end of the Sea (Fig. 8). The images corresponding to the wind peak of May 27 were impaired by clouds and are not shown in Fig. 8. A temperature gradient in the wind direction can be observed on May 28. The morning and evening median surface temperatures on this date are considerably cooler compared to the rest of the morning and evening median temperatures, respectively.

Fish kill on July 30, 2001

Strong sustained southeast winds caused a wind summation peak of 90 m/s on July 29, 2001 and a 1.8 million fish die-off was reported on July 30 (Fig. 6c). The weather was cloudy on July 29, but a temperature gradient in the southeast direction is apparent on the temperature anomaly maps of July 30 (Fig. 9). The colder patch on these maps, in the vicinity of the New River, coincides with the area of the fish kill.

Discussion

Surface temperature maps derived from MODIS/Terra images of the Salton Sea capture the occurrence of upwellings by displaying a temperature gradient in the direction of the wind. The coolest water was at the upwind end of the Sea, and the location of the coolest water coincided with the reported location of fish kills. The upwelling zones were large, often encompassing hundreds of square kilometers (Fig. 7). This is consistent with the large fetch of the Salton Sea and the low temperature difference that typifies its summer-stratified period. Examination of MODIS/Terra images for periods outside times of known fish kills, when the wind direction was not sustained and the wind magnitude not high, revealed no evidence of upwelling.

Temperature contours approximately parallel to the shoreline are observable in most of the surface temperature maps. These contours are more pronounced on days of low accumulated winds, extend over a distance of about 3 km from the shoreline, and show a temperature increase toward the shoreline in the morning images and a decrease in the evening ones. Such gradients can be observed in real lakes due to differential cooling and heating in shallow areas. The near-shore gradient observed in this study, however, is in large part the result of the temperature interpolation of mixed land-water pixels along the shoreline, which are hotter than pure water pixels in the morning and colder at night.

The implications of windstorms for the aquatic life in the Salton Sea depend on the complex interaction of many environmental and ecological factors, such as water quality conditions, the fish health, population and spatial distribution within the Sea. It should be reiterated that the upwellings or the cool water they bring to the surface do not kill fish. Rather they are the causal link that transports potentially lethal concentrations of dissolved oxygen, hydrogen sulfide, and ammonium throughout the entire water column, thereby eliminating any refuge for the fish. If the conditions that produced low dissolved oxygen and high concentrations of ammonium and hydrogen sulfide in the hypolimnion of the Salton Sea did not exist, then the upwellings would not present a threat to fish.

No attempt was made to develop a quantitative relationship between wind patterns, the extent of upwellings, and the magnitude of fish kills since that would require knowledge of the vertical thermal stratification in the Sea which is not available. A numerical modeling approach could also yield information on the extent of upwelling, although data

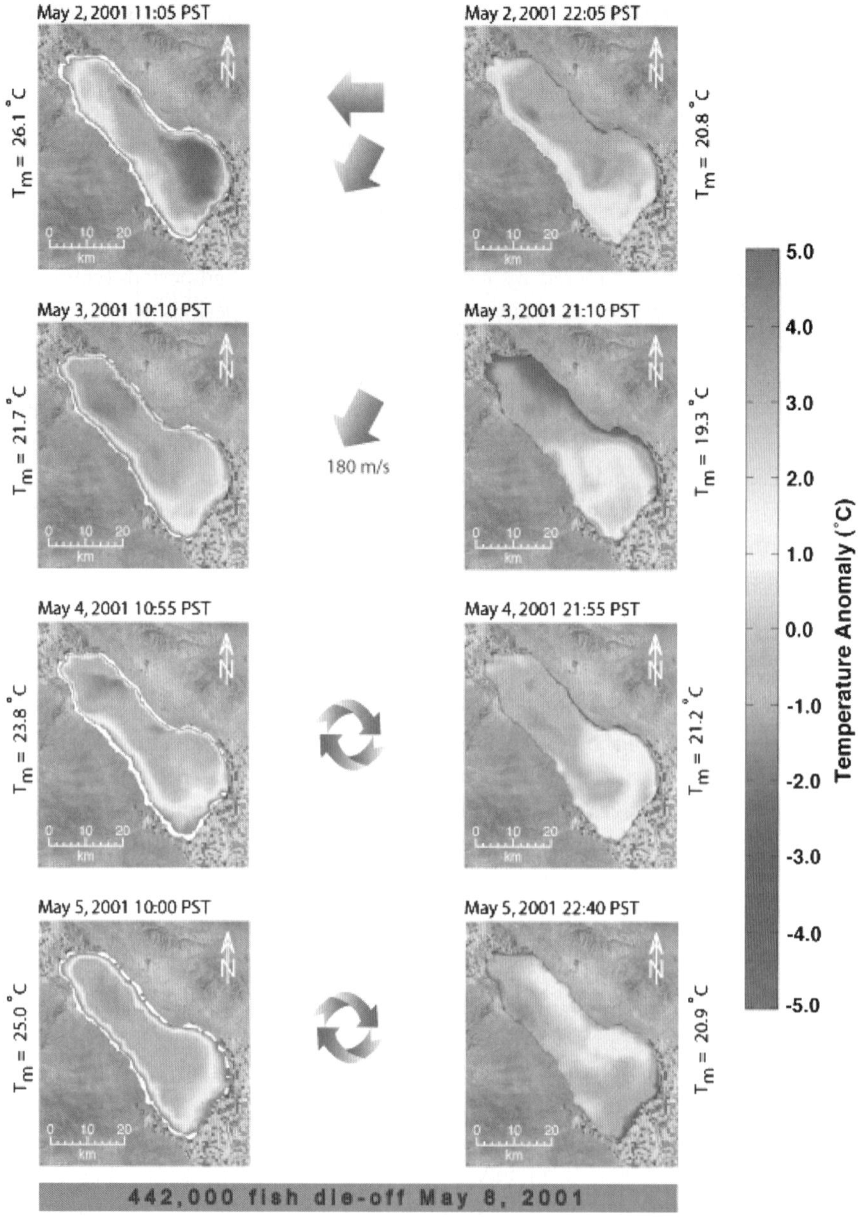

Fig. 7 Salton Sea surface temperature map. Upwelling is evident on May 2 and May 3. Fish kill reported on May 8. The median skin temperature (T_m) is shown next to each image. T_m is the Sea surface pixel temperature equal to or larger than half of the pixel temperatures and equal to or smaller than half of the pixel temperatures. Arrows indicate the wind direction or a circulating wind field

would again be required to calibrate and validate a model and were not available for the study period.

It is likely that there will be major changes within the Salton Sea. Water diversions will result in less water flowing into the Sea, and a concomitant decrease in water depth and increase in salinity. While the increase in salinity is arguably the greatest factor affecting the existing fish population, it is interesting to speculate on the effect of these changes on upwellings. A shallower Sea would tend to be less stratified; hence the buildup of harmful compounds

would be reduced. Therefore toxic upwelling conditions would tend to occur less frequently and would have a smaller affect on fish (presuming they still survive in the saltier Sea).

Hydraulic infrastructures, such as dykes and dams, have been proposed to reduce the volume of the Sea while maintaining current water levels and reducing salinity. A shorter Sea would decrease the intensity of upwellings as their amplitude would be reduced in proportion to the length. While this may appear to be a positive factor, it must be borne in mind that

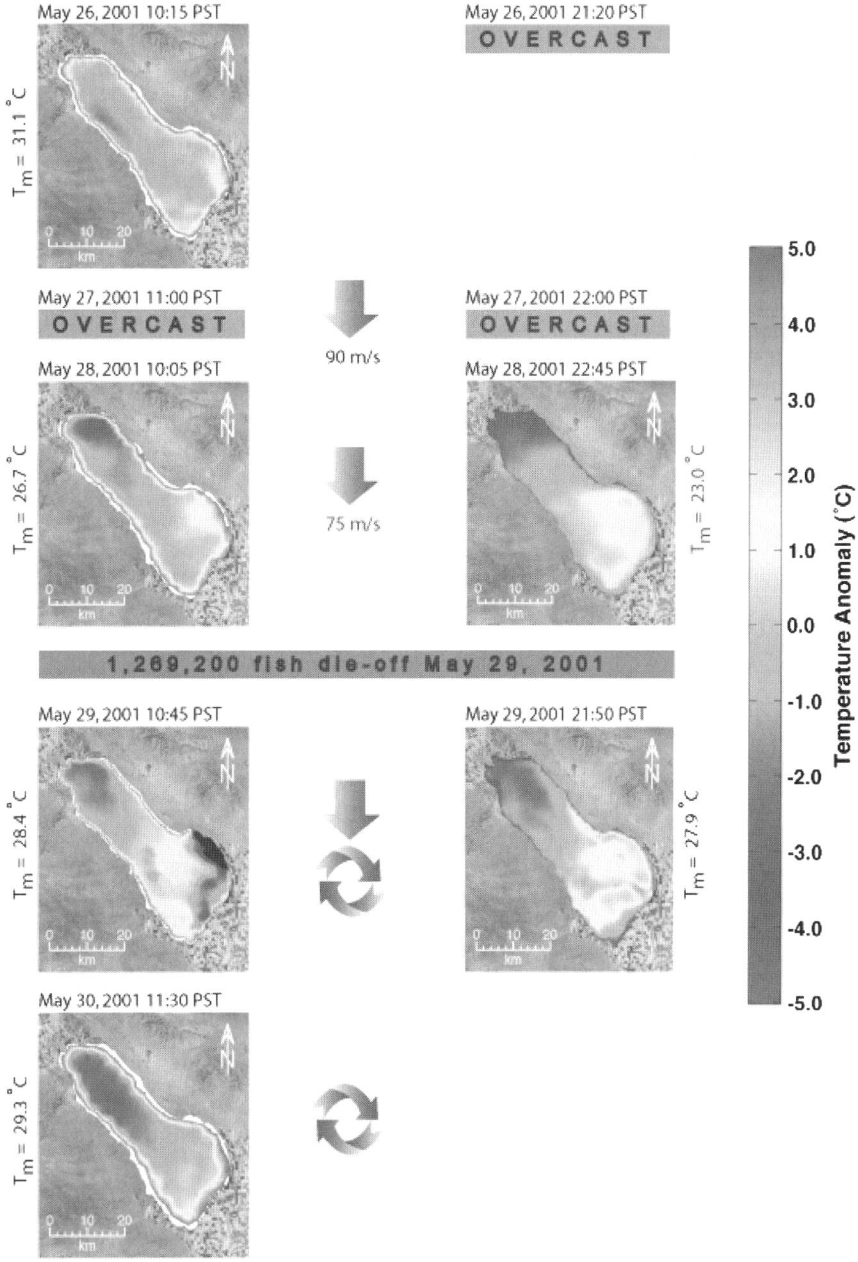

Fig. 8 Salton Sea surface temperature map. Upwelling is evident on May 28 and May 3. Fish kill reported on May 29. The median skin temperature (T_m) is shown next to each image. T_m is the Sea surface pixel temperature equal to or larger than half of the pixel temperatures and equal to or smaller than half of the pixel temperatures. Arrows indicate the wind direction or a circulating wind field

May 26, 2001 10:15 PST

$T_m = 31.1\,°C$

May 26, 2001 21:20 PST

OVERCAST

May 27, 2001 11:00 PST

OVERCAST

90 m/s

May 27, 2001 22:00 PST

OVERCAST

May 28, 2001 10:05 PST

$T_m = 26.7\,°C$

75 m/s

May 28, 2001 22:45 PST

$T_m = 23.0\,°C$

1,269,200 fish die-off May 29, 2001

May 29, 2001 10:45 PST

$T_m = 28.4\,°C$

May 29, 2001 21:50 PST

$T_m = 27.9\,°C$

May 30, 2001 11:30 PST

$T_m = 29.3\,°C$

Temperature Anomaly (°C)

5.0
4.0
3.0
2.0
1.0
0.0
-1.0
-2.0
-3.0
-4.0
-5.0

upwellings serve to transfer material from the hypolimnion to the surface of the lake. A reduction in upwelling by necessity will reduce this transfer. Therefore an areally smaller Sea, with similar water quality conditions to the current Sea, may simply replace a multitude of upwelling events that kill some of the fish population episodically throughout the summer with a single, massive release of hydrogen sulfide and ammonium when the smaller Sea undergoes its fall turnover. While upwelling events provide refugia for fish at the downwind end of the Sea, the release of toxins as part of the seasonal turnover will simultaneously affect the entire Sea.

Conclusions

Wind-driven upwellings in the Salton Sea during periods of thermal stratification are directly linked to the occurrence of fish kills. The upwellings occur

Fig. 9 Salton Sea surface temperature map. Upwelling is evident on July 30. Fish kill reported on July 30. The median skin temperature (T_m) is shown next to each image. T_m is the Sea surface pixel temperature equal to or larger than half of the pixel temperatures and equal to or smaller than half of the pixel temperatures. Arrows indicate the wind direction or a circulating wind field

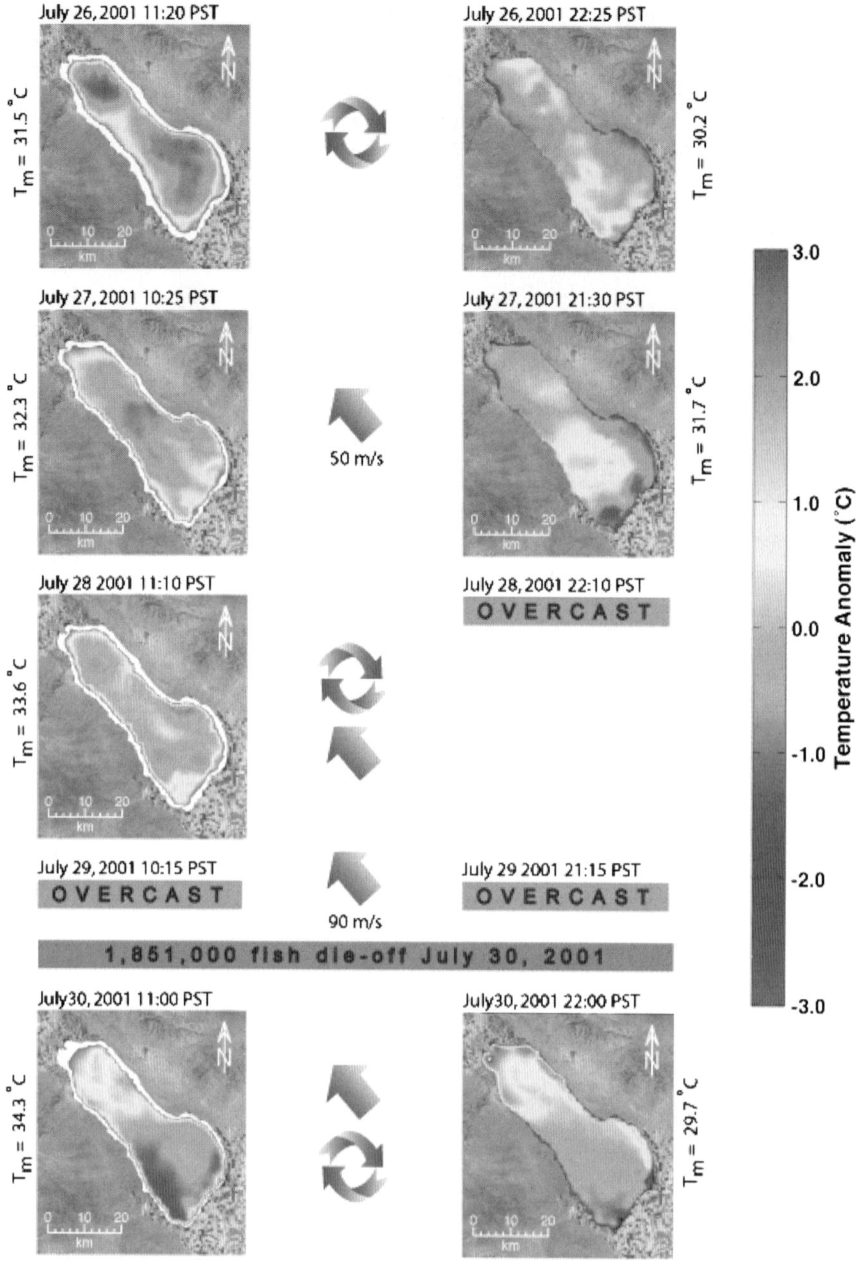

July 26, 2001 11:20 PST — $T_m = 31.5\,°C$

July 26, 2001 22:25 PST — $T_m = 30.2\,°C$

July 27, 2001 10:25 PST — $T_m = 32.3\,°C$

July 27, 2001 21:30 PST — $T_m = 31.7\,°C$

50 m/s

July 28 2001 11:10 PST — $T_m = 33.6\,°C$

July 28, 2001 22:10 PST — OVERCAST

July 29, 2001 10:15 PST — OVERCAST

July 29 2001 21:15 PST — OVERCAST

90 m/s

1,851,000 fish die-off July 30, 2001

July30, 2001 11:00 PST — $T_m = 34.3\,°C$

July30, 2001 22:00 PST — $T_m = 29.7\,°C$

Temperature Anomaly (°C): 3.0 2.0 1.0 0.0 -1.0 -2.0 -3.0

when wind is sustained in magnitude and direction such that the 24-h accumulated wind magnitude exceeds a critical threshold. Remotely sensed thermal infrared imagery has confirmed that the upwellings do occur at the times of these winds and that they are located precisely where the fish kills are observed.

Acknowledgments Funding for this research was provided by the Colorado River Basin Regional Water Quality Control Board under Contract SWRCB No. 01-265-170-0. The research described in this paper was carried out in part at the Jet Propulsion Laboratory, California Institute of Technology, under a contract with the National Aeronautics and Space Administration as part of the Earth Observing System Mission to Planet Earth Program. Numerous people have contributed to this work. In particular we would like to thank Ron Alley at JPL for the help to reduce the MODIS data.

References

Bowen, M. M., W. J. Emery, J. L. Wilkin, P. C. Tildesley, I. J. Barton & R. Knewtson, 2002. Extracting multiyear

surface currents from sequential thermal imagery using the maximum cross-correlation technique. Journal of Atmospheric and Oceanic Technology 19(10): 1665–1676.

Brown, O. B. & P. J. Minnett, 1999. MODIS Infrared Sea Surface Temperature Algorithm: Algorithm Theoretical Basis Document, Version 2.0. University of Miami, NAS5-31361.

Cook, C. B., G. T. Orlob & D. W. Huston, 2002. Simulation of wind-driven circulation in the Salton Sea: implications for indigenous ecosystems. Hydrobiologia 473: 59–75.

Farrow, D. E. & C. L. Stevens, 2003. Numerical modelling of a surface-stress driven density-stratified fluid. Journal of Engineering Mathematics 47: 1–16.

Holdren, G. C. & A. Montaño, 2002. Chemical and physical characteristics of the Salton Sea, California. Hydrobiologia 473: 1–21.

Hook, S. J., F. J. Prata, R. E. Alley, A. Abtahi, R. C. Richards, S.G. Schladow & S. O. Palmarsson, 2003. Retrieval of lake bulk and skin temperatures using along-track scanning radiometer (ATSR-2) data: a case study using Lake Tahoe, California. Journal of Atmospheric and Oceanic Technology 20: 534–548.

Hook, S. J., R. G. Vaughan, H. Tonooka & S. G. Schladow, 2007. Absolute radiometric in-flight validation of mid infrared and thermal infrared data from ASTER and MODIS on the Terra spacecraft using the Lake Tahoe, CA/NV, USA, automated validation site. IEEE Transactions Geoscience and Remote Sensing 45: 1798–1807.

Kuperman, B. I., V. E. Matey & S. H. Hurlbert, 2001. Parasites of fish from the Salton Sea, California, USA. Hydrobiologia 466: 195–208.

MacIntyre, S., 1993. Vertical mixing in a shallow, eutrophic lake: possible consequences for the light climate of phytoplankton. Limnology and Oceanography 38: 798–817.

MacIntyre, S., 1998. Turbulent mixing and resource supply to phytoplankton. In Imberger, J. (ed.), Physical Processes in Lakes and Oceans. American Geophysical Union, Washington DC: 561–590.

MacIntyre, S. & R. Jellison, 2001. Nutrient fluxes from upwelling and enhanced turbulence at the top of the pycnocline in Mono Lake, California. Hydrobiologia 466: 13–29.

Mitch, W. J. & J. G. Gosselink, 2000. Wetlands. Van Nostrand Reinhold, New York.

Monismith, S. G., 1985. Wind-forced motions in stratified lakes and their effect on mixed-layer shear. Limnology and Oceanography 30: 771–783.

Monismith, S. G., 1986. An experimental study of the upwelling response of stratified reservoirs to surface shear stress. Journal of Fluid Mechanics 171: 407–439.

Patten, M. A., G. McCaskie & P. Unitt, 2003. Birds of the Salton Sea. Status, Biogeography, and Ecology. University of California Press, Berkeley, California.

Preisendorfer, R. W., 1988. Principal Component Analysis in Meteorology and Oceanography. Elsevier, New York.

Reifel, K. M., M. P. McCoy, T. E. Rocke, M. A. Tiffany, S. H. Hurlbert & D. J. Faulkner, 2002. Possible importance of algal toxins in the Salton Sea, California. Hydrobiologia 473: 275–292.

Salomonson, V. V., W. L. Barnes, P. W. Maymon, H. E. Montgomery & H. Ostrow, 1989. MODIS: advanced facility instrument for studies of the earth as a system. IEEE Transactions on Geoscience and Remote Sensing 27: 145–153.

Shuford, W. D., N. Warnock, K. C. Molina & K. K. Sturm, 2002. The Salton Sea as critical habitat to migratory and resident waterbirds. In Barnum, D. A., J. F. Elder, D. Stephens & M. Friend (eds), The Salton Sea. Hydrobiologia 473: 255–274.

Stevens, C. & J. Imberger, 1996. The initial response of a stratified lake to a surface shear stress. Journal of Fluid Mechanics 312: 39–66.

Tiffany, M. A., S. B. Barlow, V. E. Matey & S. H. Hurlbert, 2001. Chattonella marina (Raphidophyceae), a potentially toxic alga in the Salton Sea, California. Hydrobiologia 466: 187–194.

Watts, J. M., B. K. Swan, M. A. Tiffany & S. H. Hurlbert, 2001. Thermal, mixing, and oxygen regimes in the Salton Sea, California, 1997–1999. Hydrobiologia 466: 159–176.

Yamaguchi, Y., A. B. Kahle, H. Tsu, H. T. Kawakami & M. Pniel, 1998. Overview of Advanced Spaceborne Thermal Emission and Reflection Radiometer (ASTER). IEEE Transactions on Geoscience and Remote Sensing 36: 1062–1071.

Hydrobiologia (2008) 604:97–110
DOI 10.1007/s10750-008-9308-1

Properties and distribution of sediment in the Salton Sea, California: an assessment of predictive models

M. A. Anderson · L. Whiteaker · E. Wakefield ·
C. Amrhein

Abstract The Salton Sea is the largest lake, on a surface area basis, in California (939 km^2). Although saline (>44 g/l) and shallow (mean depth approximately 9.7 m), it provides valuable habitat for a number of endangered species. The distribution of sediments and their properties within the Salton Sea are thought to have significant influence on benthic ecology and water quality. Sediment properties and their distribution were quantified and compared with predicted distributions using several sediment distribution models. Sediment samples ($n = 90$) were collected using a regular staggered-start sampling grid and analyzed for water content, organic carbon (C), calcium carbonate, total nitrogen (N), total phosphorus (P), organic phosphorus, and other properties. Water content, total N, and total and organic P concentrations were all highly correlated with organic C content. The organic C concentration showed a non-linear increase with depth, with low organic C contents (typically 1–2%) present in sediments found in depths up to 9 m, followed by a strong increase in organic C at greater depths (to about 12% at 15 m depth). The models of Hakanson, Rowan et al., Blais and Kalff, and Carper and Bachmann yielded very different predicted critical depths for accumulation (10.5–22.8 m) and areas of accumulation (0–49.5%). Hakanson's dynamic ratio model more reasonably reproduced the observed zone of elevated organic C concentrations in the Salton Sea than either exposure- or slope-based equations. Wave theory calculations suggest that strong winds occurring less than 1% of the time are sufficient to minimize accumulation of organic matter in sediments that lie at depths less than 9 m in this system.

Keywords Sediment resuspension ·
Distribution · Dynamic ratio · Wave theory ·
Salton Sea

Guest editor: S. H. Hurlbert
The Salton Sea Centennial Symposium. Proceedings of a Symposium Celebrating a Century of Symbiosis Among Agriculture, Wildlife and People, 1905–2005, held in San Diego, California, USA, March 2005

M. A. Anderson (✉) · L. Whiteaker · C. Amrhein
Department of Environmental Sciences, University of California, Riverside, CA 92521, USA
e-mail: michael.anderson@ucr.edu

E. Wakefield
Wildermuth Environmental, Inc, 23692 Birtcher Drive, Lake Forest, CA 92630, USA

Introduction

The distribution of sediments and their properties within a lake have significant influence on benthic and pelagial ecology, water quality, and paleolimnological interpretations. For example, the transport, distribution, biological exposure, and burial of particle-associated contaminants such as PCBs, DDT, and

many trace elements are intimately related to the distribution of organic C in lakes (Forstner & Wittman, 1979; Karickhoff et al., 1979). The distribution, resuspension and focusing of sediments in lakes is also important in interpreting sediment cores and inferring past conditions (e.g., Lehman, 1975). Internal nutrient recycling (Holdren & Armstrong, 1980; Sondegaard et al., 2003), sediment oxygen demand (Hatcher, 1986), macrophyte and benthic community distributions (Duarte & Kalff, 1986; Peeters et al., 2004), and other properties are also strongly dependent upon sediment properties.

The distribution and resuspension of sediment within a lake basin are generally recognized to be a function of wind speed, fetch, water depth, basin slope, particle size, sediment cohesiveness, and other factors (Lehman, 1975; Hakanson, 1982; Rowan et al., 1992a; Bloesch, 1995). A number of studies have quantified the lake area subject to erosion and/or deposition or accumulation and yielded simple predictive models using common morphometric parameters (Hakanson, 1982; Rowan et al., 1992a; Blais & Kalff, 1995). Using effective fetch, exposure, or lake surface area as surrogates for wave energy and related parameters, these studies provide a means to delineate the critical depths that separate zones of erosion (i.e., the zone of regular resuspension and no significant net sediment deposition), transportation (where sediment accumulation is interrupted by infrequent periods of resuspension), and accumulation (where no resuspension or further focusing occurs) (Hakanson & Jansson, 1983; Blais & Kalff, 1995).

In an early study, Hakanson (1982) developed the dynamic ratio concept, where lake area (a) and mean depth ($\bar{D}$) could be used to reasonably estimate the area of erosion + transportation (a_{E+T}), and therefore also the area of sediment accumulation within a lake (a_A) by the equation:

$$a_{E+T} = 100 - a_A = 25\frac{\sqrt{a}}{\bar{D}}41^{0.061\bar{D}/\sqrt{a}} \tag{1}$$

This approach was more physically sound and more accurate when applied to available data (Hakanson, 1982) than his earlier energy–topography approach (Hakanson, 1981). The distribution of the accumulation zone within a lake could, in turn, be determined using hypsographic data (Hakanson & Jansson, 1983).

In a later study, Rowan et al. (1992a) used wave theory to develop an equation based upon exposure (the circular integral of fetch) to predict the so-called mud energy boundary depth (mud EBD), taken as the upper limit to the depth (m) that separates erosional from transitional and depositional zones, given by:

$$mud\ EBD = 2.685E^{0.305} \tag{2}$$

where E is exposure (km^2). At low basin slopes (<3%), the mud EBD is equivalent to the so-called mud deposition boundary depth (mud DBD) (Rowan et al., 1995). Erosional sites are thus found at depths less than the mud DBD, depositional sites at depths greater than 1.34 times the mud DBD, and transitional or transportational sites between the mud DBD and 1.34 times * mud DBD (Rowan et al., 1995). Rowan et al. (1992a, 1995) included an alternate formulation that explicitly allowed for slope effects.

Slope effects were in fact found to be the strongest correlate with the percent accumulation area of anthropogenic Pb in the study by Blais & Kalff (1995), with percent area occupied by the accumulation zone (%a_A) following:

$$\%a_A = 49.94(\pm3.73) - 2.5(\pm0.31)\alpha'_p \tag{3}$$

where α'_p is the mean basin slope. This study also differed from earlier studies in that a highly sorptive contaminant (anthropogenic Pb) was used to define sediment zones within the lakes, rather than water content or particle size.

Sediment resuspension has been reasonably predicted in a number of studies with relationships that use wind speed, wind direction, fetch, and depth to sediment to infer loci and extent of resuspension (e.g., Carper & Bachmann, 1984). It has been shown that resuspension and erosion of fine-textured bottom sediment occurs when deep-water waves enter water shallower than one-half the wave length (Bloesch, 1995). The wavelength, L, of a deepwater wave is related to its period, T, by the relation:

$$L = \frac{gT^2}{2\pi} \tag{4}$$

where g is the gravitational constant (Martin & McCutcheon, 1999). A wave's period can be estimated using the empirical equation developed by the US Army Coastal Engineering Research Center (Carper & Bachmann, 1984) that states:

$$T = \frac{2.4\pi U \tanh\left[0.077\left(\frac{gF}{U^2}\right)^{0.25}\right]}{g} \quad (5)$$

where U is the wind speed and F is the fetch. In addition to allowing predictions about areas of resuspension within a lake, wave theory has been directly used to predict the sediment distribution pattern in Lake Ontario (Gilbert, 1999).

Thus, several different models that vary in their physical basis and their complexity were developed to predict sediment distribution in lakes, although independent tests of these models have been limited. The objectives of this study, then, were to quantify sediment properties within the Salton Sea, and to evaluate the ability of these different models to predict observed sediment distribution.

Methods

Study site

The Salton Sea is the largest lake (on an area basis) in California, with a surface area exceeding 900 km^2 (Table 1). The Salton Sea formed in 1905 following failure of an agricultural diversion on the Colorado River. The Colorado River flowed for approximately 16 months into the Salton Sink, a low-lying remnant of the Gulf of California. Since that time, the Salton Sea has been maintained by agricultural wastewaters from the extensive irrigation network and vast agricultural operations in the surrounding Imperial and Coachella Valleys.

Table 1 Morphometric properties of the Salton Sea

Property	Value
Latitude (°N)	33°20′
Longitude (°W)	115°51′
Surface elevation (m)[a]	−69.5
Surface area (km^2)[b]	939
Volume (km^3)[b]	9.12
Maximum depth (m)[b]	15.4
Mean depth (m)[b]	9.7
Mean basin slope (%)	0.16

[a] USGS (2006)

[b] USBR (2000) as modified from Ferrari & Weghorst (1997)

The Salton Sea receives an average annual inflow of approximately 1.665 km^3, with greater than 75% of that flow delivered to the southern end of the basin via the Alamo (0.765 km^3/yr) and New Rivers (0.540 km^3/yr) (Fig. 1) (Salton Sea Authority/USBR, 2000). Approximately 0.10 km^3/yr are delivered to the north end of the Salton Sea via the Whitewater River (Fig. 1), while another 0.13 km^3/yr flow from agricultural drains directly to the basin (Salton Sea Authority/USBR, 2000). These flows deliver very large loads of total P and total N (1,385 and 14,300 tonnes/yr) that help maintain eutrophic conditions there (Holdren & Montaño, 2002). A substantial amount of suspended and dissolved solid are also imported with these flows (468,100 and 3,434,000 tonnes/yr, respectively) (Holdren & Montaño, 2002). As a result of the large total dissolved solids inputs and the intense evaporative demand in the region (1.8 m/yr), the Salton Sea has increased in salinity over time, to levels exceeding 44 g/l (Holdren & Montaño, 2002). The Salton Sea is in approximate hydrologic steady-state, however, with the surface elevation varying less than about 0.5 m over a year and only about 1 m over the past 25 years (Cook et al., 2002).

The physical limnology of the Salton Sea has also been investigated by a number of researchers (Watts et al., 2001; Cook et al., 2002; Holdren & Montaño, 2002). Profile measurements of temperature, dissolved oxygen and other properties reveal strong summer heating, with stratification and infrequent mixing during the summer and more frequent mixing of the water column in the fall and winter (Watts et al. 2001; Holdren & Montaño, 2002). Cook et al. (2002) developed a 3-D hydrodynamic model of the Salton Sea based upon RMA-10 (King, 1993). Wind forcing dominated the circulation and mixing, setting up a strong counter-clockwise gyre in the south basin, with predicted velocities periodically reaching 20 cm/s or more. Velocities were particularly high in the comparatively shallow waters in the southernmost end of the Salton Sea as a result of conservation of momentum, where a large volume of water moving at some velocity must accelerate when forced into shallower conditions. Conversely, the model predicted much weaker circulation in the north basin that was less organized and somewhat chaotic (Cook et al., 2002). Acoustic Doppler current profiler measurements made at selected locations and times confirmed these general findings (Cook et al.,

Hydrobiologia (2008) 604:97–110

2002). These findings are broadly consistent with observations of Arnal (1961), who also reported high velocities generally oriented in a counter-clockwise direction in the south basin.

Field sampling and laboratory methods

Sediments from the lake were sampled on a regular, staggered-start sampling grid (Fig. 1). A total of 90 sites on the Salton Sea were sampled in the summer of 2001. Locations of sediment samples and lake margin were recorded using GPS. Depth to sediments was also recorded. Sediment samples were collected using a Ponar grab that sampled the uppermost 5–7 cm of sediment; the sediments were then briefly homogenized by mixing with a stainless steel spoon, and approximately 800 g wet-weight sediment were then transferred to 500 ml glass wide-mouthed jars. The jars were filled to the top with no headspace, capped with Teflon-lined lids and stored on ice for transport back to the laboratory, where they were stored at 4°C until analysis. The samples were later frozen for archival storage.

Sediment was promptly homogenized and sub-sampled for sediment characterization and porewater analysis. Water content was determined on subsamples that were heated overnight at 105°C. This higher temperature, commonly used in soil analyses, was chosen over the lower temperatures (60–80°C) used by Hakanson and others because of the high amount of hydrous clays and low organic matter content found in these sediments. Particle size was determined on samples mechanically dispersed in sodium hexametaphosphate without $CaCO_3$ or organic matter removal using the hydrometer method (Gee & Bauder, 1986). Total C, N, and S were measured by dry-combustion methods using a Carlo-Erba NA 1500 CNS analyzer (Carlo-Erba Instruments, Milan, Italy) (Nelson & Sommers, 1982). Inorganic C and $CaCO_3$ were determined manometrically following Loeppert & Suarez (1996). Organic C was taken as the difference between total C and inorganic C. Total

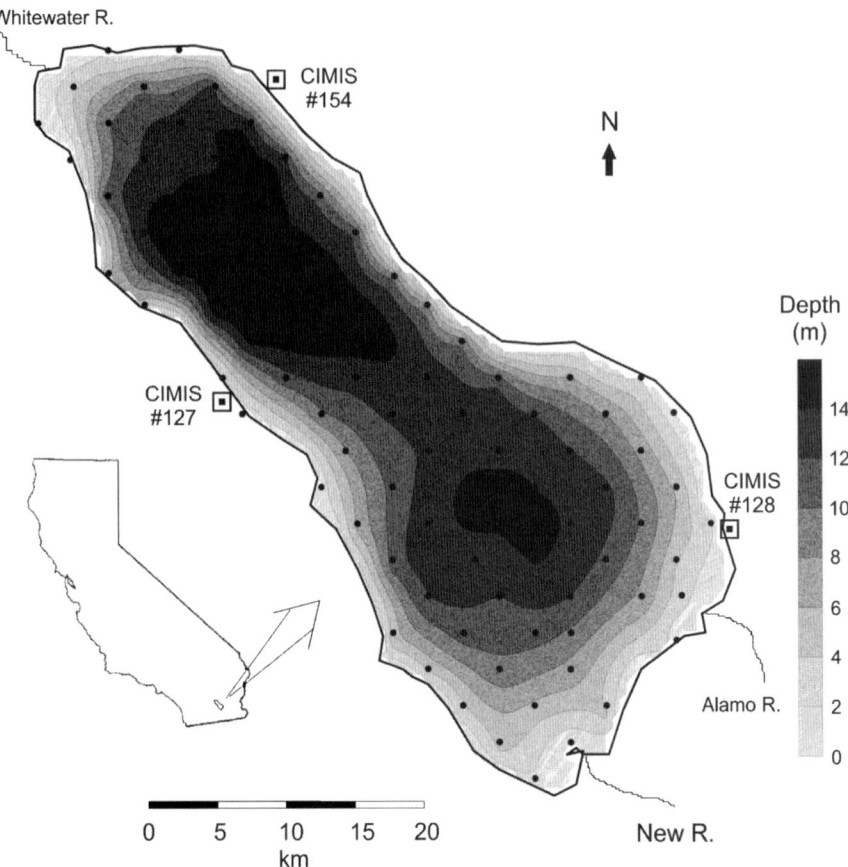

Fig. 1 Study site showing sediment sampling and CIMIS station locations and bathymetric map for the Salton Sea

and inorganic P was determined following Aspila et al. (1976). Organic P was taken as the difference between total and inorganic P (Aspila et al., 1976; Berner & Rao, 1994). Porewater was extracted by centrifugation, filtered through 0.45 μm polycarbonate filters fitted to disposable plastic syringes, and acidified with reagent-grade H_2SO_4. Porewater NH_4-N and total dissolved orthophosphate (TDP) and extracted P were analyzed using an Alpkem autoanalyzer (Astoria–Pacific International, Clackamas, OR) following standard methods (APHA, 1998).

The sediment sampling locations and depths, location of the lake margin, and other data were used to construct a bathymetric map for the lake (Fig. 1). The lake-wide distributions of various sediment properties were developed from latitude-longitude-property data using the kriging algorithm within Surfer^{TM} software (Golden Software, Inc., Golden, CO). The hypsographic data developed by the USBR (2000) as modified from Ferrari and Weghorst (1997) was used where needed in subsequent model calculations (Fig. 2).

Model calculations

Sediment distribution was predicted using the dynamic ratio model of Hakanson (1982) (Eq. 1), the exposure model of Rowan et al. (1992a, 1995) (Eq. 2), the slope model of Blais & Kalff (1995) (Eq. 3), and wave theory following Carper & Bachmann (1984) (Eqs. 4–5). Equations 1–3 require only morphometric data. Equation 3 uses mean basin slope (α'_p) to predict the percent area subject to

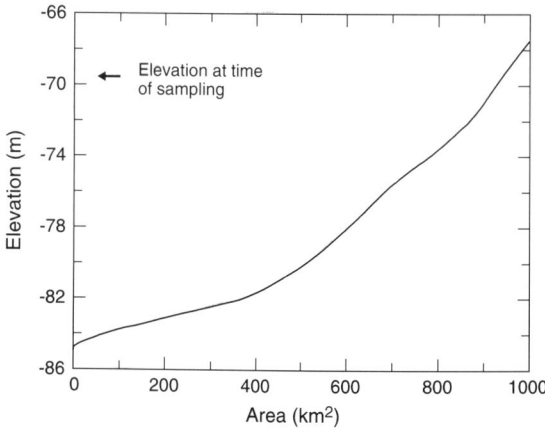

Fig. 2 Area vs. depth for the Salton Sea

deposition. The mean basin slope (Table 1) was calculated following Blais & Kalff (1995) as:

$$\alpha'_p = (l_0/2 + l_1 + l_2 + \cdots + l_{n-1} + l_n/2)Z_{max}/10nA_L \tag{6}$$

where l_0 is the shoreline length (km), l_i are the lengths of the contour lines (km), Z_{max} is the maximum depth in the lake (m), n is the total number of contour lines and A_L is total lake surface area (m^2).

Solution to the wave theory equations (Carper & Bachmann, 1984) (Eqs. 4–5) requires meteorological information; wind speeds for 2001 were taken from the California Irrigation Management Information System (CIMIS, 2004) meteorological stations #127, 128 and 154, located on the western, northeastern and southeastern shores of the Salton Sea, respectively (Fig. 1). The wind field acting on the Salton Sea is complex (Huston, 2000), but to simplify the calculations, an arithmetic average of the three stations was applied to estimate the average wind conditions acting over the whole lake surface.

Results

Sediment properties and their distribution

The "average" sediment in the Salton Sea had a particle size distribution of 24.1% sand, 45.4% silt, and 30.5% clay, although considerable variation within the sediments was found (e.g.,% sand ranged from 0.6 to 92.5%) (Table 2). Organic C within the sediments ranged from 0.13 to 12.5% and averaged 4.17% (on a dry-weight basis). Calcium carbonate ($CaCO_3$) within the sediments ranged from 5.1 to 75.2% and averaged 23.8%, while total N ranged from 0.06 to 1.30% and with a mean N content of 0.43%. Total P averaged 641 mg/kg with approximately 70% of that P in an inorganic form extractable with 1 M HCl (Aspila et al., 1976). Porewater concentrations of NH_4-N were generally quite high, averaging 15.5 mg/l and ranging from 3.6 to 57.1 mg/l. The mean porewater total dissolved P (TDP) concentration was 0.77 mg/l, with concentrations ranging from 0.02 to 5.19 mg/l. Large standard deviations were present in all measured properties, however, reflecting the wide site-to-site variation across the lake.

Table 2 Sediment properties ($n = 90$)

Property	Mean ± s.d.	Range
Sediment		
Water content (%)	51.8 ± 14.6	22.6–77.5
Sand (%)	24.1 ± 26.3	0.6–92.5
Silt (%)	45.4 ± 17.9	0.0–68.1
Clay (%)	30.5 ± 10.4	7.2–50.3
Total C (%)	7.02 ± 3.83	1.05–15.7
Organic C (%)	4.17 ± 3.08	0.13–12.5
$CaCO_3$ (%)	23.8 ± 11.4	5.1–75.2
Total N (%)	0.43 ± 0.28	0.06–1.30
Total S (%)	2.25 ± 1.11	0.23–6.12
Total P (mg/kg)	641 ± 195	201–1132
Inorganic P (mg/kg)	442 ± 153	131–924
Organic P (mg/kg)	203 ± 162	0–885
Porewater		
TDP (mg/l)	0.77 ± 0.71	0.02–5.19
NH_4-N (mg/l)	15.5 ± 9.3	3.61–57.1

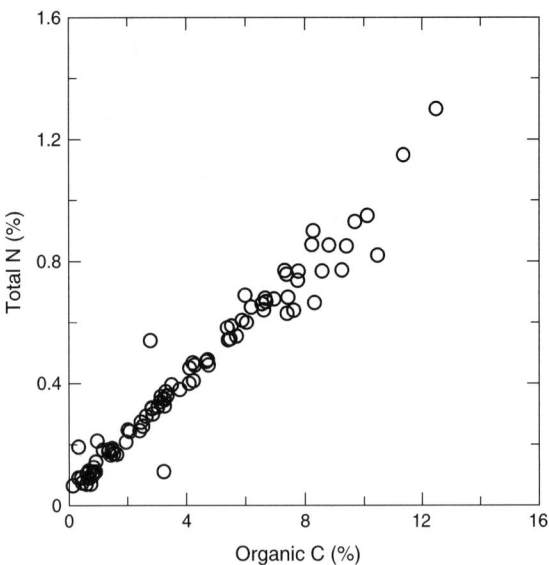

Fig. 3 Total N versus organic C concentrations in the sediments of the Salton Sea

Not unexpectedly, a number of the sediment properties were highly correlated. For example, organic C and total N within the sediments was very strongly correlated ($r = 0.98$) (Figs. 3, 4) with an average organic C:total N ratio slightly greater than 10:1. In addition to the very high correlation between total N, and organic C, water content, silt, clay, total and organic P, and $CaCO_3$ were also significantly correlated with organic C levels (at $P < 0.001$). Sediment properties were generally inversely correlated with% sand content. Excluding% sand and porewater NH_4-N and TDP, sediment properties were also positively correlated with depth.

The distribution of many of the measured sediment properties across the basin varied in a consistent way, with, e.g., higher organic C, total N, and organic P levels found in the north basin ($Z_{max} = 15.4$ m) when compared to levels found in the slightly shallower south basin ($Z_{max} = 14.3$ m) (Fig. 1).

Although linear regression indicated a strong correlation between organic C (and other properties) and depth, the data were in fact non-linear, with generally low organic C concentrations (typically 1–2%) present out to approximately 9 m (Fig. 5). Linear regression of the data over this smaller depth interval yielded a slope value not significantly different from 0, thus meeting the criterion of the erosional zone, i.e., no substantial sediment

accumulation. Beyond this depth, the organic C content of the sediments increased strongly with depth, to levels exceeding 10% (slope of 0.95%/m, significant at $P < 0.01$). The empirical data thus suggests that the erosional zone, subject to regular resuspension, extends out to approximately 9 m depth in the Salton Sea, while the transportational zone (where sediment accumulation is interrupted by infrequent periods of resuspension) extends from 9 m to near the maximum depth of the lake (Fig. 5). A clearly-defined depositional zone (where no resuspension and focusing occurs) would yield a region on Fig. 5 where organic C concentrations remain unchanged at some maximal level at large depth. Such a region is not readily apparent in Fig. 5, implying that the critical depth for accumulation is near the maximum depth of the Sea (Table 3).

Wind data

Winds out of the NNW (essentially down the long axis of the lake) occurred about 10% of the time, although winds from the SE also occurred with some regularity (as measured at CIMIS station #127) (Fig. 6). Wind records at the other 2 CIMIS stations differed somewhat, with especially strong winds out of the SW measured at CIMIS station #154 (data not

Fig. 4 Distribution of sediment (**a**) organic C, (**b**) total N and (**c**) organic P within the Salton Sea

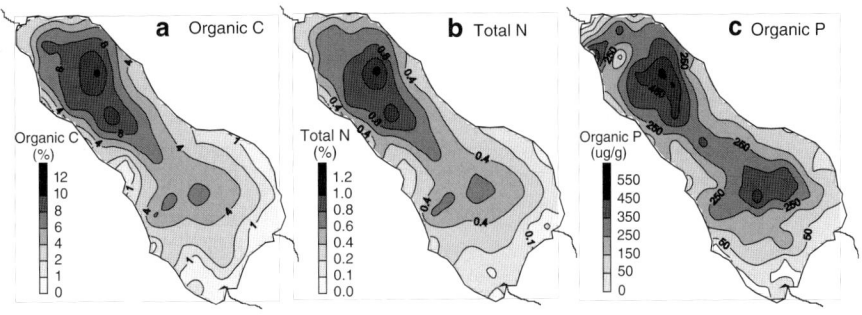

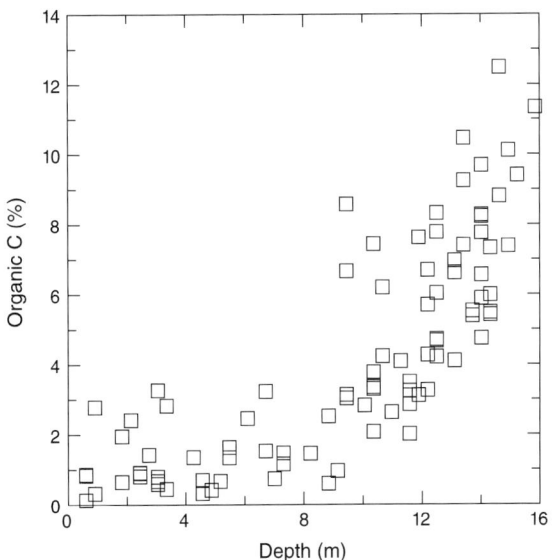

Fig. 5 Distribution of sediment organic C as a function of depth

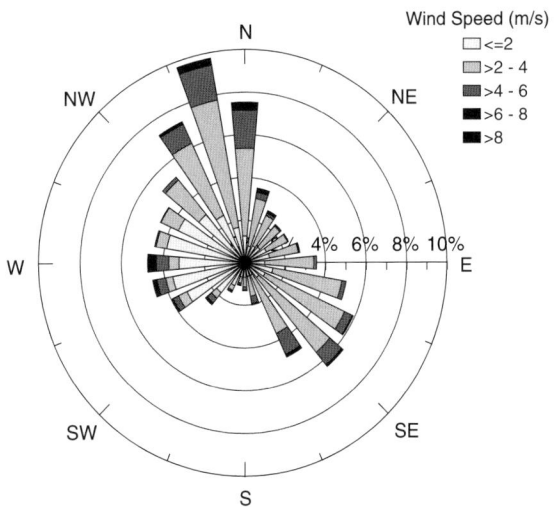

Fig. 6 Wind rose showing hourly average wind speeds and directions during 2001 measured at CIMIS station #127

Table 3 Predicted% areas and critical depths for sediment erosion and accumulation

Source	Erosion		Accumulation	
	Depth (m)	Area (%)	Depth (m)	Area (%)
Hakanson (1982)	na	na	14.3	12.1
Rowan et al. (1992a)	17.0	100.0	22.8	0
Blais & Kalff (1995)	na	na	10.5	49.5
Carper & Bachmann (1984)	na	na	12.8	30.4
Observed	9.0	58.4	≥15.0	≤4.9

and therefore also wave lengths and wave-mixed layer depths (or critical depths) are related to wind speed (Eqs. 4–5), we can expect that mixing to significant depths occurs regularly at the Salton Sea.

Predicted sediment distribution

Of the four models evaluated in this study, only the model of Rowan et al. (1992a, 1995) explicitly delineates the critical depth for erosion (Eq. 2). That model was thus used to predict from exposure the critical depth of erosion or, in their nomenclature, the mud DBD, i.e., depth that separates erosion from transitional and depositional zones. The calculations were done on a site-by-site basis to map the predicted distribution of sediment (Rowan et al., 1995). The mean exposure for all 90 sites was 441 ± 142 km². Mud DBD values were calculated from Eq. 2 for each

shown). Wind speeds were typically about 2 m/s, although periods of higher wind speeds, often greater than 8 m/s, were also witnessed with some frequency (Fig. 6). The average maximum hourly wind speed for the 3 stations was 11.9 m/s. Since wave periods

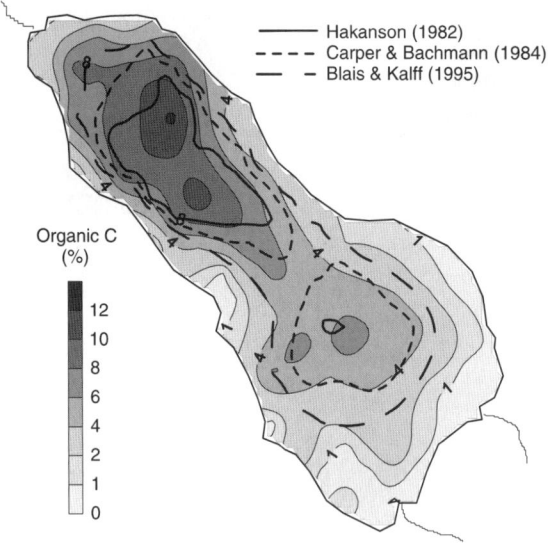

Fig. 7 Predicted areas of sediment accumulation against measured organic C distribution. The model of Rowan et al. (1992a, 1995) predicts an area of accumulation of 0% and so is not depicted in this figure

site (Fig. 1) and used to calculate the average mud DBD (17.0 ± 1.9 m). This critical depth for erosion was much greater than the 9-m depth identified based upon sediment analyses (Fig. 5, Table 3), and in fact exceeded the maximum depth of the Sea (Table 1).

All of the models provided a means to predict the critical depth delineating the boundary between erosion + transportation and accumulation, and the corresponding area for sediment accumulation. Quite divergent predictions resulted from application of these different models, however (Table 3, Fig. 7).

The dynamic ratio model of Hakanson (1982) required only mean depth and lake surface area to estimate the area of accumulation (Eq. 1), and predicted that 12.1% of the Salton Sea is accumulating sediments and not subject to erosion or transportation (Table 3). These area estimates were then used in conjunction with hypsographic data (Fig 2) (Ferrari & Weghorst, 1997; USBR, 2000) to determine the critical depth for accumulation (Hakanson & Jansson, 1983); with 12.1% of total surface area or 118 km^2 of the Salton Sea accumulating sediments, this places the critical depth for deposition (14.3 m) near the maximum depth present in the Sea (15.4 m).

Following Rowan et al. (1992a), the critical depth for accumulation from their exposure-based model was

taken as 1.34 times the mud DBD or 22.8 ± 2.5 m, and used with the hypsographic data (Fig. 2) to estimate the% area of accumulation for direct comparison with Hakanson (1982) (Table 3). Since their predicted critical depth for accumulation exceeds the maximum depth of the Salton Sea (Table 1), the% area of accumulation is 0% (i.e., no area within the Salton Sea is stable against regular resuspension).

The basin slope-based model of Blais & Kalff (1995) has a y-intercept of 49.92% and slope term of 2.50 (Eq. 3), indicating that at low basin slopes the predicted accumulation areas approach about 50%. Since the Salton Sea has a low mean basin slope (0.16%), the predicted depositional area is slightly less than 50% (Table 3) and encompassed a large portion of both the north and south basins of the lake (Fig. 7).

Wave theory (Carper & Bachmann, 1984) (Eqs. 4–5) was used to calculate the wave length and wave period for the maximum average wind speed of 11.9 m/s reported for 2001, and the average effective fetch of 21 km, taken as the square-root of the average exposure since these two terms are nearly identical mathematically (Rowan et al., 1992b). Under these conditions, wave theory predicts a critical depth of 12.8 m, corresponding to a sediment accumulation area of 30.4% (Table 3). The predicted zone of sediment accumulation using this approach is intermediate between the predicted zones of Hakanson (1982) and Blais & Kalff (1995) and includes much of the north basin and a smaller area within the south basin of the lake (Fig. 7).

Discussion

Sediment properties

The properties measured in this study indicate that the composition of the sediments in the Salton Sea have evolved over the past approximately 45 years since Arnal (1961) completed his sampling of the basin in January 1955. In that comprehensive study of the Salton Sea's limnology, sedimentation and microbiology, he collected sediment and water column samples at about 100 sites in a series of transects across the basin (Arnal, 1961). His sampling scheme included a larger number of near-shore sites and also excluded a large region on the west side of the south

basin, since it was within the confines of Salton Sea Test Base that was used in the 1940s and 1950s as a sea level ballistics range to obtain performance data on inert nuclear weapons prototypes.

Arnal (1961) reported organic matter contents of the sediments calculated from measured organic C concentrations multiplied by 1.7. Thus reported organic matter concentrations that ranged from <1 to 6% organic matter (Arnal, 1961) correspond to <1 to 3.5% organic C; these values are below the mean of 4.19% and well below the upper range of organic C concentrations measured on samples we collected in 2001 (Table 2). These sediment concentrations are generally much higher than found in the soils in the region, however. For example, Wu et al. (unpubl. data) recently reported average organic C contents of 0.8 and 0.5% for agricultural and native soils, respectively, approximately 35 km southwest of the Salton Sea near Holtville, California. Autochthonous algal production thus increased organic C contents relative to surface soils eroded by water and wind and transported to the Salton Sea. Contributing to the increased organic C concentrations in the surface sediments found in this study relative to those reported by Arnal (1961) is the higher level of productivity presently in the Sea. For example, Holdren & Montaño (2002) reported Secchi depth values of 0.4–1.4 m, while Arnal (1961) reported a mean Secchi depth of approximately 2 m with values as high as 3.05 m. The relative distribution of organic C within the basin has not changed, however; similar to our findings, Arnal (1961) found the highest concentrations in the north basin with much lower levels near the margins and in the south basin that he attributed to differences in local water currents.

The sediments of the Salton Sea are also enriched in $CaCO_3$ relative to soils in the region. Calcium carbonate concentrations averaged 11.7% in both native and irrigated agricultural soils (Wu et al., unpubl. data) and similar to the 10% content of suspended solids load in the Colorado River (Sykes, 1937). In contrast, Arnal (1961) reported a mean $CaCO_3$ content of the sediments of 18.6%, while sediments sampled in 2001 yielded an average concentration of 23.8% (Table 2). Precipitation of $CaCO_3$ due to chemical and biological processes contributes substantially to the $CaCO_3$ content of the sediments, with evidence for increasing in situ precipitation as the Sea has continued to increase in salinity.

These levels are similar to other saline lakes in the western US. For example, Walker Lake, a saline closed basin lake near the Nevada-California line, has surficial $CaCO_3$ contents of approximately 17–21% (Yuan et al., 2006), levels broadly comparable with those found in the Salton Sea. The water column of the Salton Sea is supersaturated with respect to $CaCO_3$ (Holdren & Montaño, 2002), so precipitation within the water column and deposition is considered a major source of $CaCO_3$ in the sediments. Sulfate reduction and associated alkalinity production within the sediments may also form $CaCO_3$ in situ (Anderson & Amrhein, unpubl. data).

The high amount of $CaCO_3$ present in the sediments is thought to account for the comparatively large amount of P associated with inorganic phases (approximately 70%) (Table 2) due to coprecipitation of P with $CaCO_3$ (Rodriguez et al., this issue; Danen-Louwerse et al., 1995), adsorption onto $CaCO_3$ surfaces (Freeman & Rowell, 1981), and/or formation of apatite or other discrete Ca-phosphate phases. Interestingly, porewater NH_4-N was not correlated with total N in the sediments, although porewater total dissolved P was correlated with total sediment P. The distribution of organic C, total N and organic P were all broadly similar, with comparatively low levels in the sediments near the margins of the Salton Sea, especially in the south basin, and higher levels in the deeper waters, particularly in the north basin. This is in accord with the known circulation within the Salton Sea, where very high water velocities (often 20 cm/s at mid-depths) and a strong counter-clockwise gyre are present in the south basin, with much lower velocities and less well-organized flows in the north basin (Cook et al., 2002).

Sediment distribution models

Estimates of the erosional area within the Salton Sea were limited to the model of Rowan et al. (1992a). Application of their exposure-based model indicated that regular resuspension of sediment occurs throughout the Sea (i.e., the erosional area is 100%). This appears inconsistent with the field sampling results (Fig. 4a, 5) however, where almost 60% of the bottom sediments of the Sea have little organic C (<3% or so) with no measurable increase with depth, and thus appear to experience regular resuspension (Table 3).

The predicted distribution of accumulating sediments within the lake basin varied substantially among the four models. The model of Hakanson (1982) correctly identified the center of the north basin as the principal location of sediment accumulation, delineating a region 118 km^2 in area, with only a very small region of net deposition in the south basin (Fig. 7). The predicted critical depth of 14.3 m coincided with about the 8% organic C contour, although the agreement was not perfect. Water contents have traditionally been used as a simple way to characterize sediments, with Hakanson (1977) using the 75% water content as the threshold for the depositional zone, and 50% as the threshold for the transportational zone. The 75% water content corresponded to about 9.7% organic C in sediments from Lake Ekoln (Hakanson & Jansson, 1983), while the median organic C content corresponding to 75% water content from 3 lakes in southern Ontario, Canada was similar, at about 9% organic C (Cyr, 1998).

The model of Rowan et al. (1992a) predicted no distinct depositional zone (i.e., the average mud DBD or critical depth that separates erosional and transitional zones, and the critical depth that separates the transitional and depositional zones, were both greater than the maximum depth of the Sea). Thus, the exposure-based model indicates that the entire Salton Sea basin would effectively be subject to erosion. As noted above, however, this is not consistent with the observed organic C distribution with depth (Fig. 5), because erosional zones are considered areas where there is no net deposition of fine materials (Hakanson and Jansson, 1983). An increase in organic C content with depth is clearly present at depths greater than about 9 m (Fig. 5).

In contrast, Blais & Kalff (1995) predicted a much larger area of deposition (482 km^2) based upon their empirical slope model developed from a characterization of sediments from lakes that ranged in average basin slope from 1.1 to 18.0% and surface areas of 0.04–17.9 km^2. That is, almost 50% of the area of the Salton Sea was thought to be accumulative, with the critical depth of 10.5 m following somewhat the 3% organic C contour line (Fig. 7). This is thought to be a substantial overestimate of the depositional zone, however, since up to 30% sand and low organic C and water contents were often found within samples putatively within the depositional

zone (data not shown). The average basin slope (0.16%) and the surface area (939 km^2) of the Salton Sea fall outside of the values used to develop this empirical model, so it appears that this model is not suited for large shallow lakes.

Interestingly, the non-linear increase in organic C with depth (Fig. 5) followed the trend noted for lakes subject to type II focusing as defined by Blais & Kalff (1995) in lakes with large basin slopes (5–17%). As noted, however, the Salton Sea has a very small average basin slope (0.16%) (Table 1), and even the highest slope found on the west side of the north basin ($\sim$0.5%) (Fig. 1) negligibly reduces the predicted area of accumulation using Eq. 3. Thus increases in organic C and related properties also occur in relatively shallow lakes with large erosional and/or transportational zones and small depositional zones.

Wave theory calculations using the average effective fetch for the lake and the maximum observed wind speed yielded a critical depth for accumulation that was somewhat less than that predicted from the dynamic ratio model of Hakanson (1982), and thus predicted a larger area of accumulation (Table 3; Fig. 7).

This analysis considers only the average effective fetch, although more careful geometric analyses (e.g., Hakanson & Jansson, 1983; Carper & Bachmann, 1984), shows effective fetch lengths longer than the average value when winds blow down the long axis of the Salton Sea. The maximum effective fetch (F_{eff}) was calculated following Hakanson & Jansson (1983) as:

$$F_{eff} = \sum F_i \cos \alpha \Big/ \sum \cos \alpha \qquad (7)$$

where F_i is the distance to the shore along radials at $\alpha = 0$–42° in 6° intervals from the long axis of the lake.

Using Eq. 7, one calculates a maximum fetch length to deep water of 26 km. Substituting this distance into Eq. 5 and using the maximum wind speed directed down the long axis of the lake (10.9 m/s) (Fig. 6), one predicts a wave-mixed layer depth of 12.7 m. Sites beyond this depth would lay within the depositional zone and would not be subject to resuspension. This depth is quite close to the critical depth calculated using the average effective fetch of 21 km and the average maximum wind speed of 11.9 m/s (Table 3), although as noted above, is

Springer

somewhat lower than the field-sampling based estimate (≥ 15 m) and the critical depth for accumulation of 14.3 m predicted using Hakanson's dynamic ratio model (1982).

Since the relative frequency of varying wind speeds can be extracted from meteorological records (Fig. 6), it is nevertheless informative to also calculate the relative frequency of mixing and resuspension. The wind speeds were averaged across the 3 stations, sorted and ranked to determine wind speeds corresponding to exceedance probabilities of 1, 2, 5, 10, 25, 50 (median), and 75% (Table 4). Thus, the median (50% exceedance probability) hourly wind speed at the Salton Sea was 2.1 m/s (based upon the averages from stations #127, 128, and 154), while 10% of the winds exceeded 4.2 m/s, and 2% exceeded 7.2 m/s (Table 4). These wind speeds were then used to calculate the wave periods, wave lengths, critical depths, and areas subject to erosion and/or resuspension. Wave periods increased from 0.95 to 3.26 s over these wind speeds, while wave lengths increased from 1.4 to 16.6 m, and critical depths for mixing increased from 0.7 to 8.3 m (Table 4). Under median wind conditions and average effective fetch at the Salton Sea, wave action resuspends fine organic bottom sediments only to a depth of 1.3 m; thus only 73 km^2 or 7.5% of the bottom sediments are disturbed. At higher, albeit less frequent, wind speeds, resuspension due to oscillatory horizontal motion immediately above the sediments occurs to greater depths and over larger areas. This wind-mixed depth, taken as one-half the wavelength, L (Martin & McCutcheon, 1999) increases to 3.6 m at average wind speeds of 4.2 m/s that occur 10% of the time (for an erosional area equal to 182 km^2 or 18.7% of the basin). The wind-mixed critical depth increases to 5.1 m approximately 5% of the time

(affecting 24.9% of the lake area), and further increases to a depth of 8.3 m (affecting 374 km^2) when average wind speeds reach 8.2 m/s (Table 4).

The strong increase in organic C (and total N and organic P) levels beyond about 9 m (Fig. 5) reflects net deposition to these depths, and as previously noted delineates the empirical depth boundary between the erosional and transportational zones. Setting the critical depth at 9 m and solving for wind speed, one calculates a critical wind speed of 8.8 m/s; analysis of the meteorological record found this hourly average wind-speed occurred only 0.5% of the time during 2001. Thus wave theory calculations, combined with empirical data, together suggest that high wind speeds, exceeding 8.8 m/s and occurring less than 1% of the time, are sufficient to keep organic matter from accumulating sediment up to 9 m depth in the Salton Sea.

It bears noting that the models used in this study were developed and tested for freshwater lakes, although it can be shown that only minor differences would be expected for saline systems, even with salinities as high as the Salton Sea.

First of all, since it is wind shear on the water surface that is chiefly responsible for water currents, as well as surface and internal waves, it is useful to calculate the wind shear acting on the surface of the Salton Sea for comparison with that to a typical freshwater lake. The wind shear velocity (u_*) can be calculated by (Martin and McCutcheon, 1999):

$$u_* = \sqrt{\frac{C_D \rho_a U_w^2}{\rho_w}} \qquad (8)$$

where ρ_w is the density of water, C_D is the drag coefficient, ρ_a is the density of air, and U_w is the wind

Table 4 Predicted wave properties and areas of erosion-resuspension based upon the average effective fetch (21 km) and the average 2001 windspeeds from CIMIS stations #127, 128 and 154

Wind speed (m/s)	Exceedance probability (%)	Wave period (s)	Wavelength (m)	Critical depth (m)	Area (km^2)	Area (%)
1.4	75	0.95	1.4	0.7	40	4.1
2.1	50	1.31	2.7	1.3	73	7.5
2.9	25	1.66	4.3	2.2	118	12.2
4.2	10	2.15	7.2	3.6	182	18.7
5.5	5	2.55	10.2	5.1	242	24.9
7.2	2	3.02	14.2	7.1	321	33.0
8.2	1	3.26	16.6	8.3	374	38.4
11.9	(max)	<0.1	4.05	25.6	12.8	677
69.6						

speed. The density of water, ρ_w, is function of both temperature and salinity, and can be calculated from (APHA, 1998; Martin and McCutcheon, 1999):

$$\rho = \rho_T + \Delta\rho_s \qquad (9)$$

where ρ_T is the density of water at temperature T and $\Delta\rho_s$ is the density change due to dissolved solids. Values of ρ_T and $\Delta\rho_s$ can in turn be calculated from polynomial equations relating water density to temperature and salinity (APHA, 1998). Assuming for this comparison a temperature of 25°C and salinities of 800 and 45,000 mg/l, one calculates water densities of 997.7 and 1029.4 kg/m³, respectively (APHA, 1998). The wind shear velocity is also a function of the density of air (Eq. 8). The Salton Sea is unusual in that it lies well below sea level and therefore has an air density greater than lakes at more typical elevations. The density of air (ρ_a) can be estimated from its elevation, temperature and humidity, and is calculated to decrease from a nominal value of 1.236 kg/m³ at the surface elevation of the Salton Sea to 1.193 kg/m³ at a hypothetical 300 m surface elevation.

Substituting these values into Eq. 8 and assuming a drag coefficient (C_D) of 1.3×10^{-3} (Martin and McCutcheon, 1999), we see that the wind shear velocity acting on the surface of the Salton Sea at any given windspeed U_w is within 0.2% of that calculated for a typical freshwater lake. Thus we see that the net effect of these differences in water and air densities is a minimal increase in wind shear velocity acting on the surface of the Salton Sea relative to a typical freshwater lake in the region. These differences are not expected to alter the predicted sediment distributions reported herein.

There are a number of environmental and ecological implications to these findings. First of all, sediment oxygen demand and internal nutrient loading can be expected to be greater in the higher organic content sediment within the transportational and depositional zones relative to the low organic content sediments of the erosional zone, at least under similar temperature and dissolved oxygen regimes (Hargrave, 1972). These zones should also be the loci of accumulation of pesticides and trace elements. Sediments from deep sites (>13 m) were, in fact, substantially elevated in concentrations of DDE, phenol, ρ-cresol and dimethyl naphthalenes than sediments collected from shallow or moderate depths (Schroeder et al., 2002). Higher levels of Mo, Se and

U were also found in the deeper water relative to other areas of the Sea (Vogl & Henry, 2002). Elevated levels of these contaminants may exert ecotoxicological effects on benthic organisms inhabiting these substrates. Moreover, the benthic organism distribution and the resulting food web are expected to vary regionally across the Salton Sea, e.g., with higher levels of bacteria in the organic rich sediments found at greater depths relative to the low organic content sediments of the erosional zone (Lenhard et al., 1962; Jones, 1980). Benthic invertebrate populations and distributions have been found to vary with depth and dissolved oxygen concentrations in the Salton Sea (Detwiler et al., 2002). Resuspension of bottom sediments may also deliver particulate and dissolved forms of nutrients to the water column (Reddy et al., 1996; Nagid et al., 2001); given the frequency and extent of such events, this may be an important mechanism for internal nutrient loading to the Sea.

Conclusions

The sediments in the Salton Sea varied strongly in particle size and in the levels of organic C, $CaCO_3$, total N, total S, total P, inorganic, and organic P, and porewater nutrient concentrations. Most properties were significantly positively correlated with depth. High concentrations of organic C, total N and organic P were present in the north basin, with levels exceeding 10%, 1.0%, and 450 mg/kg, respectively. Lower concentrations were found in the south basin and near the lake margins.

The sediment distribution models of Hakanson (1982), Rowan et al. (1992a, 1995), Blais & Kalff (1995), and Carper & Bachmann (1984) yielded very different results. Based on application of these models to the Salton Sea, it appears that Hakanson's dynamic ratio model may most reasonably predict the distribution of high organic C sediments in shallow lakes with simple basin shapes without well-developed littoral macrophyte communities. Calculations using wave theory and local meteorological data provided additional insight into the observed sediment distribution, indicating that net sediment accumulation, even in relatively deep waters within the erosion zone of the Salton Sea, is inhibited by winds with speeds of 8.8 m/s, occurring less than 1%

 Springer

of the time. These winds were unable to resuspend fine organic sediments at water depths greater than 9 m and thus resulted in net sediment accumulation, although wind speeds exceeding 10–11 m/s, occurring only a few times a year, were apparently able to resuspend some sediment at depths as great as 12.8 m or more. Given the importance of sediment resuspension and the resulting distribution of sediment within lakes, additional work is needed to refine sediment distribution models, including parameterization to allow prediction of erosional areas in lakes.

Acknowledgments This research was sponsored by Salton Sea Authority, La Quinta, California. Special thanks to Ed Betty for his assistance in the field sampling.

References

APHA, 1998. Standard Methods for the Examination of Water and Wastewater (20th Ed.). American Public Health Association, Washington, DC.

Arnal, R. E., 1961. Limnology, sedimentation, and microorganisms of the Salton Sea, California. Geological Society of America Bulletin 72: 427–478

Aspila, K. I., H. Agemian & A. S. Y. Chau, 1976. A semiautomated method for determination of inorganic, organic and total phosphate in sediments. Analyst 101: 187–197.

Berner, R. A. & J. L. Rao, 1994. Phosphorus in sediments of the Amazon River and estuary: implications for the global flux of phosphorus to the sea. Geochimica Cosmochimica Acta 58: 2333–2339.

Blais, J. M. & J. Kalff, 1995. The influence of lake morphometry on sediment focusing. Limnology and Oceanography 40: 582–588.

Bloesch, J., 1995. Mechanisms, measurement and importance of sediment resuspension in lakes. Marine and Freshwater Research 46: 295–304.

California Irrigation Management Information System (CIMIS). 2004. California Dept. of Water Resources. www.cimis.water.gov.

Carper, G. L. & R. W. Bachmann, 1984. Wind resuspension of sediments in a prairie lake. Canadian Journal of Fisheries and Aquatic Sciences 41: 1763–1767.

Cook, C. B., G. T. Orlob & D. W. Huston, 2002. Simulation of wind-driven circulation in the Salton Sea: implications for indigenous ecosystems. Hydrobiologia 473:59–75.

Cyr, H., 1998. Effects of wave disturbance and substrate slope on sediment characteristics in the littoral zone of small lakes. Canadian Journal of Fisheries and Aquatic Sciences 55: 967–976.

Danen-Louwerse, J. H., L. Lijklema & M. Coenraats, 1995. Coprecipitation of phosphate with calcium carbonate in Lake Veluwe. Water Research 29: 1781–1785.

Detwiler, P. M., M. F. Coe & D. M. Dexter, 2002. The benthic invertebrates of the Salton Sea: distribution and seasonal dynamics. Hydrobiologia 473: 139–160.

Duarte, C. M. & J. Kalff, 1986. Littoral slope as a predictor of the maximum biomass of submerged macrophyte communities. Limnology and Oceanography 31: 1072–1080.

Ferrari, R. L. & P. Weghorst, 1997. Salton Sea 1995 Hydrographic GPS Survey. Bureau of Reclamation, Water Resources Services, Technical Service Center, Denver, CO.

Förstner, U & G. T. W. Wittmann, 1979. Metal Pollution in the Aquatic Environment. Springer-Verlag, Berlin.

Freeman, J. S. & D. L. Rowell, 1981. The adsorption and precipitation of phosphate onto calcite. Journal of Soil Science 32: 75–84.

Gee, G. W. & J. W. Bauder, 1986. Particle-size analysis. In Klute A. (ed.), Methods of Soil Analysis. Part 1. 2nd Ed. Agronomy Monographs 9. ASA and SSSA, Madison, WI: 383–411.

Gilbert, R., 1999. Calculated wave base in relation to the observed patterns of sediment distribution in northeastern Lake Ontario. Journal of Great Lakes Research 25: 883–891.

Hakanson, L., 1977. The influence of wind, fetch, and water depth on the distribution of sediments in Lake Vanern, Sweden. Canadian Journal of Earth Sciences 14: 397–412.

Hakanson, L., 1981. On lake bottom dynamics – the energy-topography factor. Canadian Journal of Earth Sciences 18: 899–909.

Hakanson, L., 1982. Lake bottom dynamics and morphometry: the dynamic ratio. Water Resources Research 18: 1444–1450.

Hakanson, L. & M. Jansson, 1983. Principles of Lake Sedimentology. Springer-Verlag, New York, NY.

Hargrave, B. T., 1972. Aerobic decomposition of sediment and detritus as a function of particle surface area and organic content. Limnology and Oceanography 17: 583–596.

Hatcher, K. J. 1986. Sediment Oxygen Demand: Processes, Modeling & Measurement. Institute of Natural Resources, Univ. of Georgia, Athens, GA.

Holdren, G. C. & D. E. Armstrong, 1980. Factors affecting phosphorus release from intact lake sediment cores. Environmental Science and Technology. 14: 79–87.

Holdren, G. C. & A. Montaño, 2002. Chemical and physical characteristics of the Salton Sea, California. Hydrobiologia 473: 1–21.

Huston, D. W., 2000. Application of a Wind Field Analysis to a Three-Dimensional Hydrodynamic Model of the Salton Sea, California. M.S. Thesis, Univ. of California, Davis.

Jones, J. G., 1980. Some differences in the microbiology of profundal and littoral zone sediments. Journal of General Microbiology 117: 285–292.

Karickhoff, S. W., D. S. Brown & T. A. Scott, 1979. Sorption of hydrophobic pollutants on natural sediments. Water Research 13: 241–248.

King, I. P. 1993. A finite element model for three-dimensional density stratified flow. Australian Water and Coastal Studies Report, April.

Lehman, J. T., 1975. Reconstructing the rate of accumulation of lake sediment: the effect of sediment focusing. Quaternary Research 5: 541–550.

Lenhard, G., W. R. Ross & A. duPlooy, 1962. A study of methods for the classification of bottom deposits of natural waters. Hydrobiologia 20: 223–240.

Loeppert, R. H. & D. L. Suarez, 1996. Carbonate and gypsum. In Sparks, D. L. (ed.), Methods of Soil Analysis. Part 3.

3rd Ed. Agronomy Monographs 9. ASA and SSSA. Madison, WI: 437–474.

Martin, J. L. & S. C. McCutcheon, 1999. Hydrodynamics and Transport for Water Quality Modeling. Lewis Publ., Boca Raton, FL.

Nagid, E. J., D. E. Canfield & M. V. Hoyer, 2001. Wind-induced increases in trophic state characteristics of a large (27 km^2), shallow (1.5 m mean depth) Florida lake. Hydrobiologia 455: 97–110.

Nelson, D. W. & L. E. Sommers, 1982. Total carbon, organic carbon, and organic matter. In: Page, A. L., R. H. Miller & D. R. Keeney (eds), Methods of Soil Analysis, Part 2. 2nd ed. Agronomy Monographs 9. ASA and SSSA., Madison, WI: 539–580.

Peeters, E. T. H. M., R. Gylstra & J. H. Vos, 2004. Benthic macroinvertebrate community structure in relation to food and environmental variables. Hydrobiologia 519: 103–115.

Reddy, K.R., M. M. Fisher & D. Ivanoff, 1996. Resuspension and diffusive flux of nitrogen and phosphorus in a hype-reutrophic lake. Journal of Environmental Quality 25: 363–371.

Rowan, D. J., J. Kalff & J. B. Rasmussen, 1992a. Estimating the mud deposition boundary depth in lakes from wave theory. Canadian Journal of Fisheries and Aquatic Sciences 49: 2490–2497.

Rowan, D. J., J. Kalff & J. B. Rasmussen, 1992b. Profundal sediment organic content and physical character do not reflect lake trophic status, but rather reflect inorganic sedimentation and exposure. Canadian Journal of Fisheries and Aquatic Sciences 49: 1431–1438.

Rowan, D. J., J. B. Rasmussen & J. Kalff, 1995. Optimal allocation of sampling effort in lake sediment studies. Canadian Journal of Fisheries and Aquatic Sciences 52: 2146–2158.

Salton Sea Authority/USBR, 2000. Draft Salton Sea Restoration Project Environmental Impact Statement/ Environmental Impact Report. Report prepared for the Salton Sea Authority and U.S. Department of the Interior, Bureau of Reclamation, by Tetra Tech, Inc.

Schroeder, R. A., W. H. Orem & Y. K. Kharaka, 2002. Chemical evolution of the Salton Sea, California: nutrient and selenium dynamics. Hydrobiologia 473: 23–45.

Sondegaard, M., J. P. Jensen & E. Jeppesen, 2003. Role of sediment and internal loading in shallow lakes. Hydrobiologia 506–509: 135–145.

Sykes, G., 1937. The Colorado delta. American Geographic Society Special Publicaion 19, pp. 132–133.

USBR. 2000. Salton Sea Restoration Project, Draft Alternatives Appraisal Report. Bureau of Reclamation, Lower Colorado Region, Boulder City, NV, 67 pp.

USGS. 2006. National Water Information System. Water surface elevation data for the Salton Sea. USGS station #10254005, Salton Sea near Westmorland, CA. http://www.waterdata.usgs.gov.

Vogl, R. A. & R. N. Henry, 2002. Characteristics and contaminants of the Salton Sea sediments. Hydrobiologia 473: 47–54.

Watts, J. M., B. K. Swan, M. A. Tiffany & S. H. Hurlbert, 2001. Thermal, mixing, and oxygen regimes of the Salton Sea, California, 1997–1999. Hydrobiologia 466: 159–176.

Yuan, F., B. K. Linsley & S. S. Howe, 2006. Evaluating sedimentary geochemical lake-level tracers in Walker Lake, Nevada, over the last 200 years. Journal of Paleolimnology 36: 37–54.

Hydrobiologia (2008) 604:111–121
DOI 10.1007/s10750-008-9322-3

SALTON SEA

Geochemistry of iron in the Salton Sea, California

**Jason P. de Koff · Michael A. Anderson ·
Christopher Amrhein**

Abstract The Salton Sea is a large, saline, closed-basin lake in southern California. The Sea receives agricultural runoff and, to a lesser extent, municipal wastewater that is high in nutrients, salt, and suspended solids. High sulfate concentrations ($4\times$ higher than that of the ocean), coupled with warm temperatures and low-redox potentials present during much of the year, result in extensive sulfate reduction and hydrogen sulfide production. Hydrogen sulfide formation may have a dramatic effect on the iron (Fe) geochemistry in the Sea. We hypothesized that the Fe(II)-sulfide minerals should dominate the iron mineralogy of the sediments, and plans to increase hypolimnetic aeration would increase the amount of Fe(III)-oxides, which are strong adsorbers of phosphate. Sequential chemical extractions were used to differentiate iron mineralogy in the lake sediments and suspended solids from the tributary rivers. Iron in the river-borne suspended solids was mainly associated with structural iron within silicate clays (70%) and ferric oxides (30%). The iron in the bottom sediments of the lake was associated with silicate minerals (71% of the total iron in the sediments), framboidal pyrite (10%), greigite (11%), and amorphous FeS (5%). The ferric oxide fraction was <4% of the total iron in these anaerobic sediments. The morphological characteristics of the framboidal pyrite as determined using SEM suggest that it formed within the water column and experiences some changes in local redox conditions, probably associated with alternating summer anoxia and the well-mixed and generally well-aerated conditions found during the winter. The prevalence of Fe(II)-sulfide minerals in the sediments and the lack of Fe(III)-oxide minerals suggest that the classic model of P-retention by Fe(III)-oxides would not be operating in this lake, at least during anoxic summer conditions. Aeration of the hypolimnion could affect the internal loading of P by changing the relative amounts of Fe(II)-sulfides and Fe(III)-oxides at the sediment/water interface.

Keywords Pyrite · Anoxia · Saline lake · Sulfate reduction

Guest editor: S. H. Hurlbert
The Salton Sea Centennial Symposium. Proceedings of a Symposium Celebrating a Century of Symbiosis Among Agriculture, Wildlife and People, 1905–2005, held in San Diego, California, USA, March 2005

Electronic supplementary material The online version of this article (doi:10.1007/s10750-008-9322-3) contains supplementary material, which is available to authorized users.

J. P. de Koff · M. A. Anderson · C. Amrhein (✉)
Department of Environmental Sciences, University of California, Riverside, CA 92521, USA
e-mail: christopher.amrhein@ucr.edu

Introduction

The Salton Sea is a large, hypereutrophic, closed-basin lake (70 m below sea level) located 50-km north of the

U.S.–Mexico border in southern California. It initially formed between 1905 and 1907 when a diversion on the Colorado River failed and the River turned westward, filling the Salton Sink. The primary inflows to the Sea are agricultural and municipal wastewaters, mostly from use of Colorado River water. The salinity of the Sea in 1999 was approximately $44 \pm 3 \text{ g l}^{-1}$, with similar Mg, Na + K, and Cl concentrations as ocean water, but reduced Ca and HCO_3 levels and highly elevated SO_4 concentrations (Holdren & Montaño, 2002). Three main rivers flow into the Sea, with 80% of the flow from the New and Alamo Rivers at the southern end of the Sea. These rivers have high concentrations of plant nutrients, salts, and suspended solids (Holdren & Montaño, 2002). The total suspended solid load to the Sea exceeds $460,000 \text{ Mg y}^{-1}$, predominately soil clay and silt, with over $8,000 \text{ Mg y}^{-1}$ of total Fe (Holdren & Montaño, 2002). The combination of high temperatures (>30°C during the summer), intense sunlight, and high nutrient loading results in extensive algae blooms, low dissolved O_2 concentrations, and fish kills in the Sea (Watts et al., 2001). The anaerobic decomposition of organic matter via sulfate reduction produces H_2S that has been detected more than 160-km away.

The sediments of the Sea are characteristically black or dark greenish-black, indicative of anaerobic conditions and the likely presence of pyrite (FeS_2), greigite (Fe_3S_4), mackinawite (FeS), and amorphous FeS. The geochemistry of the sediments is thought to play an important role in the overall productivity of the Sea, since nutrient recycling can be a dominant and long-term source of nutrients in many lakes that have received historically high levels of external inputs (Sondegaard et al., 1999). Early studies by Mortimer (1941, 1942) linked phosphate release from sediments to reductive dissolution of Fe(III)-oxides that was associated with anoxia above the sediments. It is on this basis that aeration and oxygenation are commonly used lake management strategies for limiting internal recycling of phosphate (Cooke et al., 1993). There are geochemical conditions, however, where this classical model breaks down (Caraco et al., 1989, 1993; Ingall et al., 2005). Specifically, high rates of sulfate reduction and precipitation of Fe(II)-sulfides limit the potential for Fe(III)-oxide formation in sediments; this in turn affects phosphorus solubility by reducing P-adsorption onto Fe(III)-oxides (Caraco et al., 1993; Ingall et al., 2005).

A common method for determining the forms of metal in sediments and soils involve sequentially reacting the samples with a series of chemical extractants designed to recover different phases (Tessier & Campbell, 1988). Although problems relating to efficiency and selectivity of the various reagents have been identified with these sequential extraction procedures, they remain useful for comparing the relative reactivity across a series of samples (Tessier & Campbell, 1988; Poulton & Canfield, 2005). The assignment of particular mineral phases to each extractant is based on chemical reactivity, but we recognize that the differentiation is imprecise (Nirel & Morel, 1990; Tack & Verloo, 1995; Gleyzes et al., 2002). This is particularly true for the spectrum of iron sulfide minerals that have a wide range of reactivity from amorphous FeS to well-crystallized pyrite (Cornwell & Morse, 1987; Raiswell et al., 1994).

Due to the high iron loading from the muddy inflow waters, evidence for intense sulfate reduction within the sediments, and the role that Fe plays in regulating phosphorus release from sediments, we were interested in quantifying the Fe phases present within the sediment, including the amount of pyrite and degree of pyritization present in the sediments of this relatively new body of water (~ 100 years since formation). We hypothesized that the predominate forms of iron in the sediments would be Fe(II)-sulfide minerals, which have a low reactivity toward phosphate. If this were the case, increasing hypolimnetic aeration would increase the amount of Fe(III)-oxides in the sediments and the amount of adsorbed P. In this article, we report the distribution of iron minerals in Salton Sea sediments, the relative degree of pyritization of the iron in the sediments, and an estimate of the average annual iron sulfide mineral precipitation rate. This information is necessary to understand the potential changes to the sediment mineralogy and P-cycling under forced-air hypolimnetic aeration.

Materials and methods

Sample collection

Sediment samples (0–10 cm) were collected with a Ponar dredge at 81 sites across the Salton Sea in the summer of 2001 (Fig. 4a) and frozen at −10°C until

analyzed. Prior to use, the sediment samples were thawed in a cold room at 4°C for about 7 days. The samples were mixed to homogenize the sediment and individual portions were taken from each sample and dried at 105°C to determine water content. A second aliquot of wet sample from each site was weighed into 50-ml centrifuge tubes and corrected for water and salt contents to give a dry weight of 0.25 g. The centrifuge tubes were immediately flushed with N_2 gas and extracted following procedures described later. All sediment samples were extracted in duplicate.

Triplicated suspended solid samples from the Alamo River and New River were isolated from water samples collected on February 20, 2003 using a 4 L Van Dorn sampler. Samples were collected from ~0.5 to 1 m below the water surface and suspended solids collected using filtration of the river water and allowed to air dry. Efforts were not made to limit exposure to O_2 since these materials were derived from surface soils and transported in an oxic flowing river.

The chemical extraction procedures of Lord (1982) and Raiswell et al. (1994) were used to determine the amounts of pyrite and other iron phases in the Salton Sea sediments and river suspended solids.

Sequential extraction procedures

The procedure of Lord (1982) was used for each of the samples whereby three consecutive extractions were conducted to separate the (1) reducible-iron(III) oxides, (2) silicate minerals, and (3) pyrite (Fig. 1). The first extractant is commonly used to remove iron(III)-oxides from soils, and studies have shown that it also dissolves the amorphous FeS (Raiswell et al., 1994). Ten milliliters of a buffer solution containing 78 g l^{-1} sodium citrate and 9.3 g l^{-1} sodium bicarbonate was added to each centrifuge tube and the tubes placed in a water bath at 75°C. At 5-min intervals, the centrifuge tubes were individually shaken. After 10 min in the water bath, 0.25 g of sodium dithionite was added to each tube and the tubes returned to the water bath. To ensure complete reduction of the Fe(III)-oxides, sodium dithionite was added to the tubes a total of four times. The samples were then centrifuged at 7000 RCF for 10 min. The

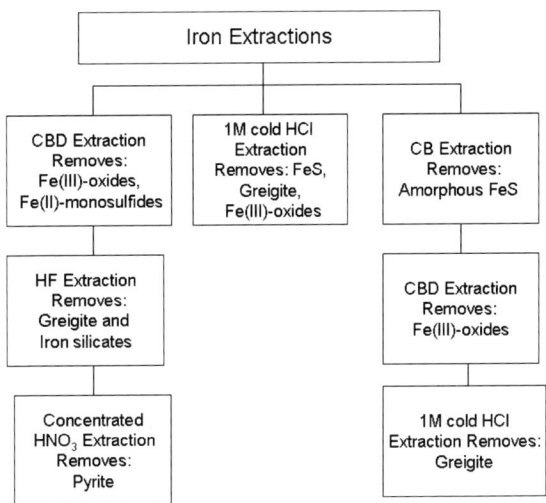

Fig. 1 Flow diagram of iron extractions

supernatant was collected and the remaining sediment washed once with 10-ml of the citrate–bicarbonate buffer and then with 10-ml distilled water. The supernatant from these washings was added to the original supernatant. This cumulative supernatant sample is called the CBD extraction (citrate, bicarbonate, dithionite).

The sediment residue from the CBD extraction was then treated with an HF solution which dissolves the silicate minerals but does not react with the pyrite (Lord, 1982). For this extraction, 20 ml of distilled water plus 10-ml concentrated HF were added to the residue from the first extraction. These suspensions were shaken overnight, after which 5 g of boric acid was added to each tube, and the samples shaken for another 24 h. The samples were centrifuged at 8,150 RCF for 20 min and the supernatants collected. The remaining sediment was washed twice with 5 ml of boiling distilled water and this wash water added to the HF supernatant.

For the third extraction in the sequence, 10 ml of concentrated nitric acid was added to the remaining sediment in the tubes to dissolve the pyrite. The samples were shaken for 2 h and then diluted with 15 ml of distilled water. The supernatant was collected by suction filtration using glass fiber filters.

In a separate extraction (Fig. 1), ~1 g of wet sample (equivalent to ~0.3 g dry sample) was added to each centrifuge tube followed by 30 ml of 1 M HCl. The tubes were shaken for 16 h, centrifuged, and the supernatants collected by decanting. This

extraction dissolves iron(III)-oxides, Fe(II) monosulfides, greigite, and small amounts of iron silicates (Cornwell & Morse, 1987; Raiswell et al., 1994).

A final series of extractions (Fig. 1) were performed using 9 sediment samples chosen for their different locations in the Salton Sea as well as their dissimilar pyrite concentrations (determined from the previous sequential extractions). This series of extractions was a modification and combination of both the Lord (1982) and Raiswell et al. (1994) procedures. The samples were set up the same way as the first series of extractions, however the extracts varied. The first extraction, a citrate–bicarbonate extraction, (CB) followed the same procedure as the CBD extractions, minus the sodium dithionite additions. This CB extraction removed any amorphous FeS present in the samples but without the reductant, did not solubilize the Fe(III)-oxides. The CB extractant was tested on laboratory-prepared amorphous FeS to verify the efficiency of extraction. The CBD extraction, already outlined above, followed the CB extraction and removed the iron(III)-oxides.

Following the CBD extract, the sediments were extracted with 30 ml of 1 M cold HCl with agitation on a shaker for 24 h. This procedure was originally suggested by Raiswell et al. (1994) but was not tested at that time. It is likely that 1 M cold HCl solubilized siderite and greigite because Cornwell & Morse (1987) reported that this extractant dissolved 40–67% of a synthetic greigite after 45 min. Siderite ($FeCO_3$) is not considered a likely phase in Salton Sea sediments due to the high sulfide concentrations found in the sediment pore water and hypolimnion (Watts et al., 2001); so this extraction is believed to solubilize predominately greigite.

Chemical analysis

The total soluble iron from the extractions was measured using a Perkin Elmer 3000DV inductively coupled argon plasma, optical emission spectrophotometer. Standard QA/QC protocol was followed which included matrix-matched standards, spikes, blanks, and duplicates. A set of analyses was considered acceptable if the spike recoveries and duplicates agreed to better than 15%.

The spatial distribution of extractable Fe across the Sea was then calculated and plotted using the kriging algorithm within SURFER, Version 7 (Golden Software).

A scanning electron microscope (SEM, Philips Model XL30-FEG), equipped with both secondary electron (SE) and backscattered electron (BSE) detectors, was used to image the sediments. The SEM was operated at 30 kV and elemental composition determined by energy dispersive X-ray (EDX) analysis.

Results

Sequential extraction of Salton Sea sediments and suspended solids isolated from the New and Alamo River following Lord (1982) yielded Fe solid phase speciation that varied across these three sediment sources (Table 1). The suspended solids entering the Sea from the New and Alamo Rivers were dominated by the silicate-Fe fraction (20.1 ± 0.3 and 18.3 ± 1.6 mg Fe g^{-1} solid, respectively) that accounted for 70% of the total extractable Fe in the samples. The remaining Fe was principally recovered in the Fe(III)-oxide fraction (28% of the total extractable Fe), with 2% or less recovered in the pyrite extraction (Table 1). Bottom sediments from the Sea had an average silicate-Fe iron concentration very similar to that found in the river-borne suspended solids (18.1 ± 6.8 mg Fe g^{-1} sediment), but a much lower Fe level recovered in the Fe(III)-oxide fraction (1.9 ± 0.7 mg Fe g^{-1} sediment) and an elevated average FeS_2 content (2.1 ± 1.0 mg Fe g^{-1} sediment) relative to the suspended solids isolated from the New and Alamo Rivers (Table 1). Of the total iron in the Salton Sea sediments, an average of 9.5% was in pyrite.

These extraction results suggest that Fe transformations are occurring within the water column and bottom sediments of the Sea, with reduction of Fe(III)-oxides and the formation of Fe-sulfide phases. It is important to note that these analyses were done on sediment samples collected during the summer when the hypolimnion was most intensely anaerobic and much of the lake had <1 mg l^{-1} dissolved oxygen. It is likely that during the winter months, when the Sea is better oxygenated, there would be some oxidation of the Fe-sulfides back to Fe(III)-oxides. Further work on seasonal variations seems warranted.

Table 1 Extractable-Fe phases in river-borne suspended solids and Salton Sea sediments following Lord (1982)

	mg Fe g^{-1} dry-weight sediment		
	Alamo River ($n = 3$)	New River ($n = 3$)	Salton Sea ($n = 81$)
Iron(III)-oxides	7.5 ± 0.4	8.2 ± 0.4	1.9 ± 0.7
Silicate-Fe	18.3 ± 1.6	20.1 ± 0.3	18.1 ± 6.8
FeS$_2$	0.40 ± 0.04	0.60 ± 0.04	2.1 ± 1.0

Errors are ±one standard deviation. River samples are average of three replicates collected on 1 day at one place on the rivers; Salton Sea samples are average of replicates of 81 samples collected over the whole Sea

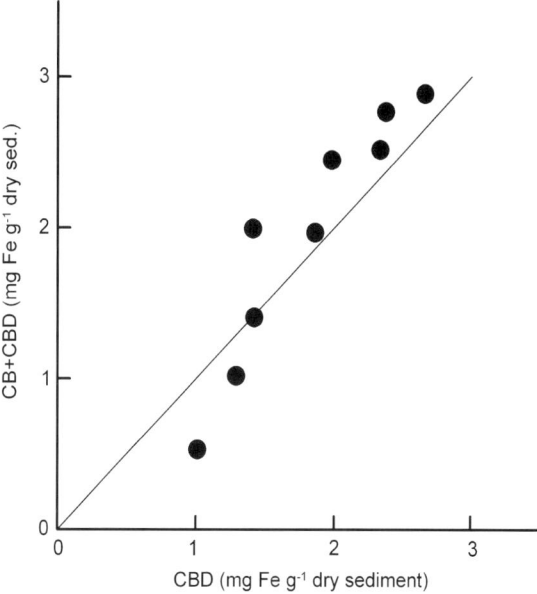

Fig. 2 Comparison of extractable Fe using the two-step citrate–bicarbonate (CB) plus the citrate–bicarbonate–dithionite (CBD) extractions and the single citrate–bicarbonate–dithionite (CBD) extraction. The solid line represents the 1:1 line

The average total extractable Fe concentration of the Sea sediments was lower than that in the suspended solids from the rivers due to variations in deposition of suspended solids within the lake and dilution effects from organic matter and CaCO$_3$ accumulation in the sediments. The average CaCO$_3$ content of the bottom sediments is 24% by weight, with values as high as 50% in the deeper parts of the Sea (Anderson et al., 2008). Virtually all CaCO$_3$ is formed within the lake and not input via the rivers. In addition, the average organic matter content of the sediments is 17% (dry weight basis) due to algal and bacterial accumulations (Anderson et al., 2008). Both these additions dilute the Fe from river-borne suspended solids.

The sequential extraction process of Lord (1982) fails to distinguish between dissimilar phases that are solubilized in the extractants, e.g., Fe(III)-oxides and amorphous FeS in the CBD extract, and silicate-Fe and greigite (Fe$_3$S$_4$) in the HF extract. As a result, the 1.0 M cold HCl method of Raiswell et al. (1994) was used on all the sediment samples, and a modification of the CBD extraction of Lord (1982) was applied to 9 sediment samples collected from across the Sea (Fig. 1). The sum of CB + CBD-extractable Fe (two sequential steps) adequately reproduced the CBD-extractable Fe (Lord, 1982) (Fig. 2). This additional extraction thus provides a means by which amorphous FeS contents can be split out from the lumped Fe(III)-oxide + FeS fraction of the Lord procedure.

The ratio of amorphous FeS, recovered in the CB extract, to the Fe recovered as Fe(III)-oxide + FeS in the CBD extract increased linearly with depth, with lower concentrations of amorphous FeS present at

shallow depths (CB/(CB + CBD) = 0.3) and higher ratios at greater depth (CB/(CB + CBD) = 0.7) (Fig. 3). Thus, a greater relative amount of reduced

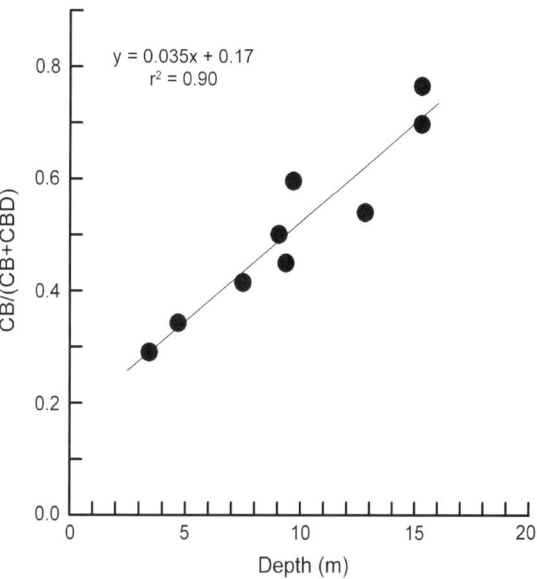

Fig. 3 Ratio of amorphous FeS (as measured with the citrate–bicarbonate (CB) extraction) to the dithionite reducible iron + CB amount (CB/(CB + CBD)) versus water depth

Table 2 Distribution of Fe minerals in Salton Sea sediments using the modified extraction procedure

Fe-Phase	mg Fe g^{-1} dry-weight sediment		
	Mean ($n = 81$)	Standard deviation	Range
Amorphous FeS	1.0	0.5	0.1–2.4
Iron(III)-oxides	0.9	0.5	0.3–2.6
Fe$_3$S$_4$	2.4	1.0	0.3–6.8
Silicate-Fe	15.7	6.2	3.9–45
FeS$_2$	2.1	1.0	0.2–5.3

Fe, as amorphous FeS, was positively correlated with depth ($r^2 = 0.90$). Using this relationship we were able to estimate amorphous FeS and Fe(III)-oxide concentrations for the remaining Salton Sea bottom sediments (Table 2).

The bathymetry of the Salton Sea and the sampling site locations (Fig. 4a) show the two deeper basins, where organic matters tend to accumulate (Anderson et al., 2008) and the most reducing conditions are expected. The distribution of Fe(III)-oxides varied across the Sea but tended to be associated with suspended solids input near the Alamo River and San Felipe Creek on the west side of the south basin (>2 mg Fe g^{-1} sediment), while concentrations were quite low (<0.7 mg Fe g^{-1} sediment) in the deep portions of the lake (Fig. 4b).

Conversely, the concentrations of amorphous FeS were low near the river inlets and margins of the Sea, and highest near the center of the south basin (Fig. 4c). Interestingly, the highest FeS concentrations were not located in the deep north basin, previously found to have the highest organic C concentrations (Anderson et al., 2008). It appears that proximity to the large riverine inputs of Fe in the southern basin have promoted local increases in amorphous FeS levels there.

Following the CBD extraction, reaction of the residual material with HF provided a measure of the Fe-silicates + Fe$_3$S$_4$ (Lord, 1982). This fraction dominated the Fe within the bottom sediments of the Sea and the river-borne suspended solids (Table 1), although the method does not distinguish between the two phases. Cornwell & Morse (1987) and Raiswell et al. (1994) previously demonstrated that cold 1 M HCl effectively extracts most Fe(III)-oxides, amorphous FeS, and Fe$_3$S$_4$ while dissolving only small amounts of Fe-silicates. As a result, the Fe

extracted in cold 1 M HCl was used with CB + CBD and HF extractions result to estimate Fe$_3$S$_4$ and Fe-silicate levels by difference (Table 2, Fig. 4d and e).

Silicate-Fe concentrations were elevated in a band across the south basin and lower in the north basin, except near the inlet of the Whitewater River where high concentrations were found (Fig. 4d). The silicate-Fe which is assumed to be derived from the suspended solids deposited by the rivers and dust deposition, do not appear to be prone to extensive redistribution; for example, the Whitewater River inputs were confined to sites near the inlet. San Felipe Creek is an ephemeral, flash-flood stream, draining non-agricultural lands, appears to be an important source of silicate-Fe and Fe(III)-oxides. The counter clockwise circulation in the southern basin, which has greater velocities than in the gyre in the northern basin (Cooke et al., 2002), has redistributed the silicate-Fe to a greater extent, compared to the north basin (Fig. 4d). The sediment loads in the New River and Alamo River comprise 88% of the measured suspended solids input to the Sea (Holdren & Montaño, 2002). The amount of suspended solids input by San Felipe Creek is unknown because of its flash flows, but appears to be significant.

Greigite (Fe$_3$S$_4$) was the second most abundant Fe phase with a mean concentration of 2.4 ± 1.0 mg Fe g^{-1} sediment (Table 2). High concentrations were present near the Whitewater River inlet, while somewhat lower concentrations were also present in a band across the south basin (Fig. 4e). Greigite levels deviated quite strongly from the observed amorphous FeS distribution, where high concentrations were found near the center of the south basin (Fig. 4c). Since the 1.0 M HCl extraction is known to recover some portion of Fe from Fe-silicates, and the broad similarity in their distributions (Fig. 4d) suggests that these concentrations represent an upper bound on Fe$_3$S$_4$ contents within the bottom sediments.

Pyrite (FeS$_2$) is a well-crystallized phase that is quite inert to the extractants used to recover the Fe(III)-oxide, amorphous FeS, Fe$_3$S$_4$, and silicate-Fe phases (Lord, 1982; Raiswell et al., 1994). As a result, concentrated HNO$_3$ was required to solubilize residual FeS$_2$ within the sediments (Lord, 1982). Pyrite comprised on average about 10% of the total recoverable Fe within the bottom sediments (2.1 ± 1.0 mg Fe g^{-1} sediment) (Table 2), with high

Fig. 4 (**a**) Bathymetry and sediment sampling locations. Distribution of: (**b**) Fe(III)-oxides, (**c**) amorphous FeS, (**d**) silicate iron, (**e**) greigite (Fe_3S_4), and (**f**) pyrite (FeS_2) in Salton Sea sediments

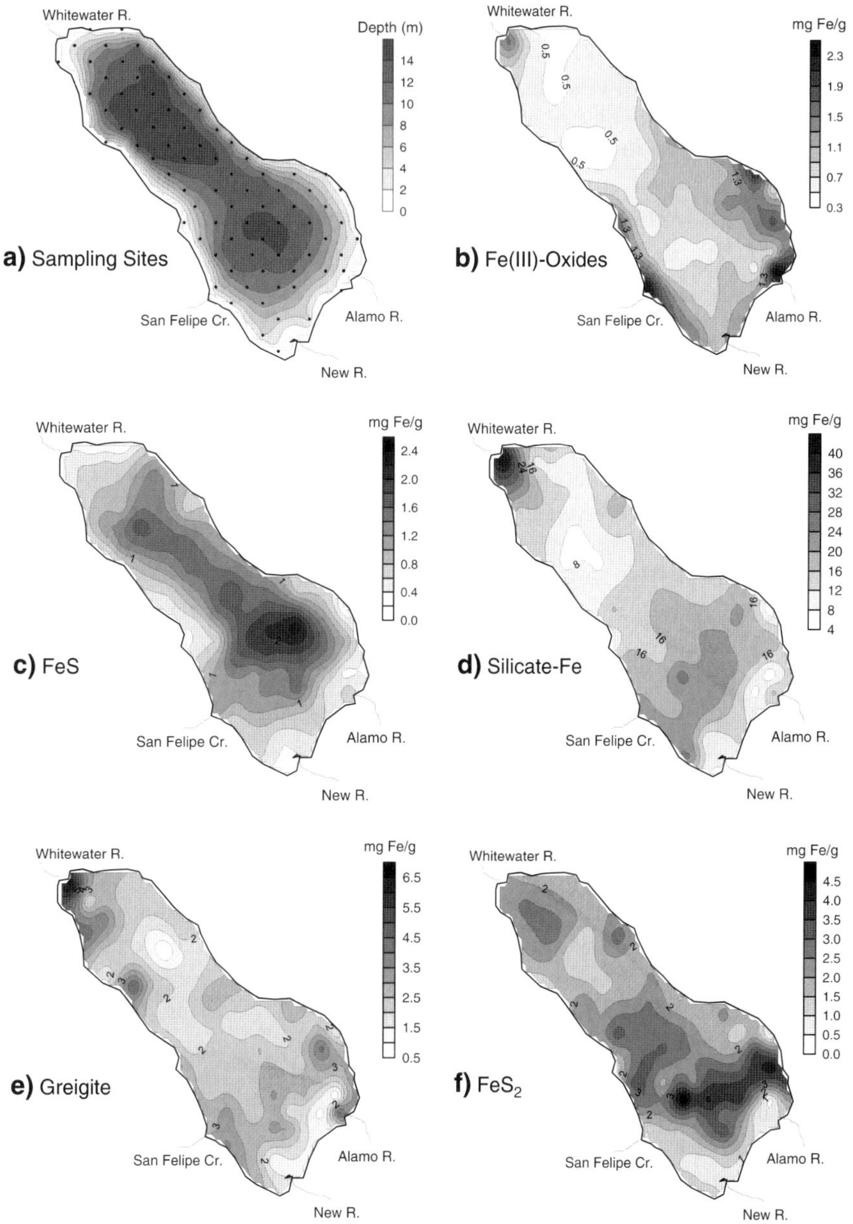

concentrations (>3.5 mg Fe g^{-1} sediment) found within a narrow band near the middle of the south basin extending eastward to the margin of the Sea (Fig. 4f). This region overlaps with the higher silicate-Fe contents found there; in fact, pyrite was most closely correlated with silicate-Fe ($r = 0.44$, significant at $P < 0.01$), but was less strongly correlated with Fe_3S_4 and amorphous FeS ($r = 0.27$ and 0.22, significant at $P < 0.05$, respectively) and weakly negatively correlated with Fe(III)-oxides

($r = -0.09$, not significant). The highest concentrations of FeS_2 and amorphous FeS were found in the southern basin, apparently due to the inputs of readily reducible Fe(III)-oxides and slower-reacting Fe-silicates via the Alamo and New Rivers and San Felipe Creek (Carroll, 1958; Grossman et al., 1979).

Since the sequential extractions provide convenient but indirect measures of Fe phases present in sediments, direct SEM images and EDAX spectra of sedimentary material were also collected. SEM

Hydrobiologia (2008) 604:111–121

analyses confirmed the presence of well-developed spherical, framboidal pyrite in all the samples inspected (e.g., see Electronic supplementary material—Appendix 1a). EDAX analyses confirmed the elemental composition of the particles as FeS_2, with only small amounts of Al, Si, and other impurities present (see Electronic supplementary material—Appendix 1b).

Discussion

Pyrite formation

The mechanism of pyrite formation is a topic that has been debated for many years (Rickard et al., 1995; Schoonen, 2004). It is well recognized that amorphous FeS, greigite, and pyrite can coexist in sediments, although the rates of formation of each species is dependent on many factors including the availability of reactive Fe and metabolizable organic matter, pH, temperature, concentrations of H_2S, elemental sulfur, and soluble polysulfides (Wilkin & Barnes, 1997a; Hurtgen et al., 1999; Schoonen, 2004). It is generally recognized that the formation of pyrite (FeS_2) occurs via the transformation of amorphous FeS and/or greigite (Sweeney & Kaplan, 1973; Schoonen, 2004). Wilkins & Barnes (1997a) suggested that greigite is an important precursor to framboidal pyrite because greigite crystals are magnetic, which causes them to aggregate into larger particles. After the greigite framboids are formed, they can then be converted to pyrite either by the addition of sulfur (Sweeney & Kaplan, 1973) or the loss of iron (Furukawa & Barnes, 1995).

Another possible mechanism for the formation of framboidal pyrite has been described by Rickard (1997) and Butler & Rickard (2000) whereby amorphous FeS reacts with H_2S with the generation of H_2 gas. This mechanism takes place in the absence of any magnetic or biological intermediates. Pyrite formation can occur in either the pore water or the water column depending upon the location of the oxic–anoxic boundary. Pyrite framboids formed in the water column (syngenetic pyrite) are typically small (~ 4–6 μm) with a narrow size range. In oxic waters, the formation of pyrite occurs in the sediment pore water just below the sediment/water interface. Pyrite framboids formed in sediments are generally larger and more variable in size than those formed in the water column (Wilkin et al., 1996, 1997; Wilkin & Barnes, 1997b).

The framboids we observed by SEM were composed of uniformly sized microcrystals 0.4–0.6 μm in diameter. The framboidal aggregates were all of comparable sizes and relatively small (~ 5 μm), suggesting that nucleation and aggregation occurred in the water column rather than within the pore water of the sediments (Wilkin et al., 1996).

Many of the pyritic framboids from the sediments were found with a smooth outer surface suggesting chemical alteration (see Electronic supplementary material—Appendix 1a). It appears that the framboids may have been partially dissolved due to changes in redox state. The redox state of the water column and surficial sediments are known to undergo strong seasonal changes, with high dissolved oxygen levels and high-redox potentials present during the cool winter months, and very low dissolved oxygen concentrations and strongly reducing conditions (e.g., $E_h < -200$ mV) present during the summer (Watts et al., 2001; Holdren & Montaño, 2002). We noticed that upon exposure to air, the dark black sediments (Munsell color 5Y 2.5/1) turned into a light-olive-brown color (Munsell color 2.5Y 5/4) within a few hours. We also observed a rapid color change when laboratory-prepared amorphous FeS was left open to the air for a few hours, indicating that oxidation of amorphous FeS could be occurring in the sediments during the cool winter months when dissolved oxygen levels are relatively high. The conversion of amorphous FeS to Fe(III)-oxides could result in the formation of a phosphate-adsorbing layer at the sediment/water interface during the well-aerated winter months, thereby reducing P loading to the water column.

Degree of pyritization (DOP)

The extent to which potentially reactive iron species have been transformed to pyrite can be expressed as the degree of pyritization or DOP (Berner, 1970; Raiswell et al., 1988). DOP is the ratio between the weight percent pyritic iron, and the sum of pyritic iron, and acid-soluble iron on a weight percent basis. The 1.0 M cold HCl extract Fe was used in the DOP calculation and is considered to be the reactive form

of iron in these sediments. DOP ratios less than 0.4 are typically found in well-aerated marine waters with low amounts of organic matter and low H_2S; those between 0.5 and 0.7 indicate oxic to suboxic environments that also have pore water H_2S; and those greater than 0.7 are usually associated with euxinic conditions (H_2S in pore water and overlying water) and low input rates of available iron. Values around 1.0 usually indicate iron deficiency and are found in ancient, carbon-rich sediments (Berner, 1970; Leventhal & Taylor, 1990; Raiswell et al., 1994; Lyons, 1997; Raiswell & Canfield, 1998). The DOP for the Salton Sea sediments averaged 0.45 ± 0.13 and ranged from 0.11 to 0.85, suggesting that the formation is occurring under conditions intermediate between oxic and suboxic states. This is logical considering the hypolimnetic water is not permanently euxinic and there are high inputs of available iron. In addition, only the surface sediments were sampled and higher DOP values would be expected in older, deeper sediments.

It is instructive to consider the potential amount of pyrite forming within the Sea. We observed a decrease in the Fe(III)-oxide fraction in the bottom sediments of the Sea relative to that input in the form of river-borne suspended solids (Table 1). This Fe(III)-oxide fraction in the riverine suspended solids is considered "readily reducible iron" and can be used to estimate the amount of sulfidic minerals that are forming. Using the suspended solids' loading rates from Holdren & Montaño (2002) for the New, Alamo, and Whitewater Rivers, and assuming that all the readily reducible iron that enters the Sea is converted to pyrite, we estimate that approximately 7,700 metric tons of pyrite could be formed each year. However, we recognize that this readily reducible Fe may be converted to Fe-sulfide intermediates as well and the diagenetic transformation of amorphous FeS and greigite to pyrite is an ongoing process in the lake.

The conversion of Fe(III)-oxides to Fe-sulfide minerals is thought to dominate the iron geochemistry and thereby influence the phosphorus cycle in the Sea. That is by sequestering most of the available Fe in low solubility sulfidic phases, little Fe can form P-sorbing Fe(III)-oxides phases (Ingall et al., 2005). Moreover, the high porewater sulfide concentrations present throughout the year (unpublished data) will limit reoxidation of reduced forms

of Fe even when the water column is relatively well mixed. Thus, Fe control on P release is not thought to be a major mechanism in the biogeochemical cycling of P in the Salton Sea. Artificial hypolimnetic aeration would change the relative distribution of Fe(II)-sulfides and Fe(III)-oxides at the sediment/water interface, thereby increasing the adsorption of P.

Conclusions

Sequential extractions demonstrated that reducible Fe(III)-oxides from riverine suspended solids are being converted to amorphous FeS, Fe_3S_4, and FeS_2 in the sediments of the Salton Sea. Direct imaging of the sediment with SEM and spectroscopic analyses by EDAX confirm the presence of FeS_2. These reduced forms of Fe constitute $\sim 1\%$ of the sediment by weight. Based on morphology, the pyrite appears to be forming in the water column and not in the sediment pore water. The high deposition rates of both organic matter and reducible-Fe minerals, the intense sulfate reduction rates during the summer months, the oxic conditions in the water column during the winter, and the relatively young age of the surface sediments result in "degree of pyritization" (DOP) values near 0.5.

The strongly reducing conditions that are present during much of the year promote reduction and precipitation of Fe as Fe(II)-sulfides, thereby limiting Fe(II) solubility and thus also limiting the availability of Fe(III)-oxides for adsorption of phosphate. On this basis, then, the classic model for P retention by ferric oxides if present would not operate until the cool winter months when there is a lower oxygen demand, enhanced vertical mixing, and increased dissolved oxygen levels in the water column which convert amorphous FeS to Fe(III)-oxides. Rapid internal recycling of phosphate would thus be expected for most of the year, although high dissolved Ca^{2+} concentrations and high alkalinity production rates associated with sulfate reduction may provide an alternative P sequestration process within the Salton Sea (Rodriguez et al., 2008). The introduction of artificial hypolimnetic aeration via force-air or propeller mixers would increase the Fe(III)-oxides and decrease the internal loading of P.

 Springer

References

Anderson, M. A., L. Whiteaker, E. Wakefield & C. Amrhein, 2008. Properties and distribution of sediment in the Salton Sea, California: An assessment of predictive models. Hydrobiologia (this issue).

Berner, R. A., 1970. Sedimentary pyrite formation. American Journal of Science 268: 1–23.

Butler, I. B. & D. Rickard, 2000. Framboidal pyrite formation via the oxidation of iron (II) monosulfide by hydrogen sulphide. Geochimica et Cosmochimica Acta 64: 2665–2672.

Caraco, N., J. J. Cole & G. E. Likens, 1989. Evidence for sulphate-controlled phosphorous release from sediments of aquatic systems. Nature 341: 316–318.

Caraco, N. F., J. J. Cole & G. E. Likens, 1993. Sulfate control of phosphorus availability in lakes. A test and re-evaluation of Hasler and Einsele's model. Hydrobiologia 253: 275–280.

Carroll, D., 1958. Role of clay minerals in the transportation of iron. Geochimica et Cosmochimica Acta 14: 1–28.

Cook, C. B., G. T. Orlob & D. W. Huston, 2002. Simulation of wind-driven circulation in the Salton Sea: Implications for indigenous ecosystems. Hydrobiologia 473: 59–75.

Cooke, G. D., E. B. Welch, S. A. Peterson & P. R. Newroth, 1993. *Restoration and Management of Lakes and Reservoirs*, 2nd ed. Lewis Publishers, Boca Raton, FL.

Cornwell, J. C. & J. W. Morse, 1987. The characterization of iron sulfide minerals in anoxic marine sediments. Marine Chemistry 22: 193–206.

Furukawa, Y. & H. L. Barnes, 1995. Reactions forming pyrite from precipitated amorphous ferrous sulphide. In Vairavamurthy, M. A. & M. A. A. Schoonen (eds), *Geochemical Transformations of Sedimentary Sulphur*. American Chemical Society, 194–204.

Gleyzes, C., S. Tellier & M. Astruc, 2002. Fractionation studies of trace elements in contaminated soils and sediments: A review of sequential extraction procedures. Trends in Analytical Chemistry 21: 451–467.

Golden Software, Inc., 1999. Surfer: A surface mapping system. Version 7. Golden, CO.

Grossman, R. H., R. S. Liebling & H. S. Scherp, 1979. Chlorite and its relationship to pyritization in anoxic marine environments. Journal of Sedimentary Petrology 49: 611–613.

Holdren, G. C. & A. Montaño, 2002. Chemical and physical characteristics of the Salton Sea, California. Hydrobiologia 473: 1–21.

Hurtgen, M. T., T. W. Lyons, E. D. Ingall & A. M. Cruse, 1999. Anomalous enrichments of iron monosulfide in euxinic marine sediments and the role of H_2S in iron sulfide transformations: Examples from Effingham Inlet, Orca Basin and the Black Sea. American Journal of Science 299: 566–588.

Ingall, E., L. Kolowith, T. Lyons & M. Hurtgen, 2005. Sediment carbon, nitrogen and phosphorus cycling in an anoxic fjord, Effingham Inlet, British Columbia. American Journal of Science 305: 240–258.

Leventhal, J. S. & C. Taylor, 1990. Comparison of methods to determine degree of pyritization. Geochimica et Cosmochimica Acta 54: 2621–2625.

Lord, C. J., III, 1982. A selective and precise method for pyrite determination in sedimentary materials. Journal of Sedimentary Petrology 52: 664–666.

Lyons, T. W., 1997. Sulfur isotopic trends and pathways of iron sulfide formation in upper Holocene sediments of the anoxic Black Sea. Geochimica et Cosmochimica Acta 61: 3367–3382.

Mortimer, C. H., 1941. The exchange of dissolved substances between mud and water in lakes (Parts I and II). Journal of Ecology 29: 280–329.

Mortimer, C. H., 1942. The exchange of dissolved substances between mud and water in lakes (Parts III and IV). Journal of Ecology 30: 147–201.

Nirel, P. M. & F. M. M. Morel, 1990. Pitfalls of sequential extractions. Water Research 24: 1055–1056.

Poulton, S. W. & D. E. Canfield, 2005. Development of a sequential extraction procedure for iron: Implications for iron partitioning in continentally derived particulates. Chemical Geology 214: 209–221.

Raiswell, R., F. Buckley, R. A. Berner & T. F. Anderson, 1988. Degree of pyritization of iron as a paleoenvironmental indicator of bottom-water oxygenation. Journal of Sedimentary Petrology 58: 812–819.

Raiswell, R. & D. E. Canfield, 1998. Sources of iron for pyrite formation in marine sediments. American Journal of Science 298: 219–245.

Raiswell, R., D. E. Canfield & R. A. Berner, 1994. A comparison of iron extraction methods for the determination of degree of pyritisation and the recognition of iron-limited pyrite formation. Chemical Geology 111: 101–110.

Rickard, D., 1997. Kinetics of pyrite formation by the H_2S oxidation of iron(II) monosulfide in aqueous solutions between 25 and 125°C: The rate equation. Geochimica et Cosmochimica Acta 61: 115–134.

Rickard, D., M. A. A. Schoonen & G. W. Luther, 1995. Chemistry of iron sulfides in sedimentary environments. In Vairavamurthy, M. A. & M. A. A. Schoonen (eds), *Geochemical Transformations of Sedimentary Sulfur*. Washington D.C., American Chemical Society Symposium Series 612, 168–193.

Rodriquez, I. R., C. Amrhein & M. A. Anderson, 2008. Laboratory studies on the coprecipitation of phosphate with calcium carbonate in the Salton Sea, California. Hydrobiologia (this issue).

Schoonen, M. A. A. 2004. Mechanisms of sedimentary pyrite formation. In Amend, J. P., K. J. Edwards & T. W. Lyons (eds), *Sulfur Biogeochemistry: Past and Present*. The Geological Society of America Special Paper 379, 117–134.

Sondegaard, M., J. P. Jensen & E. Jeppesen, 1999. Internal phosphorus loading in shallow Danish Lakes. Hydrobiologia 29: 664–686.

Sweeney, R. E. & I. R. Kaplan, 1973. Pyrite framboid formation: Laboratory synthesis and marine sediments. Economic Geology 68: 618–634.

Tack, F. M. G. & M. G. Verloo, 1995. Chemical speciation and fractionation in soil and sediment heavy metal analysis: A review. International Journal of Environmental Analytical Chemistry 59: 225–238.

Tessier, A. & P. G. C. Campbell, 1988. Partitioning of trace metals in sediments. In Kramer, J. R. & H. E. Allen (eds), *Metal Speciation: Theory, Analysis and Application*. Lewis Publishers, Chelsea, MI, Chapter 9.

Watts, J. M., B. K. Swan, M. A. Tiffany & S. H. Hurlbert, 2001. Thermal, mixing, and oxygen regimes of the Salton Sea, California, 1997–1999. Hydrobiologia 466: 159–176.

Wilkin, R. T., M. A. Arthur & W. E. Dean, 1997. History of water-column anoxia in the Black Sea indicated by pyrite framboid size distributions. Earth and Planetary Science Letters 148: 517–525.

Wilkin, R. T. & H. L. Barnes, 1997a. Formation processes of framboidal pyrite. Geochimica et Cosmochimica Acta 61: 323–339.

Wilkin, R. T. & H. L. Barnes, 1997b. Pyrite formation in an anoxic estuarine basin. American Journal of Science 297: 620–650.

Wilkin, R. T., H. L. Barnes & S. L. Brantley, 1996. The size distribution of framboidal pyrite in modern sediments: An indicator of redox conditions. Geochimica et Cosmochimica Acta 60: 3897–3910.

Hydrobiologia (2008) 604:123–135
DOI 10.1007/s10750-008-9319-y

SALTON SEA

Transport and distribution of trace elements and other selected inorganic constituents by suspended particulates in the Salton Sea Basin, California, 2001

Lawrence A. LeBlanc · Roy A. Schroeder

Abstract In order to examine the transport of contaminants associated with river-derived suspended particles in the Salton Sea, California, large volume water samples were collected in transects established along the three major rivers emptying into the Salton Sea in fall 2001. Rivers in this area carry significant aqueous and particulate contaminant loads derived from irrigation water associated with the extensive agricultural activity, as well as wastewater from small and large municipalities. A variety of inorganic constituents, including trace metals, nutrients, and organic carbon were analyzed on suspended material isolated from water samples collected at upriver, near-shore, and off-shore sites established on the Alamo, New, and Whitewater rivers. Concentration patterns showed expected trends, with river-borne metals becoming diluted by organic-rich algal particles of lacustrine origin in off-shore stations. More soluble metals, such as cadmium, copper, and zinc showed a more even distribution between sites in the rivers and off-shore in the lake basin. General distributional trends of trace elements between particulate and aqueous forms were discerned by combining metal concentration data for particulates from this study with historical aqueous metals data. Highly insoluble trace metals, such as iron and aluminum, occurred almost entirely in the particulate phase, while major cations and approximately 95% of selenium were transported in the soluble phase. Evidence for greater reducing conditions in the New compared to the Alamo River was provided by the greater proportion of reduced (soluble) manganese in the New River. Evidence of bioconcentration of selenium and arsenic within the lake by algae was provided by calculating "enrichment" concentration ratios from metal concentrations on the algal-derived particulate samples and the off-shore sites.

Keywords Trace metals · Transport · Suspended sediments · Salton Sea

Guest editor: S. H. Hurlbert
The Salton Sea Centennial Symposium. Proceedings of a Symposium Celebrating a Century of Symbiosis Among Agriculture, Wildlife and People, 1905–2005, held in San Diego, California, USA, March 2005

Roy A. Schroeder—Retired.

L. A. LeBlanc (✉)
University of Maine, 5741 Libby Hall, Orono, ME 04469-5741, USA
e-mail: Lawrence.leblanc@umit.maine.edu

R. A. Schroeder
U.S. Geological Survey, 304 North Sierra Avenue, Solana Beach, CA 92075, USA

Introduction

The Salton Sea is a large (1,000 km^2) shallow (mean depth = 8 m) saline lake; its shoreline is at an elevation of about 70 m below sea level. It was formed as a freshwater lake between 1905 and 1907 as the

result of uncontrolled flooding from the Colorado River and since has been maintained largely by agricultural irrigation drainage, which presently is about 1.7 km^3 per year (Schroeder et al., 2002). Irrigation water in this area is derived from the Colorado River, which is already high in dissolved salts due to prior irrigation activity occurring along the river before it reaches Imperial County and the Salton Sea watershed, where it is once again used for irrigation. The present salinity in the Salton Sea measured as conductivity is approximately 55,000 µS/cm, roughly equivalent to 45 parts per thousand (Setmire & Schroeder, 1998; Schroeder et al., 2002) and is steadily increasing, due to evaporation in the closed-basin lake.

Declining environmental quality in this area is of great concern, due in large part to the importance of the Salton Sea in providing over-wintering habitat for several species of migrating birds. There are also human health concerns centering on potential exposure to toxicant-laden dust, as changing water use patterns have the potential to expose more lake bed sediments to the atmosphere. The situation is often compared to that of the dry lakebed of Owens Lake in Owens Valley, California, where windblown dust emissions and metal cycling have been the subject of numerous investigations (Tyler et al., 1997; Rya et al., 2002; Gillette et al., 2004). The dust from Owens Valley is considered to be a human health hazard, due to the presence of toxic trace metals (Gill et al., 2002). Respiratory ailments arising from windblown dust have been noted at other drying saline lakes, such as the Aral Sea in Uzbekistan (Kunii et al., 2003; Wiggs et al., 2003).

An in-depth examination of the transport of essential nutrients and toxic trace metals and cycling of these constituents within the lake basin was performed by Schroeder et al. (2002). Major findings were: (1) Nitrogen and selenium (and other inorganic species such as molybdenum and uranium) enter the sea as soluble oxyanions, which become reduced and at least temporarily sequestered in anoxic bottom sediments. Nitrogen can be reintroduced into the water column, through diffusion of ammonia from sediment porewaters, as well as by oxygenation of reduced N-species during sediment resuspension; (2) Dissolved phosphorus concentrations remain fairly constant, due to the sequestering of P in bones of fishes incorporated in bottom sediment and to coprecipitation with calcite in the bottom sediment; and (3)

Mass-balance calculations indicate that at least 78% of the organic carbon in the Salton Sea is derived from autochthonous production.

Other studies have examined various aspects of the importance of metal inputs into the Salton Sea basin. Cooke and Bruland (1987) examined the dynamics of selenium cycling within the lake and provided evidence of loss of selenium through gaseous efflux of methylated Se species. The authors also suggested incorporation of selenium by resident fish that are currently in the lake is an important component affecting selenium speciation. A recent review by Moreau et al. (2007) demonstrated detectable concentrations of arsenic and selenium in resident tilapia (an *Oreochromis mossambicus* Peters x *O. urolepis hornorum* (Trewavas) hybrid), and used them to estimate safe fish consumption levels for humans. An extensive survey of bottom sediments from throughout the Salton Sea conducted by Vogl and Henry (2002) demonstrated elevated levels of cadmium, copper, molybdenum, nickel, zinc, and selenium.

In 2001, as part of a larger study designed to examine the role of suspended sediments as a mechanism of transport of pesticides to sediments in the Salton Sea, suspended material was isolated from water samples collected in the lake, as well as from the three major rivers discharging into the lake basin. The purpose of this was twofold. One objective was to examine the role of suspended material in transporting different inorganic constituents into the lake basin. This was accomplished by examining the partitioning of variety of elements between aqueous and particulate phases within the rivers. The second objective was to look for evidence of bioaccumulation of metal species in the Salton Sea at the base of the food chain. Although monitoring trace-element concentrations in biological tissues from species representing all trophic levels is necessary to fully evaluate the toxic threat posed to wildlife, a strong argument can be made that the most important and useful biological data comes from chemical analyses at the base of the food chain. It is here that some three orders of magnitude enrichment (bioconcentration) from aqueous to tissue concentration occurs (Bowie et al., 1996). This is especially important for selenium because bacteria, algae, and fungi have the capability of converting inorganic oxidized forms of selenium (selenate or selenite) that typically predominate in water into the organic selenium that is subsequently incorporated into higher plants and animals (Bottino et al.,

1984; Besser et al., 1993). It is difficult and time consuming to obtain sufficient material on which to perform chemical analyses from microorganisms. However, sufficient suspended material, which comprises microbial and detrital biomass, can be isolated from the water column and analyzed, with ratios of concentrations then employed to ascertain the level of enrichment for various trace elements. We hypothesized that bioaccumulable metal species would be enriched in suspended material collected from the lake when compared to riverine particulate material and that examination of authigenic suspended material is a reasonable approach to assessing the danger of bioaccumulation of toxic metal compounds by higher organisms.

Methods

Sample collection methods

Sampling occurred during October 20–29, 2001. Transects consisting of three sampling sites each for the three major rivers in the Salton Sea Basin were established to examine the transport of inorganic constituents to the lake (Fig. 1). The in-river sites are

located approximately 1.6 km upstream of the mouths of the New and Alamo rivers and 6 km upstream of the mouth of the Whitewater River. The near-shore sites are located within the lake close to the river mouths and the off-shore sites are located further from the shoreline in water 4.6-m deep. The coordinates for each site are listed in Table 1. Water temperature and conductivity were measured in the field using a handheld digital thermometer (VWR Scientific, Westchester, Pennsylvania.) and an Orion conductivity meter (Orion Instruments, San Jose, California.), respectively. Fifty-ml water aliquots were collected during the pumping of the large volume water samples (described below) and sent to the Organic Chemistry Laboratory at the USGS California District office in Sacramento for more accurate determination of pH and conductivity.

Details of the collection of large volume (up to several hundred liters) water samples and subsequent isolation and dewatering of the suspended particulates via flow-through centrifugation and high-speed centrifugation can be found in LeBlanc et al. (2004a). Briefly, large volume water samples were collected using a large peristaltic pump powered by a portable generator and equipped with a stainless steel and Teflon inlet hose. Sample water was pumped directly

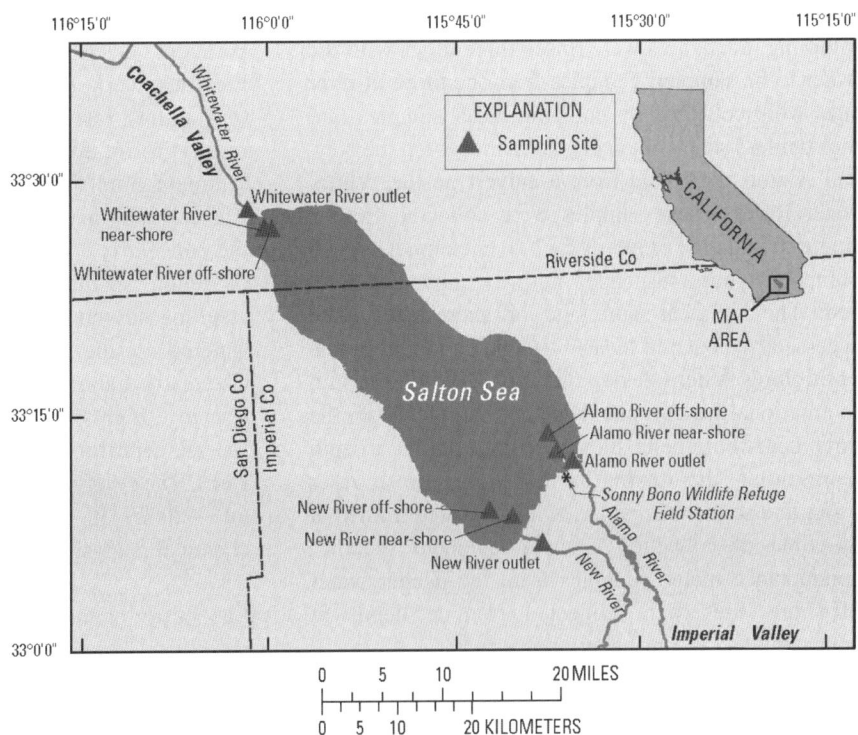

Fig. 1 Location of study area and of sampling sites in the Salton Sea Basin, California

Table 1 List of sampling sites and their coordinates

Station Name	Latitude	Longitude	Description[a]
Alamo River upriver	33° 11′ 56″	115° 35′ 46″	Bridge at Garst Road, 1.6 km upstream from river mouth river width: 25 m, river depth 3.5 m, total discharge, 903 ft³/s
Alamo River near-shore	33° 12′ 42″	115° 37′ 14″	Within Salton Sea, 250 m from river mouth approx depth: 1.5 m
Alamo River off-shore	33° 14′ 00″	115° 38′ 00″	Within Salton Sea, 2.9 km from river mouth approx depth 5 m
New River upriver	33° 05′ 59″	115° 38′ 56″	Bridge at Lack Road, 1.6 km upstream from river mouth river width: 26 m, river depth ~ 1 m, mean total discharge, 606 ft³/s
New River near-shore	33° 08′ 03″	115° 41′ 40″	Within Salton Sea, 100 m from river mouth depth ~ 1 m
New River off-shore	33° 08′ 35″	115° 43′ 45″	Within Salton Sea, 3.5 km from river mouth approx depth 5 m
Whitewater River upriver	33° 31′ 29″	116° 04′ 44″	Culvert at Lincoln Road, 6 km upstream from river mouth river width 20 m, river depth ~ 5 m, mean total discharge, 67 ft³/s
Whitewater River near-shore	33° 30′ 06″	116° 03′ 15″	Within Salton Sea approximately 500 m from river mouth approx depth 1.5 m
Whitewater River off-shore	33° 29′ 58″	116° 02′ 35″	Within Salton Sea approximately 1.5 km from river mouth approx depth 5 m

[a] Distances from river mouth are approximations, discharge measurements from U.S. Geological Survey (2001)

into pre-cleaned 20-l stainless steel containers for transport to processing facilities nearby. The stainless steel containers were cleaned by rinsing twice with approximately 100-ml distilled deionized water, followed by 100 ml rinses of methanol, acetone, and deionized water and allowed to air dry. Prior to sampling, containers were rinsed three times with the water to be sampled. Samples from the three in-river sites were collected by pumping water samples into the stainless steel containers from bridges on the New and Alamo rivers and from a culvert on the White-water River. These samples were collected from a single point at the center of each river channel using a pump inlet hose suspended midway above the river bed. The six near-shore and off-shore sites were accessed by boat and located using a handheld global positioning system device (Garmin GPS 12, Garmin International Inc., Olathe, Kansas). Samples from the three near-shore delta sites were collected at a depth representing the mixing zone between fresh river water and saline lake water, which was determined at the time of collection using a handheld specific-conductance meter. Samples from the deeper water off-shore sites were collected from a depth of approximately 0.5 m below the surface.

Additional samples were collected to measure suspended-sediment concentrations and percent fine-grained sediments. For these samples, water was pumped into pre-cleaned 500-ml clear glass bottles during the large volume sample collection.

Suspended sediments for chemical analysis were isolated by pumping the water through a flow-through centrifuge (Westphalia model KA-2, Westfalia Corporation, Odele, Federal Republic of Germany). Previous work (Horowitz et al., 1989) has shown that a flow rate of 2 l/min through the centrifuge results in the efficient capture of a wide variety of sediment grain sizes, from 2 μm and larger. Sediment was carefully and completely removed from each of the concentric collecting bowls using a combination of teflon spatulas and sample rinse water. The resulting suspended-sediment slurry was further de-watered in the USGS California district Organic Chemistry Laboratory in Sacramento by centrifugation in a high-speed refrigerated centrifuge (Sorvall RC-5B centrifuge, Du Pont Company, Wilmington, Del.). The resulting solid samples were stored frozen (at −20°C) in pre-cleaned glass containers until chemical analysis.

Laboratory methods

Suspended sediments were analyzed for organic carbon and nitrogen content using a Perkin Elmer CHNS/

O analyzer (Perkin Elmer Corporation, Norwalk, Connecticut) as described by LeBlanc et al. (2004a). Percent fines were determined on suspended-sediment samples by the USGS laboratory in Marina, California, using standard methods (Guy, 1969). The percentage of fines is defined as the fraction of sediment that passes through a 62-μm sieve.

Suspended sediments collected during the fall 2001 sampling period were analyzed by the USGS National Water Quality Laboratory in Denver, Colorado, using methods described by Arbogast (1996), and Briggs & Meier (1999). Instrumental analysis was performed by inductively coupled argon-plasma emission spectrometry (ICP) for a suite of trace metals and inorganic constituents.

Trace element phase distribution and partitioning

Particulate metal concentration data obtained by this study in 2001 were used along with aqueous dissolved concentrations obtained in 1986 (Setmire et al., 1990) and in 1988–1989 (Schroeder et al., 1993) to apportion element contributions between these two phases at the Alamo River in-river site. Similar calculations were not done for the Whitewater River because it was not monitored during the 1980s studies. For the New River, only two constituents, manganese and selenium, were considered due to the confounding influence of the large increase in population and industrial activity in Mexicali, a large Mexican city on the border that discharged municipal wastewaters into the New River between the 1980s and 2001.

Mean dissolved concentrations from as few as 1 to as many as 10 analyses of 22 chemical constituents were used. The chemical contribution associated with the particulate phase in the water column was calculated by multiplying particulate metal concentration data by suspended-sediment concentration at the in-river sites. A value of "1" indicates equal contribution from each phase; ratios less than 1 indicate greater contribution from the soluble constituent.

Calculation of enrichment ratios

In order to discern whether metals were concentrated in off-shore particulates, enrichment ratios described by Schroeder (1985, 1995) and Schroeder et al.

(1988) were calculated. This was done by normalizing element concentrations to a continental crust-derived element (aluminum) and calculating the ratio of off-shore particulate concentrations to in-river particulate concentrations for each constituent as follows:

$$ER = \frac{\frac{[E]_{off-shore}}{[Al]_{off-shore}}}{\frac{[E]_{outlet}}{[Al]_{outlet}}} \tag{1}$$

where ER = the enrichment ratio, $[E]_{off-shore}$ = element concentration in off-shore suspended particulates, $[Al]_{off-shore}$ = aluminum concentration in off-shore suspended particulates, $[E]_{in-river}$ = element concentration in suspended particulates collected at in-river sites, and $[Al]_{in-river}$ = aluminum concentration in suspended particulates collected at in-river sites

The resulting value is a measure of the degree to which metals and other constituents are enriched in the off-shore suspended particulates and, by inference, bioconcentrated by algae.

Results

River flow and water chemistry

Daily mean stream flows ranged from 24.5 to 28.6 m³/s (872–1,010 ft³/s) at the Alamo and from 13.9 to 18.8 m³/s (490–665 ft³/s) at the New River in-river site during the 2001 sampling event (October 20, 2001 to October 28, 2001) and are similar to the mean daily flows for the 42-year period of record (USGS, 2001). Stream flows for the Whitewater River were approximately 40% lower than the 42-year average mean flows and ranged from 1.6 to 1.9 m³/s (57–67 ft³/s) during the fall 2001 sampling event.

The gradient of specific conductance (measured as a surrogate for salinity) was extremely high in all three river-to-lake transects and varied by more than a factor of 10 (Table 2). Salinity in the Salton Sea has been discussed in depth by Schroeder et al. (2002), Setmire et al. (1993), and Setmire & Schroeder (1998). Greater influence of saline Salton Sea water can be seen in the high conductivity measured at the Whitewater River near-shore site relative to the near-shore sites from the other two rivers; a result that is consistent with the lower flows measured in the Whitewater River.

Table 2 Stream discharge rates and water properties at sampling sites in the Salton Sea Basin, California during the October 2001 sampling event

	In-river	Near-shore	Off-shore
Alamo River: 24.5–26.8 m^3/s[a]			
Sampling date	10/21	10/20	10/23–10/24
Sample depth (m)	0.5	0.5	0.5
Temp[b] (°C)	28.9	28.0	29.8
pH	7.4	7.8	8.1
Conductance[c] (μS/cm)	3,080	11,500	53,300
DOC[d] (mg/l)	6.3	11.6	50
New River: 13.9–18.8 m^3/s[a]			
Sampling date	10/23	10/21	10/25
Sample depth (m)	0.5	0.3	0.5
Temp (°C)	26.1	26.4	29.5
pH	7.8	7.6	8.1
Conductance (μS/cm)	3,800	34,500	53,800
DOC (mg/l)	6.6	33.6	53
Whitewater River: 1.6–1.9 m^3/s[a]			
Sampling date	10/26	10/28	10/27–10/28
Sample depth (m)	0.5	1	0.5
Temp (°C)	22.0	25.3	31.9
pH	7.7	7.8	7.8
Conductance (μS/cm)	1,900	53,200	53,200
DOC (mg/l)	5.2	52	52

[a] Range of daily mean sample flow during the sampling period; [b] Temp, temperature; [c] Conductance, specific conductance; [d] DOC, dissolved organic carbon; pH, specific conductance, and DOC were measured in the laboratory, and temperature was measured in the field

Measurements of pH indicated a slight trend of increasing pH from the in-river sites to the off-shore sites (Table 2). Dissolved organic carbon (DOC) concentrations were high at all sites, and increased from a minimum of 5.2 mg/l at in-river sites to a maximum of 53 mg/l at the off-shore sites. The influence of Salton Sea water closer to shore was also seen at the Whitewater River near-shore site, where DOC concentration was the same as that at the off-shore site. In contrast, DOC concentrations at the Alamo and New river near-shore sites were intermediate between the in-river and lake off-shore sites.

Suspended matter

Concentrations of suspended sediments were much higher in the Alamo and New Rivers than in the much smaller Whitewater River (Table 3), and decreased in all three rivers from the in-river to the off-shore sites. However, the percentage of fine-grained material did not change appreciably from in-river to off-shore sites. These trends are in agreement with historical trends described by Setmire et al. (1990, 1993).

Organic carbon concentrations in suspended sediment collected from the in-river sites ranged from 0.93 to 4.0% (Table 3). Suspended sediment at all the off-shore sites consisted of a high percentage of algae, which is reflected in extraordinarily high percent organic carbon concentrations (28–44%). The Whitewater River near-shore site also had very high organic carbon concentrations (43%), in contrast to the much lower concentrations at the near-shore sites of the other two rivers (1.2 and 2.7% for the Alamo and New rivers) where the large flows resulted in suspended sediment still predominately of riverine origin.

Molar ratios of carbon to nitrogen in off-shore samples also indicated a biogenic origin for the suspended material from within the lake (i.e. the off-shore sites). Ratios of C/N ranged from 1:1 to 8:1 in off-shore suspended solids, values that are near, or even less than the traditional Redfield ratio for biologically derived oceanic particulate organic matter of

Table 3 Suspended sediment concentrations and percent fine-grained material in water from sampling sites along the Alamo, New and Whitewater rivers, Salton Sea watershed, California, during October 2001

Sampling site	Total		Organic
	SS[a] (mg/l)	Fines[b] (%)	Carbon (%)
Alamo River in-river	418	80	0.93
Alamo River near-shore	255	88	1.2
Alamo River off-shore	35	76	28.4
New River in-river	245	97	1.74
New River near-shore	145	48	2.71
New River off-shore	43	86	na[c]
Whitewater River in-river	72	82	3.99
Whitewater River near-shore	39	66	42.7
Whitewater River off-shore	21	79	43.7

[a] SS, suspended sediments; [b] Fines, percent particles smaller than 62 microns; [c] na, not analyzed

approximately 7:1 (Chester, 1990). Visual observations of intense red and green algal blooms during sample collection and the appearance and characteristic odor of the dewatered material also supported the assertion that suspended sediment in the Salton Sea consists of organic-rich material of algal origin.

Trace elements and other particle-bound constituents

Elements associated with continental crust materials such as aluminum, barium, chromium, iron, titanium, and vanadium were elevated in suspended sediments from the in-river sites by a factor of 10 or greater relative to the off-shore sites, consistent with a river-derived source which is diluted by organic-rich particulates off-shore (Table 4). Cadmium, copper, and zinc were more evenly distributed between in-river, near-shore, and off-shore sites, reflecting their greater solubility, potential for release from sediments, and redistribution in the environment (Table 4, Boehm et al., 1987; Allen, 1995). For the Whitewater River, concentration differences between in-river and the off-shore sites were often a factor of 10 or greater for both continental crust-derived elements as well as the more biologically active trace metals. Selenium (Se), which also can be remobilized from sediments, was markedly elevated in the off-shore suspended material relative to suspended sediments from the in-river sites by factors of 10 or greater for all three transects (Table 4). Arsenic was similarly elevated in off-shore lacustrine suspended particulates relative to riverine suspended sediments from in river sites, although by much less than selenium.

Trace element phase distribution and partitioning

The chemical contribution associated with the suspended-sediment phase in the water column was calculated by multiplying sediment chemical data in Table 4 by suspended-sediment concentration at the in-river sites in Table 3. Ratios of concentrations contributed by suspended sediment to soluble concentrations for 22 chemical species are listed in Table 5.

The particulate phase was the predominant contributor for 10 elements (Al, Ba, Cr, Cu, Fe, Mn, Ni, P, V, and Zn) in the Alamo River (Table 5).

Dissolved phase contributed most to transport for 12 elements (As, Cd, Ca, Li, Mg, Mo, K, Se, Na, Sr, U, and organic carbon). Selenium was transported primarily in soluble form (Table 5).

Although partitioning ratios were not calculated for the entire suite elements for the New River, calculation of manganese partitioning was made using the 1988–1989 dissolved data (mean dissolved concentration = 112 μg/l, from 9 observations ranging from 60 to 200 μg/l) and compared to the Alamo River. Results show that for manganese the soluble contribution in the New River was much more than double that in the Alamo River, with a concentration ratio of 1.5, compared to 16 for the Alamo River. In contrast, the concentration ratio of 0.058 calculated for selenium in the New River (based upon a mean dissolved selenium concentration of 4.2 μg/l, from 10 observations ranging from 4 to 5 μg/l) was similar to the ratio of 0.037 calculated for the Alamo River, consistent with a large dissolved, relatively unreactive source of selenium.

Enrichment ratios

Results show that those elements associated with aluminosilicate materials, such as the rare earth elements (e.g. lanthanum), had ratios of approximately one (i.e. no enrichment, Table 5). A number of constituents important in biological metabolism, such as calcium, potassium, magnesium, and zinc, had ratios of 6.2 to 18. Elements known as "heavy metals" such as copper, chromium, lead, nickel, and mercury, and shown to accumulate in phytoplankton (Fisher, 1986; Luoma et al., 1998), had enrichment ratios between 11 and 31. Phosphorus, an essential nutrient had a very high enrichment ratio (140), consistent with the observation that off-shore suspended particulates primarily comprise algal material. Selenium exhibited the highest enrichment ratio of the 32 elements examined in Table 5.

Discussion

Suspended particulate concentration and trends in molar C/N ratios

Suspended particulate concentrations followed the expected pattern along each transect, with the

Table 4 Element concentrations in suspended sediment for samples from the Alamo, New, and Whitewater rivers, Salton Sea watershed, California, fall 2001

Element	Alamo River			New River			Whitewater River		
	In-river	Near-shore	Off-shore	In-river	Near-shore	Off-shore	In-river	Near-shore	Off-shore
Aluminum	**7.8**	**8.4**	**0.46**	**6.7**	**6.8**	**0.14**	**7.9**	**0.13**	**0.11**
Antimony	0.71	0.82	0.2	0.76	0.62	0.02	0.62	0.02	<0.02
Arsenic	8.3	9.2	17	9.1	9.1	30	11	27	27
Barium	490	460	34	520	480	19	630	14	13
Beryllium	1.9	2.1	0.18	2	1.8	0.04	2	0.03	0.03
Bismuth	<0.005	<0.005	<0.005	0.17	0.12	<0.005	0.84	0.1	<0.005
Cadmium	0.4	0.44	0.18	0.45	0.4	0.12	0.36	0.15	0.09
Calcium	**5.5**	**5.8**	**1.2**	**5.1**	**4.8**	**0.88**	**5.5**	**0.51**	**0.54**
Cerium	63	71	3.4	67	62	1.2	85	0.7	0.8
Chromium	120	70	60	57	110	28	83	7.9	8.2
Cobalt	10	11	2.8	9.9	9.5	4.5	19	5.9	4.5
Copper	28	34	16	37	31	18	69	7.9	8
Gallium	15	17	0.91	15	14	0.27	21	0.27	0.27
Iron	**2.9**	**3.3**	**0.2**	**3**	**2.6**	**0.071**	**5.2**	**0.056**	**0.051**
Lanthanum	33	37	2.2	35	32	0.7	45	0.5	0.5
Lead	22	24	27	23	23	15	27	2.5	1.9
Lithium	44	51	22	46	43	15	44	12	15
Magnesium	**2**	**2.1**	**0.42**	**2.1**	**2**	**0.36**	**2.3**	**0.4**	**0.45**
Manganese	670	700	250	650	740	140	1,200	170	120
Mercury	0.04	0.04	0.06	0.07	0.07	0.05	0.08	0.03	0.04
Molybdenum	6.3	1.3	3	2.5	6.5	2.3	4.1	1.1	1
Nickel	51	34	22	30	43	11	36	6	6
Niobium	18	20	<2	17	12	<2	26	<2	<2
Phosphorus	**0.11**	**0.13**	**0.48**	**0.2**	**0.24**	**0.54**	**0.29**	**0.6**	**0.55**
Potassium	**2**	**2.1**	**0.42**	**2.1**	**2**	**0.36**	**2.3**	**0.4**	**0.45**
Scandium	11	12	0.9	11	9.5	<0.3	16	<0.3	<0.3
Selenium	0.7	1	8.7	1	1.4	16	1	14	11
Silver	<3	<3	<3	<3	<3	<3	<3	<3	<3
Sodium	**0.54**	**0.52**	**6.4**	**0.6**	**1.6**	**4.6**	**1.3**	**3.8**	**4.7**
Strontium	320	340	230	320	280	200	500	150	140
Thallium	0.62	0.68	0.01	0.65	0.61	<0.003	0.77	<0.003	<0.003
Thorium	11	12	0.52	11	10	0.18	15	0.13	0.09
Titanium	3,100	3,600	170	3,400	3,000	50	6,200	50	40
Uranium	3	3.6	5.7	3.3	3	4.7	4.4	1.2	1.2
Vanadium	78	90	16	80	69	23	110	17	16
Yttrium	23	25	1.6	27	30	0.5	26	<0.3	0.3
Zinc	86	100	92	120	100	62	190	27	27

Units are µg/g dry sediment (regular font) or percent (in bold)

highest concentrations found at the in-river sites. This pattern is consistent with a riverine source of particulates, derived from soil transported from surrounding agricultural fields and carried via agricultural drains to rivers and finally into the Salton Sea. This study shows, however, that the composition of particulate matter within the lake is very different from that carried down by the rivers

Table 5 Element enrichment ratios for suspended sediment from the Salton Sea Basin, California. Suspended sediment/aqueous concentration ratios from the Alamo River in-river site, along with historical aqueous concentration data used for the ratio calculation

Element	Enrichment ratio[a] Mean (range)	Concentration ratio[b]	Dissolved concentration Mean (range) (µg/l)	Number of samples
Aluminum (Al)		2,000	17 (10–20)	3
Antimony (Sb)	3 (1–5)			
Arsenic (As)	120 (35–180)	0.58	6 (5–7)	
Barium (Ba)	1.5 (1.2–1.7)	2	100	1
Beryllium (Be)	1.2 (1.0–1.6)			
Bismuth (Bi)	NC[c]			
Cadmium (Cd)	13 (7.6–18)	0.08	2	1
Calcium (Ca)	6.3 (3.7–8.3)	0.14	160,000 (150,000–180,000)	11
Cerium (Ce)	0.8 (0.7–0.9)			
Chromium (Cr)	13 (7.1–24)	>50	>50	1
Cobalt (Co)	15 (4.7–24)			
Copper (Cu)	14 (8.3–24)	2.9	4	1
Gallium (Ga)	0.9 (0.9–1.0)			
Iron (Fe)	1.0 (0.7–1.2)	530	23 (10–40)	11
Lanthanum (La)	1.0 (0.80–1.1)			
Lead (Pb)	19 (5.1–21)			
Lithium (Li)	16 (8.5–24)	0.11	160 (160–170)	4
Magnesium (Mg)	8.6 (3.6–14)	0.10	87,000 (78,000–100,000)	7
Manganese (Mn)	8 (6.3–10)	16	20 (<10–40)	11
Mercury (Hg)	32 (25–36)			
Molybdenum (Mo)	23 (8.1–44)	0.18	14 (12–21)	7
Nickel (Ni)	12 (7.3–18)	7.1	3	1
Niobium (Nb)	NC[c]			
Phosphorus (P)	110 (74– 1 40)	2.2	210	1
Potassium (K)	8.6 (3.6–14)	0.78	10,700 (10,000–12,000)	7
Scandium (Sc)	~1.3			
Selenium (Se)	590 (210–790)	0.04	6–10	11
Silver (Ag)	NC[c]			
Sodium (Na)	280 (200–370)	0.005	420,000 (400,000–600,000)	7
Strontium (Sr)	20 (12–30)	0.05	28,000 (2700–31000)	6
Thallium (Tl)	0.3			
Thorium (Th)	0.7 (0.4–0.8)			
Titanium (Ti)	0.8 (0.7–0.9)			
Uranium (U)	40 (20–68)	0.07	18	1
Vanadium (V)	9.2 (3.5–14)	1.9	33	1
Yttrium (Y)	1 (0.8–1.2)			
Zinc (Zn)	18 (10–25)	1.8	36	1

[a] Enrichment ratio (see Eq. 1 in text), [b] see Methods for a description of the concentration ratio, [c] NC, not calculated because too many concentrations were below the detection limit

and consists primarily of biogenic particles. Chemical evidence, provided by the low C/N ratios in off-shore suspended material, is consistent with visual observations of this material. Low C/N values (from 6 to 8) were reported by Schroeder et al. (2002) in bottom sediment collected off-shore in the lake, who

 Springer

also postulated that algae are the primary source of sediment organic carbon.

Suspended material collected from the near-shore site off the Whitewater River had more representative lacustrine particulate material than riverine material. This is largely due to the lower river discharge in the Whitewater River, which is also reflected in the lower values of salinity and DOC in the water from this site compared to the New and Alamo River near-shore sites.

Trace element concentration trends

Numerous trace element concentration patterns in suspended particulate samples followed spatial trends consistent with a river-derived source. It must be mentioned, that the sample collection techniques employed were intended to minimize contamination for organic analysis. Since the water was processed through stainless steel tubing, passed through a flow-through stainless steel centrifuge, and further spun down in stainless steel centrifuge tubes, minor contamination for selected metals present in stainless steel is possible. All samples were treated exactly the same however, and therefore, were subjected to the same degree of contamination. Despite this confounding factor, distributional trends were observed.

Trace element phase distribution and partitioning

In order to quantify the partitioning of metals between solid and aqueous phases, multiple datasets (dissolved concentrations and particulate concentrations) collected more than 10 years apart, were utilized. The limitations imposed by data collected at different times and by variations in suspended-sediment concentration associated with seasonality and changing river discharge are readily recognized, so the results presented here have been used only to illustrate general trends. The method applied would likely not be appropriate to most rivers but can be used here because the Alamo River exhibits a small range in water flow rates, and therefore also exhibits small changes in suspended-sediment and aqueous-chemical concentrations, which typically fluctuate with river discharge. Comparisons of total aqueous concentrations between 1986 and those reported by Holdren & Montaño (2002) from sampling performed in 1999 show variations of a factor of 2 or less for

many of the elements (data not shown). The more recent data was not used, however, since only total aqueous metal concentrations (dissolved plus particulate) were reported. The comparatively constant conditions exist because water in the Alamo River is composed almost entirely of agricultural irrigation runoff, and crops are grown year-round in the Imperial Valley. Storm runoff to the river is infrequent because annual mean rainfall is only about 7 cm.

It is known that irrigation water supplied by the Colorado River is the source of selenium in agricultural drainage in the Imperial Valley, and that selenium is transported virtually entirely as a conservative (i.e., unreactive) dissolved constituent as it moves through the system, from diversion to irrigation, to drainage and ultimately to discharge in the Salton Sea (Schroeder et al., 1993; Setmire et al., 1993). In contrast, particulate manganese contributes 16 times more manganese than dissolved concentrations (Table 5).

The differences seen between the New and Alamo Rivers in terms of manganese partitioning reflects important differences between the two rivers. The higher soluble contribution in the New River is likely caused by discharge of municipal and domestic wastewater in Mexicali, which results in more reducing conditions in the New River and that has a measurable effect even as far downstream as the in-river site. Dissolved oxygen measured in the top 1 m along the New River ranged from 0.3 to 1.1 mg/l, and 7 mg/l at the in-river site, compared to 9.4–10 mg/l along the length of the Alamo River, when measured in April, 2003 (LeBlanc et al., 2004b). The impact of wastewater on oxidized relative to reduced forms of nitrogen has been noted by Schroeder (1996). It is well-known that reduced manganese is far more soluble than oxidized manganese species (Berner, 1980). This trend would be even greater if the use of older data from 1988 to 1989 underestimated the present day dissolved concentration of manganese.

In contrast, the similarity of the concentration ratios calculated for selenium between the New and Alamo Rivers reflects the presence of a large dissolved, relatively unreactive source as mentioned above. The total selenium concentration of 4.6 µg/l reported by Schroeder et al. (1993) is similar to the concentration reported by Holdren & Montaño (2002) of 3.4 µg/l for total selenium, which is consistent

with this scenario and reflects the more constant source from irrigation drainage.

Enrichment ratios

Along with evidence demonstrating a biogenic (algal) origin for particulates within the Salton Sea, calculated enrichment ratios provided evidence for the bioaccumulation of heavy metals in phytoplankton. Especially significant is the extremely high enrichment ratio for selenium. It has long been known that selenium residues are quite high in resident invertebrates, fish, and birds relative to the low selenium concentration in water from the Salton Sea (Setmire et al., 1990; Schroeder et al., 1993; Moreau et al., 2007), and therefore has been thought that benthic invertebrates (pileworms) residing in the selenium-rich bottom sediments provide the link whereby selenium is bioaccumulated in higher trophic levels. Cook and Bruland (1987) postulated that the presence of reduced selenium species in the Salton Sea is caused by reductive incorporation by organisms. The finding herein of high selenium concentrations in lacustrine particulates (8–16 µg/g), comparable to tissue concentrations reported for tilapia in the Salton Sea by Moreau et al. (2007) suggests that selenium is already bioconcentrated at the lowest trophic level by algae in the water column. The results of this study suggest a similar conclusion for arsenic, although its enrichment is noticeably less than that of selenium.

Summary and conclusions

The concentrations of trace elements in suspended sediments followed expected trends, with continental crustal materials decreasing in concentration from the rivers to the off-shore sites due to the dilution of river-derived material with algal particulates formed within the lake. Differences in element distribution in the Whitewater River can be explained by its much lower discharge rate, causing its influence on the Salton Sea water chemistry and chemical composition of suspended particulates to extend to a lesser distance from the mouth of the river. Concentration differences between river and off-shore sites were noticeably less for more soluble trace elements, suggesting remobilization from the sediment into overlying waters.

Estimates of partitioning between particulate and soluble phases showed highly insoluble trace metals such as iron and aluminum to occur virtually entirely in the particulate phase, while soluble major cations were carried mostly as dissolved constituents. About 95% of selenium was determined to be transported in the soluble phase in the Alamo River. The greater reducing conditions in the New River favored the formation of more soluble reduced manganese, which was reflected in the smaller concentration ratio.

Enrichment ratios provided evidence of bioaccumulation of biologically important constituents such as phosphorus and calcium as well as several trace elements, most notably selenium and arsenic, in lacustrine suspended particulates. Evidence that the off-shore material was largely algae was provided by concentration ratios of elemental carbon to nitrogen in suspended particulates similar to those of attributed to living algae, along with visual evidence.

This study demonstrates the importance and utility of examining suspended solids in conducting studies of element fate and transport. Suspended solids within the lake basin were composed largely of algae, and so could be used to examine bioconcentration of toxic elements such as selenium at the base of the food web. Analyses of suspended solids should be included in future studies of pollutant transport and input into the Salton Sea and other hypersaline lakes.

Acknowledgments The authors thank G. Edward Moon for help with field sampling and boat operations. Also, many thanks go to Charlie Pelizza and the staff of the Sonny Bono Wildlife Refuge for use of their Boston Whaler and general use of the refuge facilities. This study was conducted with funds provided by the California Regional Water Quality Control Board along with matching funds provided by the U.S. Geological Survey California Water Resources Division.

References

Allen, H. E. (ed.), 1995. Metal Contaminated Aquatic Sediments. Ann Arbor Press, Ann Arbor, Michigan, 292 pp.

Arbogast, B. F. (ed.), 1996. Analytical Methods Manual for the Mineral Resource Surveys Program: U.S. Geological Survey Open-File Report 96-525, 248 pp.

Berner, R. A., 1980. Early Diagenesis, a Theoretical Approach. Princeton University Press, Princeton, New Jersey, 250 pp.

Besser, J. M., T. J. Canfield & T. W. La Point, 1993. Bioaccumulation of organic and inorganic selenium in a laboratory food chain. Environmental Toxicology and Chemistry 12: 57–72.

Boehm, P., M. Steinhauer., E. Crecelius, J. Neff & C. Tuck-field, 1987. Beaufort Sea monitoring program: analysis of trace metals and hydrocarbons from outer continental shelf (OCS) activities. Minerals Management Service 87-0072, 263 pp.

Bottino, N. R., C. H. Banks, K. J. Irgolic, P. Micks, A. E. Wheeler & R. A. Zingaro, 1984. Selenium-containing amino-acids and proteins in marine-algae. Phytochemistry 23: 2445–2452.

Bowie, G. L., J. G. Sanders, G. F. Riedel, C. C. Gilmore, D. L. Breitburg, G. A. Cutter & D. B. Porcella, 1996. Assessing selenium cycling and accumulation in aquatic systems. Water, Air and Soil Pollution 90: 93–104.

Briggs, P. H. & A. L. Meier, 1999. The determination of forty two elements in geological materials by inductively coupled plasma-mass spectrometry. U.S. Geological Survey Open-File Report 99-166, 15 pp.

Chester, R., 1990. Marine Geochemistry. Unwin Hyman, Ltd., London, 698 pp.

Cooke, T. D. & K. W. Bruland, 1987. Aquatic chemistry of selenium: evidence of biomethylation. Environmental Science and Technology 21: 1214–1219.

Fisher, N. S., 1986. On the reactivity of metals for marine phytoplankton. American Society of Limnology and Oceanography 31: 443–449.

Gill, T. E., D. A. Gillette, T. Niemeyer & R. T. Winn, 2002. Elemental geochemistry of wind-erodible playa sediments, Owens Lake, California. Nuclear Instruments and Methods in Physics Research Section B—Beam Interactions With Materials and Atoms, 189: 209–213.

Gillette, D., D. Ono & K. Richmond, 2004. A combined modeling and measurement technique for estimating windblown dust emissions at Owens (dry) Lake, California. Journal of Geophysical Research-Earth Surface 109 (F1): Art. No. F01003.

Guy, H. P., 1969. Laboratory theory and methods for sediment analysis: U.S. Geological Survey Techniques of Water-Resources Investigations, Vol. 5, Chap. C1, 58 pp.

Holdren, G. C. & A. Montaño, 2002. Chemical and physical characteristics of the Salton Sea, California. Hydrobiologia 473: 1–21.

Horowitz, A. J., K. A. Elrick & R. C. Hooper, 1989. A comparison of instrumental dewatering methods for the separation and concentration of suspended sediment for subsequent trace element analysis. Hydrological Processes 2: 163–184.

Kunii, O., M. Hashizume, M. Chiba, S. Sasaki, T. Shimoda, W. Caypil & D. Dauletbaev, 2003. Respiratory symptoms and pulmonary function among school-age children in the Aral Sea region. Archives of Environmental Heath 58: 676–682.

LeBlanc, L. A., R. A. Schroeder, J. L. Orlando & K. M. Kuivila, 2004a. Occurrence, distribution and transport of pesticides, trace elements and selected inorganic constituents into the Salton Sea Basin, California, 2001–2002. U.S. Geological Survey Scientific Investigations Report No. 2004-5117, 40 pp.

LeBlanc, L. A., J. L. Orlando & K. M. Kuivila, 2004b. Pesticide concentrations in water and in suspended and bottom sediments in the New and Alamo rivers, Salton Sea

watershed, California, April 2003: U. S. Geological Survey Data Series 104, 15 pp.

Luoma, S. N., A. van Geen, B.-G. Lee & J. E. Cloern, 1998. Metal uptake by phytoplankton during a bloom in south San Francisco Bay: implications for metal cycling in estuaries. Limnology and Oceanography 43: 1007–1016.

Moreau, M. F., J. Surico-Bennett, M. Vicario-Fisher, D. Crane, R. Gerads, R. M. Gersberg & S. H. Hurlbert, 2007. Contaminants in tilapia (Oreochromis mossambicus) from the Salton Sea, California, in relation to human health, piscivorous birds and fish meal production. Hydrobiologia 576: 127–165.

Rya, J. H., S. D. Gao, R. A. Dahlgran & R. A. Zierenberg, 2002. Arsenic distribution, speciation and solubility in shallow groundwater of Owens Dry Lake, California. Geochimica et Cosmochimica Acta 66: 2981–2994.

Schroeder R. A., 1985. Sediment accumulation rates in Irondequoit Bay, New York based on lead-210 and cesium-137 geochronology. Northeastern Environmental Science 4: 23–29.

Schroeder, R. A., 1995. Potential for chemical transport beneath a storm-runoff recharge (retention) basin for an industrial catchment in Fresno, California. U.S. Geological Survey Water-Resources Investigations Report 93-4140, 38 pp.

Schroeder, R. A., 1996. Transferability of environmental assessments in the Salton Sea Basin, California, and other irrigated areas in the western United States to the Aral Sea Basin, Uzbekistan. In Micklin, P. P. & W. D. Williams (eds), The Aral Sea Basin Proceedings of the NATO Advanced Research Workshop "Critical Scientific Issues of the Aral Sea Basin: State of Knowledge and Future Research Needs". Tashkent, Uzbekistan, May 2–5, 1994. NATO ASI Series, Partnership Sub-Series, 2. Environment, Springer Verlag Press, Berlin, 12, 121–137.

Schroeder, R. A., J. G. Setmire & J. C. Wolfe, 1988. Trace elements and pesticides in the Salton Sea area, California. In Proceedings on Planning Now for Irrigation and Drainage: Irrigation Division. American Society of Civil Engineers, Lincoln, Nebraska, July 19–21, 1988, 700–707.

Schroeder, R. A., A. M. Rivera, B. J. Redfield, J. N. Densmore, R. L. Michel, D. R. Norton, D. J. Audet, J. G. Setmire & S. L. Goodbred, 1993. Physical, chemical and biological data for detailed study of irrigation drainage in the Salton Sea area, California, 1988–90. U. S. Geological Survey Open-File Report 93-83, 179 pp.

Schroeder, R. A., W. H. Orem & Y. K. Kharaka, 2002. Chemical evolution of the Salton Sea, California: nutrient and selenium dynamics. Hydrobiologia, 473: 23–45.

Setmire, J. G. & R. A. Schroeder, 1998. Selenium and salinity concerns in the Salton Sea area of California. In Frankenberger, W. T., Jr. & Engberg, R. A. (eds), Environmental Chemistry of Selenium, Chap. 12. Marcel Dekkar Inc., New York, 205–221.

Setmire, J. G., J. C. Wolfe & R. K. Stroud, 1990. Reconnaissance investigation of water quality, bottom sediment, and biota associated with irrigation drainage in the Salton Sea area, California, 1986–87: U.S. Geological Survey Water-Resources Investigations Report 89-4102, 68 pp.

Setmire, J. G., R. A. Schroeder, J. N. Densmore, S. L. Good-bred, D. J. Audet & W. R. Radke, 1993. Detailed study of water quality, bottom sediment, and biota associated with irrigation drainage in Salton Sea area, California, 1988–90. U.S. Geological Survey Water-Resources Investigations Report 93-4014, 102 pp.

Tyler, W. W., S. Kranz, M. B. Parlange, J. Albertson, G. G. Katul, G. F. Cochran, B. A. Lyles & G. Holder, 1997. Estimation of groundwater evaporation and salt flux from Owens lake, California, USA. Journal of Hydrology 200: 110–135.

U.S. Geological Survey, 2001. National Water Information System (NWISWeb), U.S. Geological Survey, accessed March 15, 2003, at http://waterdata.usgs.gov/nwis/.

Vogl, R. A. & R. H. Henry, 2002. Characteristics and contaminants of the Salton Sea sediments. Hydrobiologia 473: 47–54.

Wiggs, G. F. S., S. L. O'Hara, J. Wegerdt, J. Van der Meer, I. Small & R. Hubbard, 2003. The dynamics and characteristics of aeolian dust in dryland Central Asia: possible impacts on human exposure and respiratory health in the Aral Sea basin. Geographical Journal 169: 142–157, Part 2.

Hydrobiologia (2008) 604:137–149
DOI 10.1007/s10750-008-9320-5

SALTON SEA

Organochlorine pesticides, polychlorinated biphenyls, metals, and trace elements in waterbird eggs, Salton Sea, California, 2004

Charles J. Henny · Thomas W. Anderson · John J. Crayon

Abstract The Salton Sea is a highly eutrophic, hypersaline terminal lake that receives inflows primarily from agricultural drainages in the Imperial and Coachella valleys. Impending reductions in water inflow at Salton Sea may concentrate existing contaminants which have been a concern for many years, and result in higher exposure to birds. Thus, waterbird eggs were collected and analyzed in 2004 and compared with residue concentrations from earlier years; these data provide a base for future comparisons. Eggs from four waterbird species (black-crowned night-heron [*Nycticorax nycticorax*], great egret [*Ardea alba*], black-necked stilt [*Himantopus mexicanus*], and American avocet [*Recurvirostra Americana*]) were collected. Eggs were analyzed for organochlorine pesticides, polychlorinated biphenyls (PCBs), metals, and trace elements, with current results compared to those reported for eggs collected from the same species and others during 1985–1993. The two contaminants of primary concern were p,p'-DDE (DDE) and selenium. DDE concentrations in night-heron and great egret eggs collected from the northwest corner of Salton Sea (Whitewater River delta) decreased 91 and 95%, respectively, by 2004, with a concomitant increase in eggshell thickness for both species. Decreases in bird egg DDE levels paralleled those in tissues of tilapia (*Oreochromis mossambicus* × *O. urolepis*), an important prey species for herons and egrets. Despite most nests of night-herons and great egrets failing in 2004 due to predation, predicted reproductive effects based on DDE concentrations in eggs were low or negligible for these species. The 2004 DDE findings were in dramatic contrast to those in the past decade, and included an 81% decrease in black-necked stilt eggs, although concentrations were lower historically than those reported in night-herons and egrets. Selenium concentrations in black-necked stilt eggs from the southeast corner of Salton Sea (Davis Road) were similar in 1993 and 2004, with 4.5–7.6% of the clutches estimated to be selenium impaired during both time periods. Because of present selenium concentrations and future reduced water inflow, the stilt population is of special concern. Between 1992

Guest editor: S. H. Hurlbert
The Salton Sea Centennial Symposium. Proceedings of a Symposium Celebrating a Century of Symbiosis Among Agriculture, Wildlife and People, 1905–2005, held in San Diego, California, USA, March 2005

C. J. Henny (✉)
U.S. Geological Survey, Forest & Rangeland Ecosystem Science Center, 3200 SW Jefferson Way, Corvallis, OR 97331, USA
e-mail: charles_j_henny@usgs.gov

T. W. Anderson
Salton Sea Authority, Sonny Bono Salton Sea National Wildlife Refuge, 906 W. Sinclair Road, Calipatria, CA 92233, USA

J. J. Crayon
California Department of Fish & Game, 78078 Country Club Drive, Suite 109, Bermuda Dunes, CA 92203, USA

and 1993 and 2004 selenium in night-heron and great egret eggs from the Whitewater River delta at the north end of the Sea decreased by 81 and 55%, respectively. None of the night-heron or egret eggs collected in 2004 contained selenium concentrations above the lowest reported effect concentration (6.0 µg/g dw). Reasons for selenium decreases in night-heron and egret eggs are unknown. Other contaminants evaluated in 2004 were all below known effect concentrations. However, in spite of generally low contaminant levels in 2004, the nesting populations of night-herons and great egrets at Salton Sea were greatly reduced from earlier years and snowy egrets (*Egretta thula*) were not found nesting. Other factors that include predation, reduced water level, diminished roost and nest sites, increased salinity, eutrophication, and reduced fish populations can certainly influence avian populations. Future monitoring, to validate predicted responses by birds, other organisms, and contaminant loadings associated with reduced water inflows, together with adaptive management should be the operational framework at the Salton Sea.

Keywords Salton Sea · Black-crowned night-heron · Great egret · Black-necked stilt · American avocet · DDE · Selenium

Introduction

The Salton Sea and surrounding areas provide critical habitat for both resident and migratory birds, including many wading and fish-eating species (Patten et al., 2003, Hurlbert et al., in press). Elevated concentrations of pesticides, certain metals, and trace elements in the area have been of concern for many years. Setmire et al. (1993), in their detailed summary, noted a high level of concern for effects of selenium in fish-eating birds, primarily at the Salton Sea, and for effects of DDE throughout all aquatic habitats in the Imperial Valley. Setmire et al. (1993) concluded that in the Salton Sea area the potential for adverse effects from selenium was high in black-crowned night-herons (*Nycticorax nycticorax* L.), great egrets (*Ardea alba* L.), black-necked stilts (*Himantopus mexicanus* Müller), and American avocets (*Recurvirostra Americana* Gmelin), and high from DDE for three of these species (excluding American avocet). Boron was of no or low concern.

Declines in fish populations between 2000 and 2002 (Riedel et al., 2002; Caskey et al., 2007; Hurlbert et al., in press) became increasingly apparent in 2003 and 2004 (Crayon, CDFG, unpublished data). Interestingly, Hurlbert et al. (in press) discuss the fish and fish-eating bird populations at Salton Sea over the last century as a cycle of "boom and bust" with a number of stressors (rising salinity, cold winter temperatures, and high sulfide levels and anoxia associated with mixing events) playing roles in crashes of fish populations. Fish-eating birds, of course, respond to fish availability. Thus, contaminants are not the only issue for birds, especially fish-eating birds, at Salton Sea. Annual colonial waterbird nesting surveys conducted at the Salton Sea from 1986 to 1999 by the U.S. Fish and Wildlife Service documented peak annual numbers of breeding pairs (Molina & Sturm, 2004). Numbers of breeding great egrets and black-crowned night-herons seemed to peak in the early to mid-1990s, and then both species declined in the late 1990s, although year-to-year numbers were highly variable. In 2004, night-herons and great egrets were found at only one small-mixed colony (about 40 pairs in ratio 30:70) nesting in a stand of dead tamarisk (*Tamarix* spp.) (Fig. 1). Nesting habitat such as this or as represented by dead cottonwoods (*Populus fremontii* S. Wats.) was

Fig. 1 Stand of dead tamarisk with nests of black-crowned night-herons (within ca. 1 m of water surface) and great egrets (>1 m above surface) near the Whitewater River delta. This was the only nesting colony on the margin of the Salton Sea in 2004 for these two species

produced by rising levels of the saline lakewater beginning in the 1970s. This inundated stands of trees at many points along the lake margin. Such snag habitat is now becoming increasingly limited as these dead trees decay and fall over time.

Recent enactment of the Quantification Settlement Agreement, a series of documents addressing water allocations on the Colorado River, and California State legislation will result in a diversion of 200,000 acre-feet (246,700,000 m³) of inflows currently received by the Salton Sea (State of California, 2003). How this reduction in inflows will affect future shorelines and habitats along the river deltas is being actively studied as part of a California Resources Agency program focused on Salton Sea restoration. Various scenarios have been proposed detailing reconfigurations of the Sea to make the best use of projected future inflows. Knowledge of the occurrence and dynamics of contaminants will contribute to informed decisions leading to successful management of this critical habitat. Thus, in 2004 we collected a series of black-crowned night-heron, great egret, black-necked stilt, and American avocet eggs in order to compare current residue concentrations with those in eggs collected between 1985 and 1993, and to provide a continuum for future evaluations. Studies of these species and others conducted in earlier years showed highly elevated concentrations of DDE in eggs of black-crowned night-herons, great egrets, snowy egrets (*Egretta thula* Molina), black skimmers (*Rynchops niger* L.), and white-faced ibis (*Plegadis chihi* Vieillot) (Ohlendorf & Marois, 1990; Audet et al., 1997; Bennett, 1998; Roberts, 2000). In addition, there was concern about selenium concentrations reported in eggs of most of the same species, with the possible exception of white-faced ibis which feed in fields. The present study summarizes residue levels for bird eggs previously collected, and compares them with levels in eggs collected from the same site and species in 2004. Most of the small, mixed night-heron and great egret colony was depredated early in the nesting cycle in 2004, and there was no follow-up of the stilt and avocet nests. Published literature was used to interpret residue concentrations reported in the eggs and to estimate effects on productivity.

Study area and methods

Sample collection and preparation

In 2004, we collected eggs from night-herons and great egrets located at the northwest end of Salton Sea (Whitewater River delta), and black-necked stilts and American avocets nesting at the southeast corner of the Sea; collections from earlier years were more widely distributed (Fig. 2). In 2004, one random egg was collected from each clutch during early incubation and the nest sites were marked. Black-necked stilts nested at numerous locations in 2004, but the collection of eggs was limited to Davis Road where eggs were collected in 1993. Night-heron and great egret eggs were first collected on March 18 with eggs of both species ranging from fresh to about 7–10 days into incubation. Additional nests with eggs of both species were found on March 29. Stilt eggs were collected on May 18 and all eggs were about 7–10 days into incubation. Fresh avocet eggs were collected on June 8. Eggs were refrigerated within 6 h after collection and were opened within a few days and prepared for chemical analyses. Egg contents were placed in chemically cleaned jars and frozen at −20°C. Organochlorine pesticide and industrial chlorinated hydrocarbon concentrations

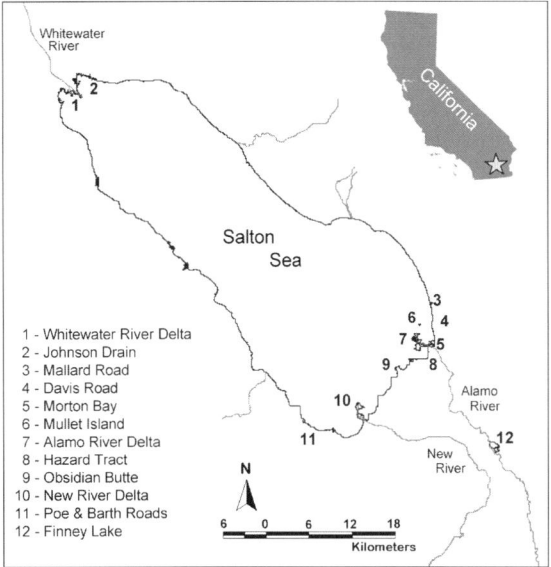

Fig. 2 Map of the Salton Sea showing egg collection sites in 2004 and in earlier years

 Springer

were expressed on a fresh wet weight (Stickel et al., 1973). Selenium and other trace elements and metals were presented on a dry-weight basis.

Eggshell measurements

Shell thickness was measured to the nearest 0.01 mm with a modified Starrett micrometer after shells had dried at room temperature for at least 6 months. The thickness measurements were taken at the equator of each egg and included both shell and shell membranes. A mean of three measurements was used for each egg. To determine normal eggshell thickness (pre-DDT era, i.e., pre-1947) we used data summarized earlier by Ohlendorf & Marois (1990) for black-crowned night-herons and great egrets, and data summarized by Henny et al. (1985) for black-necked stilts.

Analytical chemistry methods

The contents of 38 eggs and 2 regurgitated samples of prey from young night-herons were placed in chemically cleaned jars, frozen, and sent to the Great Lakes Institute of Environmental Research (GLIER) at the University of Windsor, Windsor, Ontario, Canada, for chemical analyses. Analytical chemistry methods included the use of sample blanks, replicate samples, certified reference samples, and in-house reference samples. The following is a synopsis of Standard Operating Procedures (SOPs) as described in the GLIER Quality Control Manual. Detection limits for the various compounds and elements are listed in the tables where data are first presented.

Chlorobenzenes (4; listed in Table 3), PCBs (41 congeners, but presented in this paper as sum PCBs), and organochlorine pesticides or metabolites (15, listed in Table 3) were determined by Column-Solid/Liquid Extraction and Gas Chromatography with Electron Capture Detector (GC-ECD) (Drouillard, 2005). The method includes column solid–liquid extraction with DCM-Hexane (1:1), lipid and moisture content determination, clean-up and fractionation on fluorisil column, and instrumental analysis by GC-ECD. Quality assurance (QA) procedures involved the following tests run in conjunction with batches of 6 or 14 samples: (1) comparison of

measured and expected values of chemical concentrations in a reference material (Canadian Wildlife Service's herring gull [*Larus argentatus* Pontoppidan] egg homogenate), (2) sample duplicates run every second batch of samples extracted and analyzed, and (3) surrogate Standard Recoveries (1,3,5-tribromobenzene), spiked to each sample prior to sample extraction. The data must fall within the established control limits, defined as + or −3 SD of a pre-determined reference value, for each analyte in the reference sample or in the spiked blank. The same control limits were used for other procedures defined below.

Total mercury was determined by Atomic Absorption Spectrometry Vapor Generation (Haffner, 2004a). For QA: (1) two sets of sample duplicates were run with each batch of samples analyzed, and (2) measured and expected total mercury concentrations were compared in water control sample (W-CntVG), in-house biological reference sample (BT-Cnt), and three certified reference samples (BT-Luts1, BT-Dolt3, BT-Dorm2).

Trace elements (17; listed in Table 4) were determined by microwave digestion and Inductively Coupled Plasma Optical Emission Spectrometry (Haffner, 2004b). This method involves nitric acid high temperature and pressure closed system digestion for complete sample dissolution and includes a peroxide oxidation step. For QA: (1) two sets of sample duplicates were run with the batch of samples analyzed, and (2) measured and expected concentrations were compared in two water control samples (W-CntL2, W-CntH2), an internal biological reference sample (BT-Cnt), and a certified reference sample (BT-Dolt3).

Statistical analyses

Analysis of variance (ANOVA) and t-tests were conducted using geometric means when evaluating residue concentrations. Geometric means were computed only when the contaminant was detected in at least 50% of the eggs. For those compounds, a value of one-half the detection limit was substituted for the non-detected values to permit logarithmic transformations. Geometric means are reported in this article, except in the case of eggshell thickness data for which arithmetic means are reported.

Results

DDE and selenium

The two contaminants previously found in bird eggs at the Salton Sea and of primary concern are p,p'-DDE (DDE) and selenium. DDE and selenium residue concentrations for the period 1985–1993 and for 2004 (this study) are summarized in Table 1. Egg collection sites at the Salton Sea were conveniently divided into two general regions, the north end and the south end.

Night-heron eggs collected at the north end from the Whitewater River delta had much higher concentrations in 1992 than in 2004 of both DDE (6.42 vs. 0.56 µg/g ww; $P = 0.0002$) and selenium (6.18 vs. 1.77 µg/g dw; $P = 0.0009$) (Table 1). Raw data for 1985 and 1991 were not available for statistical comparisons. Similarly, great egrets nesting at the same site as the night-herons had higher concentrations in 1993 than in 2004 for both DDE (11.90 vs. 0.56 µg/g ww; $P < 0.0001$) and selenium (6.69 vs. 3.02 µg/g dw; $P = 0.009$).

Table 1 Summary of selenium (µg/g, dry weight) and p,p'-DDE (µg/g, wet weight) geometric means (and range) in waterbird eggs from the Salton Sea

Species	Year	Location	N	Selenium	p,p'-DDE	Source
North end						
Night-heron	1985	WRD	10	5.88 (4.92–7.49)[a]	8.62 (2.5–20)	A
Night-heron	1991	WRD/JD	3	5.27 (4.6–6.5)	2.34 (1.7–3.6)	B
Night-heron	1992	WRD	10	6.18 (3.30–7.85)	6.42 (1.1–25)	C
Night-heron	2004	WRD	11	1.77 (<1.1–5.03)[b]	0.56 (0.14–11)	D
Snowy egret	1991	WRD/JD	3	3.93 (3.9–4.0)	15.7 (5.0–31)	B
Snowy egret	1993	WRD	10	4.97 (3.51–8.32)	6.33 (1.7–41)	E
Great egret	1993	WRD	5	6.69 (6.1–7.9)	11.90 (4.5–50)	E
Great egret	2004	WRD	12	3.02 (<1.1–4.90)[c]	0.56 (0.40–0.86)	D
Black skimmer	1993	JD	10	6.01 (4.61–7.19)	3.30 (2.9–5.0)	C
Black skimmer	1993	JD	2	5.32 (4.8–5.9)	0.88 (0.7–1.1)	E
Cattle egret	1991	WRD/JD	3	3.60 (2.7–5.4)	2.81 (1.6–4.8)	B
South end						
Great egret	1985	MNR	10	3.53 (2.97–4.51)[a]	24.0 (16–48)	A
Great egret	1992	MRWU	8/6[d]	4.95 (3.45–6.17)	2.00 (0.33–5.9)	C
Great egret	1993	PR	5	7.62 (6.8–9.9)	14.44 (11–17)	E
White-faced ibis	1992	FLIWA	5	0.25 (<0.1–1.2)	11.00 (1.3–23)	C
Caspian tern	1993	MI	5	2.60 (1.40–3.81)	2.00 (1.3–3.7)	C
Black skimmer	1993	MI	10	6.78 (5.10–8.17)	3.10 (0.72–9.0)	C
Black skimmer	1993	OB	10	6.35 (3.59–8.92)	2.30 (0.61–5.9)	C
Black skimmer	1993	MB	9	4.47 (3.25–8.03)	0.90 (0.31–2.2)	C
Black-necked stilt	1992	U/ML	38	6.60 (3.74–14.2)	2.02 (0.36–7.5)	E
Black-necked stilt	1993	HT	24	5.43 (3.7–8.0)	2.58 (1.0–9.2)	E
Black-necked stilt	1993	BR	3	7.05 (6.7–7.7)	1.64 (1.4–2.1)	E
Black-necked stilt	1993	DR	15	6.32 (4.0–9.0)	2.91 (0.65–23)	E
Black-necked stilt	2004	DR	12	6.16 (3.29–8.53)	0.55 (0.19–1.7)	D
American avocet	2004	DR	3	9.27 (8.13–10.89)	1.14 (0.83–2.1)	D
Black skimmer	1991	U/ML	12	4.65 (2.2–8.2)	4.90 (1.8–16.4)	B
Black skimmer	1992	U/ML	5	5.87 (5.71–6.24)	11.00 (4.2–26)	C
Black-necked stilt	1986–1990	U/ML	84	NA	2.54 (0.05–12.0)	F

Table 1 continued

Species	Year	Location	N	Selenium	p,p'-DDE	Source
Black-necked stilt	1988–1990	U/ML	127	4.30 (1.60–35.0)	NA	F
Black-necked stilt	1992	U/ML	38	6.60 (3.74–14.2)	2.02 (0.36–7.5)	E
Great blue heron	1991	U/ML	4	3.86 (2.8–5)	5.78 (2.6–10)	B
Great egret	1991	U/ML	9	4.77 (3.5–7.1)	8.36 (0.86–31)	B

Locations: WRD = Whitewater River delta, WRD/JD = Whitewater River delta/Johnson Drain, JD = Johnson Drain, MNR = Mouth New River, MRWU = Mallard Road, Wister Unit, PR = Poe Road, FLIWA = Finney Lake Imperial Wildlife Area, MI = Mullett Island, OB = Obsidian Butte, MB = Morton Bay, U/ML = Unknown or Mixed Locations, HT = Hazard Tract, BR = Barth Road, DR = Davis Road

Sources: A = Ohlendorf & Marois (1990), B = Audet et al. (1997), C = Roberts (2000), D = This study, E = Bennett (1998), F = Setmire et al. (1993). NA = Not available

Note: Detection limits for 2004: p,p'-DDE (0.02 μg/g, ww), selenium (1.05 μg/g, dw). Caspian tern (*Sterna caspia* Pallus), great blue heron (*Ardea herodias* L.)

[a] Adjusted from wet weight to dry weight by percent moisture in eggs reported by Ohlendorf and Marois (1990)

[b] Four eggs below detection limit

[c] One egg below detection limit

[d] $N = 8$ for selenium, $N = 6$ for DDE

At Davis Road at the south end of the Sea, black-necked stilt and American avocet eggs were collected in 1993 and 2004 (Table 1). Concentrations of DDE in stilt eggs were higher in 1993 than 2004 (2.91 vs. 0.55 μg/g ww; $P < 0.0001$), while selenium concentrations remained unchanged (6.32 vs. 6.16 μg/g dw; $P = 0.79$). Stilt eggs collected at other sound end locations (Hazard Tract, Barth Road) in 1993 also contained concentrations of both DDE and selenium similar to those found at Davis Road in 1993. Another series of stilt eggs collected in 1992 from several locations at the Sea also contained concentrations similar to 1993 for both DDE and selenium. The American avocet is not a regular breeder at Salton Sea; therefore, no data from the early 1990s were available to compare with the three eggs collected in 2004. Geometric means for avocet eggs were 1.14 μg/g ww for DDE and 9.27 μg/g dw for selenium (Table 1). Great egret eggs were collected at the south end of Salton Sea in 1985 and 1992–1993, but great egrets were not found nesting there in 2004.

Eggshell thickness and DDE

Many studies have documented eggshell thinning (arithmetic means) in wild birds in North America and throughout the world, with DDE being the contaminant primarily responsible for this (see review by Blus [2003]). With DDE reported in heron and egret eggs from the Salton Sea at some of the highest concentrations in the United States in 1985 and 1991–1993 (Ohlendorf & Marois 1990; Roberts, 2000), DDE, and eggshell thickness became important topics in the earlier investigations of these species. Night-herons showed substantial eggshell thinning at the Salton Sea in 1985 and 1992, but DDE levels were lower by 2004, and eggshells were thicker (Table 2). Similarly, great egret eggs showed lower concentrations of DDE in 2004 than in 1993. No pre-DDT era eggshell thickness data were available, and the lack of individual egg records on 1985 DDE concentrations in great egrets preclude a formal statistical comparison, but an 18.4% increase in eggshell thickness from 1985 to 2004 is noteworthy.

Eggshell thickness data in 2004 for black-necked stilts and American avocets are presented in Table 2. DDE concentrations in stilt eggs in the early 1990s were generally lower than those reported in heron and egret eggs, and earlier studies by Henny et al. (1985) indicated that stilts were not especially sensitive to eggshell thinning. Although no pre-2004 thickness measurements at Salton Sea were reported for stilts, those for 2004 compared favorably with pre-DDT era values for stilts in Utah.

 Springer

Table 2 A comparison of arithmetic mean shell thickness (mm) of black-crowned night-heron, great egret, black-necked stilt, and American avocet eggs collected in 1985–1993, 2004, and those collected in the pre-DDT era (before 1947)

Species (Location)	Years	N	Thickness (mm) ± SD	Percent change[a]	Source
Black-crowned night-heron					
San Francisco Bay, CA	1899–1945	23	0.280 ± 0.011	–	A
San Joaquin Valley, CA	1906–1945	29	0.266 ± 0.015	–	A
Whitewater River delta	1985	10	0.240 ± 0.010	−9.8 to −14.3[b]	A
Whitewater River delta	1992	13	0.247 ± 0.019	−7.1 to −11.8[b]	B
Whitewater River delta	2004	11	0.256 ± 0.021	−3.8 to −8.6[b]	C
Great egret					
Mouth New River	1985	11	0.244 ± 0.016	NA	A
Poe and Lack/Lindsey Rds/Whitewater River delta	1993	29	0.282 ± 0.024	NA	B
Whitewater River delta	2004	12	0.289 ± 0.012	NA	C
Black-necked stilt					
Utah	Pre-DDT	40	0.205 ± 0.013	–	D
Davis road	2004	12	0.209 ± 0.011	+2.0	C
American avocet					
Davis road	2004	3	0.220 ± 0.027	NA	C

Sources: A = Ohlendorf and Marois (1990), B = Bennett (1998), C = This Study, D = Henny et al. (1985)

NA = Not available

[a] From pre-DDT era means

[b] From both San Francisco Bay and San Joaquin Valley pre-DDT era means

Other chlorinated hydrocarbon contaminants, trace elements, and metals

Eggs collected in 2004 contained additional chlorinated hydrocarbon contaminants that included organochlorine pesticides and PCBs (Table 3). These contaminants were all found at generally low concentrations and reported on a μg/kg basis (contrast with μg/g for DDE). Some metals do not accumulate very well in eggs, while others are essential and found in fairly high concentrations. Aluminum, arsenic, cadmium, cobalt, chromium, lead, and vanadium were not detected in any of the eggs, and nickel was reported in only 2 (American avocet) of the 38 eggs (Table 4). Mercury was reported in all eggs, with a considerably higher mean concentration in black-necked stilts.

Discussion

DDE effects on reproduction

Fish-eating birds and those whose diets include at least some fish have received most of the attention when evaluating possible reproductive effects from DDE and other organochlorine pesticides. Exceptionally high mean concentrations of DDE were found in night-heron (8.62 μg/g ww) and great egret (24.0 μg/g) eggs collected at the Salton Sea in 1985 (Ohlendorf & Marois, 1990). The mixed colony of night-herons and great egrets at the Whitewater River delta failed in 2004 (destroyed by predators) during incubation (Fig. 2), except for a few nests at the periphery of the colony.

At DDE concentrations above 8 μg/g ww, clutch size and reproductive success of night-herons decrease and the incidence of cracked eggs increases (Custer et al., 1983; Henny et al., 1984). Thus, residue concentrations were used to estimate potential contaminant effects. Seven of 10 (70%) night-heron eggs collected at Salton Sea in 1985 contained DDE concentrations exceeding 8 μg/g (Ohlendorf & Marois, 1990), but by 1991–1992 only 4 of 13 eggs (31%) exceeded 8-μg/g DDE (Audet et al., 1997; Roberts, 2000, Roberts, pers. comm.). During our 2004 study, only 1 of 11 eggs (9%) contained DDE >8 μg/g. In addition to the reduced incidence rate of eggs above the critical 8-μg/g DDE concentration, the mean concentration decreased 91% from 1992 to

Table 3 Concentrations of other chlorinated hydrocarbon contaminants (geometric means, μg/kg wet weight) in black-crowned night-heron, great egret, black-necked stilt, and American avocet eggs, Salton Sea, 2004

Species (N)	% Lipid	% Moisture	QCB	HCB	OCS	TRNO	Mirex	βHCH	OXY	TRCH	ClCH	DDD	ClNO	DDT	HE	Dieldrin	ΣPCBs
Heron (11)	6.03	80.81	ND	0.73	0.15	6.83	0.26	NC[a]	1.54	0.99	NC[b]	0.97	2.85	2.74	2.86	8.87	180.27
Egret (12)	6.23	81.70	ND	NC[c]	ND	4.06	NC[d]	NC[e]	1.80	0.42	ND	1.23	1.52	2.49	2.54	3.85	146.44
Stilt (12)	11.17	73.05	0.13	1.75	NC[f]	0.78	NC[g]	1.09	1.48	0.83	NC[h]	0.82	0.20	1.40	NC[i]	3.86	2.13
Avocet (3)	10.88	74.55	0.59	1.45	ND	0.86	0.38	0.23	0.38	3.89	ND	1.81	1.06	4.37	ND	0.75	39.83
Det. Limit	NA	NA	0.04	0.03	0.05	0.05	0.07	0.02	0.01	0.02	0.03	0.08	0.02	0.02	0.02	0.02	NA

Note: All non-detections (detection limit): 1,2,4,5-TCB (0.04), 1,2,3,4-TCB (0.04), a-HCH (0.05), g-HCH (0.01). ND = not detected in any sample, NC = no mean calculated, found in <50% of samples. NA = Not applicable

QCB = pentachlorobenzene, HCB = hexachlorobenzene, OCS = octachlorostyrene, HCH = hexachlorocyclohexane, TRNO = *trans*-nonachlor, OXY = oxychlordane, TRCH = *trans*-chlordane, ClCH = *cis*-chlordane, DDD = *p,p′*-DDD, ClNO = *cis*-nonachlor, DDT = *p,p′*-DDT, HE = heptachlor epoxide

[a] One detection (1.70 μg/kg)

[b] Three detections (0.62, 0.73, and 0.90 μg/kg)

[c] Three detections (0.38, 0.19, and 0.25 μg/kg)

[d] Three detections (0.62, 0.69, and 0.32 μg/kg)

[e] Two detections (0.80 and 0.74 μg/kg)

[f] One detection (3.65 μg/kg)

[g] Three detections (0.45, 0.38, and 7.24 μg/kg)

[h] Five detections (0.47, 0.82, 0.27, 0.38, and 0.22 μg/kg)

[i] Five detections (0.68, 3.29, 7.51, 0.20, and 0.74 μg/kg)

Table 4 Concentrations of other metals (geometric means, µg/g dry weight) in black-crowned night-heron, great egret, black-necked stilt, and American avocet eggs, Salton Sea, 2004

Species (N)	Hg	Ca	Ni	Cu	Fe	Mg	K	Mn	Na	Zn
Heron (11)	0.08	2,176	ND	4.92	105.8	489.6	6746	1.42	9,952	43.82
Range	0.03–0.15	1,739–2,552	–	3.56–6.22	84.8–137.9	408.1–549.6	5762–7712	0.68–2.91	8,326–11,080	39.55–53.00
Egret (12)	0.09	2,190	ND	3.76	116.0	574.5	5890	0.72	11,287	39.75
Range	0.04–1.83	1,833–2,366	–	ND-4.96	91.8–129.3	493.8–715.4	5220–6449	0.52–1.08	10,220–12,410	35.42–46.57
Stilt (12)	0.43	3,049	ND	3.51	95.1	487.7	5138	1.28	6,356	43.81
Range	0.25–1.14	2,376–4,263	–	2.37–5.55	75.5–125.8	388.2–591.6	4614–5908	0.71–2.03	5,941–6,964	35.42–52.02
Avocet (3)	0.05	2,640	0.70	3.61	129.6	408.2	5269	1.76	5,684	43.75
Range	0.02–0.13	2,466–2,861	ND-0.91	3.08–4.87	122.2–136.4	368.5–446.8	5203–5360	1.43–2.18	5,552–5,763	42.62–45.04
Detection limit	0.02	28	0.87	1.5	0.68	4.5	61	0.04	26	1.2

Note: All non-detections (detection limit): aluminum (9.6), arsenic (3.2), cadmium (0.29), cobalt (0.31), chromium (1.5), lead (0.97), and vanadium (1.3)

ND = Not detected

2004 (Table 1) and eggshell thickness correspondingly increased (Table 2). Mean whole body DDE levels in adult tilapia collected at Salton Sea showed a similar decrease of 70% from 1986 to 2000–2001 (0.28 µg/g vs. 0.085 µg/g ww; Moreau et al., 2007). Young tilapia, recently eaten and regurgitated, were found at two night-heron nests in 2004; DDE concentrations were lower in the regurgitated tilapia (0.005 and 0.003 µg/g ww) than in the adult tilapia. Potential reproduction in only a small segment of the night-heron population in 2004 was believed to be adversely impacted by DDE. Interestingly, a diet of 0.004-µg/g ww DDE (young tilapia) in 2004 would project to a night-heron egg concentration of 0.35-µg/g ww DDE (based upon a DDE biomagnifications factor of 87 reported from fish to osprey [*Pandion haliaetus* L.] eggs [Henny et al., 2003]). The estimated 0.35-µg/g DDE in night-heron eggs is similar to the observed 0.56-µg/g DDE we reported in 2004 (Table 1).

Much as with the night-herons at the Salton Sea in 1985, great egret eggs that year contained high concentrations of DDE with a mean of 24 µg/g, and individual eggs ranged from 16 to 48 µg/g (Table 1, Ohlendorf & Marois, 1990). Great egret eggs collected at another site at the south end of the Sea in 1992 contained low DDE concentrations (all <8 µg/g), while eggs at another southern site in 1993 contained high DDE concentrations (all >8 µg/g ww) (Table 1, Bennett, 1998; Roberts, 2000). Thus, location of the nesting colony and, perhaps, associated foraging areas seemed to influence DDE egg concentrations at the Salton Sea.

The critical DDE concentration in great egret eggs required to reduce reproduction is not known, but if for snowy egrets and night-herons reproductive problems begin at ~5–8 µg/g (Findholt, 1984; Henny et al., 1985), a similar threshold may exist for great egrets. At the north end of the Sea where great egret eggs were collected in 2004, eggs were also collected in 1993, so a direct comparison was made. The five eggs collected in 1993 included four (80%) above 5-µg/g DDE and two of them (40%) above 8 µg/g, whereas in 2004 none of the eggs were above 5 µg/g (Table 1, Bennett, 1998, this study).

As with the night-herons, eggshell thickness of great egrets improved as the DDE concentrations decreased, with an 18.4% improvement from 1985 to 2004 (Table 2). This 18.4% improvement in shell

thickness (although no pre-DDT era norm available) is particularly relevant in view of the statement by Lincer (1975) that not one North American raptor population exhibiting 18% or more eggshell thinning was able to maintain a stable self-perpetuating population, and the statement by Hickey & Anderson (1968) that thinning by 18–20% or more over several years is related to population decline. Clearly, the great egret suffered from severe DDE problems at Salton Sea in earlier years. Snowy egrets nesting at the north end of Salton Sea in the early 1990s contained high DDE concentrations in eggs that were similar to night-herons and great egrets, while cattle egrets (*Bubulcus ibis* L.) (small sample) and black skimmers had generally lower DDE concentrations (Table 1).

Many black-necked stilt eggs from Salton Sea have been analyzed for DDE. A study in the Imperial Valley in 1989–1990 reported markedly different DDE concentrations in stilt eggs associated with different sites (Setmire et al., 1993). The highest mean concentrations were found in eggs collected near man-made ponds constructed on land previously used for agriculture. Stilt eggs collected in 2004 were collected at the same site (Davis Road) where eggs were collected in 1993 so that a direct comparison of residue concentrations could be made (Table 1). DDE concentrations decreased 81% from 1993 to 2004. Henny et al. (1985) reported mean DDE concentrations in stilt eggs for four consecutive years from 1980 to 1983 at Carson Lake, Nevada (3.26, 1.96, 1.94, and 1.38 µg/g ww) that also suggested a downward trend over time. Mean eggshell thickness for the four 10-egg samples from Nevada was 0.217, 0.203, 0.209, and 0.217 mm, respectively, which provided little evidence of a change in that variable. Furthermore, no marked correlation was evident between DDE and eggshell thickness of the 40 eggs (Henny et al., 1985). A pre-DDT era series of stilt eggs from Utah had similar thickness (arithmetic mean ± SD: 0.205 mm ± 0.013) based upon 40 eggs, each from a different clutch (Henny et al., 1985). Eggs collected during this study in 2004 had a mean shell thickness of 0.209 mm (Table 2), which was also similar to the pre-DDT era mean. The DDE concentrations in eggs documented in this and other studies seem not to produce eggshell thinning in stilts and the associated adverse effect on reproductive success.

Selenium effects on reproduction

Selenium concentrations in bird eggs usually average ≤ 3 µg/g dw with the maxima usually <5 µg/g when collected from study areas with "background selenium" (Skorupa & Ohlendorf, 1991; Ohlendorf et al., 1993; Ohlendorf, 2003). Heinz (1996) reported that reproductive impairment is one of the most sensitive response variables and eggs are the most reliable tissues for interpretive purposes. Fortunately, much research has been conducted with black-necked stilt eggs; Skorupa (1998) reported the selenium threshold point for hatchability effects at 6–7 µg/g dw, and the response profile in relation to egg concentration for stilt clutch viability is shown in "Guidelines for Interpretation of the Biological Effects of Selected Constituents in Biota, Water and Sediment" (NIWQP 1998). Bennett (1998) estimated that stilt reproduction was depressed 4.5% due to selenium at the Salton Sea in 1992–1993, while the same residue data subjected to the NIWQP (1998: p. 164) clutch viability profile analysis yield a similar estimate with 5.6% of the clutches impaired. Three of 12 eggs collected in 2004 were at the "no effect" background level (<6 µg/g), and 9 of the 12 eggs contained 6- to 8.5-µg/g selenium, which was in the lowest selenium effect level (range 6–15 µg/g, with 19% probability of impairment) (NIQWP, 1998). After weighting the selenium egg concentrations by the number of eggs in each selenium exposure category (0–5 and 6–15 µg/g), an estimated 16% of the clutches may have had impaired clutch viability. In areas with normal (<6 µg/g in eggs) selenium exposure, the impaired viability was only 9% (NIQWP, 1998). Therefore, an estimated 7% of nests (16–9%) were impaired by selenium in 2004. This estimate probably was high because the lowest "effect level" included 6- to 15-µg/g selenium and none of the eggs sampled in 2004 were above 8.5 µg/g. The percentage of impaired clutches at Salton Sea in 2004 was probably similar to that in 1992–1993. Furthermore, mean selenium concentrations in stilt eggs at Davis Road did not change between 1993 and 2004 (6.32 vs. 6.16 µg/g).

Few American avocets were nesting at Salton Sea in 2004, and only three eggs were collected. American avocets are more tolerant to selenium than stilts, with normal clutch viability reported at egg selenium concentrations up to 20 µg/g (NIWQP, 1998). Based upon the literature, even at the highest selenium

concentration observed in avocet eggs (10.9 μg/g), adverse effects on reproduction would not be expected.

Interestingly, selenium concentrations in night-heron and great egret eggs collected at the White-water River delta decreased from 1992 and 1993 to 2004 by 81% and 55%, respectively (Table 1). Furthermore, none of the eggs collected in 2004 contained selenium concentrations above 5.03 μg/g. This finding is important since it is below the lowest reported selenium effect level in bird eggs (6–7 μg/g: Ohlendorf, 2003). Reasons for the decrease in egg selenium levels at Whitewater River delta remain unknown. It would have been useful to know where the night-herons and egrets foraged in 2004, and if foraging locations and prey species consumed changed between 1992–1993 and 2004, night-herons in particular have a very diverse diet (Hancock & Kushlan, 1984). Selenium concentrations in snowy egret and black skimmer eggs in the early 1990s were similar to night-heron and great egret eggs in the early 1990s, and the limited number of cattle egret eggs (a field feeder like the ibis mentioned earlier) contained the lowest selenium concentrations.

Other contaminants

All the other contaminants found in the eggs were reported at low concentrations; in fact, other OCs and polychlorinated biphenyls were reported in the μg/kg range, while reproduction problems generally begin for this group in the μg/g range (see Beyer et al., 1996). Likewise, the other metals or trace elements of concern were low. The theoretical "effect level" for mercury in bird eggs is 0.5–0.8 μg/g ww (or 2.5–4.0 μg/g dw) (Heinz, 1979; Newton & Haas, 1988; Thompson, 1996). No eggs collected during this study in 2004 had mercury concentrations above 1.83 μg/g dw.

Conclusions

Data collected during the 2004 study provide useful recent information for evaluating future changes in egg residue concentrations. DDE concentrations decreased dramatically in the eggs of night-herons and great egrets at the north end of the Salton Sea and

in eggs of stilts at the south end which is a positive finding. Thus, the serious DDE-associated reproductive problems for fish-eating birds at the Salton Sea in the 1980s and early 1990s (and most likely earlier years, although no data collected) were not apparent for the species we studied in 2004. However, field-feeding white-faced ibis nesting at nearby Finney Lake, Imperial Wildlife Area (IWA) may be an exception. Ibis eat earthworms which concentrate DDE and egg residues did not decline between 1985 and 1996 at the Nevada colony which was not exposed locally to DDE, but has a population segment that stages/winters in the Imperial Valley (Henny, 1997). Ibis eggs collected at Finney Lake in 1992 contained an extremely high geometric mean of 11.00-μg/g DDE (>4 μg/g is the effect level [Henny & Herron, 1989]); eggs should be analyzed from this site in the future. Selenium concentrations decreased in night-heron and great egret eggs at the north end, but remained unchanged in stilt eggs from the south end near Davis Road where an estimated 4.5–7.65% of clutches were impaired.

Future reductions in water flow into the Salton Sea will most likely influence waterbird exposure to the contaminants measured during this study and other contaminants. Thus, it is important to establish a long-term monitoring program which includes standardized waterbird censuses and an evaluation of contaminants in waterbird eggs and other media including fish and invertebrates. The Salton Sea has a long history of "boom and bust" years for fish populations and the dependant fish-eating bird populations. Stressors affecting fish populations are cited in this paper, but discussed in detail elsewhere in this Special Issue. Thus, contaminants are just one factor that can influence the overall viability of bird populations. The DDT era and the "boom and bust" of fish populations at the Salton Sea caused difficulties for birds in the past. Future contaminant concentrations in fish and birds, associated with less water entering the ecosystem, may be modeled with existing protocols to predict contaminant levels and effects on wildlife, e.g., the stilt population is of special concern with a portion of the population already affected by selenium. Unfortunately, predictions based upon models for complicated ecosystems like the Salton Sea are susceptible to large errors. Future monitoring, to validate predicted responses from birds and other organisms with less water inflow

(including contaminant loadings), and adaptive management should be the operational framework.

Acknowledgments We thank Rey C. Stendell and Douglas A. Barnum, Salton Sea Science Office, U.S. Geological Survey, La Quinta, California, for their support in the planning of this study. Carol A. Roberts, U.S. Fish & Wildlife Service, Carlsbad, California, kindly provided unpublished information from their files. Personnel of the U.S. Fish & Wildlife Service, Sonny Bono Salton Sea National Wildlife Refuge, kindly allowed the use of their facilities and equipment.

References

Audet, D. J., M. Shaughnessy & W. Radke, 1997. Organochlorines and selenium in fishes and colonial waterbirds from the Salton Sea. Final Report. U.S. Fish and Wildlife Service, Carlsbad, CA, 20 pp.

Bennett, J., 1998. Biological effects of selenium and other contaminants associated with irrigation drainage in the Salton Sea area, California, 1992–1994. National Irrigation Water Quality Program, Information Rept. No. 4, Washington, DC, 35 pp.

Beyer W. N., G. H. Heinz & A. W. Redmon-Norwood (eds), 1996. Environmental Contaminants in Wildlife – Interpreting Tissue Concentrations. Lewis Publishers, Boca Raton, FL, 494 pp.

Blus, L. J., 2003. Organochlorine pesticides. In Hoffman, D. J., B. A. Rattner, G. A. Burton, Jr. & J. Cairns, Jr. (eds), Handbook of Ecotoxicology, 2nd edn. Lewis Publishers, Boca Raton, FL, 313–339.

Caskey, L. L., R. R. Riedel, B. Costa-Pierce, J. Butler & S. H. Hurlbert, 2007. Population dynamics, distribution, and growth rate of tilapia (*Oreochromis mossambicus*) in the Salton Sea, California, with notes on bairdiella (*Bairdiella icistia*) and orangemouth corvina (*Cynoscion xanthulus*). Hydrobiologia 576: 185–203.

Custer, T. W., G. L. Hensler & T. E. Kaiser, 1983. Clutch size, reproductive success and organochlorine contaminants in Atlantic coast black-crowned night-herons. Auk 100: 699–710.

Drouillard, K. G., 2005. Determination of chlorobenzenes (CBs), organochlorine pesticides (OCs) and PCBs in biological tissue samples (vegetals and animals) by Column-Solid/Liquid Extraction and Gas Chromatography with Electron Capture Detector (GC-ECD). SOP No. 02-001, Revision 8, Great Lakes Institute of Environmental Research, Windsor, Ontario, Canada, 88 pp.

Findholt, S. L., 1984. Organochlorine residues, eggshell thickness, and reproductive success of snowy egrets nesting in Idaho. Condor 86: 163–169.

Haffner, G. D., 2004a. Analysis of total recoverable mercury in biological tissues (vegetals and animals) by Atomic Absorption Spectrometry (AAS-VG). SOP No. 01-002, Revision 8, Great Lakes Institute of Environmental Research, Windsor, Ontario, Canada, 24 pp.

Haffner, G.D., 2004b. Analysis of total recoverable metals in biological tissues (vegetals and animals) by microwave digestion and Inductively Coupled Plasma Optical Emission Spectrometry (ICP-OES). SOP No. 01-006, Revision 2, Great Lakes Institute of Environmental Research, Windsor, Ontario, Canada, 50 pp.

Hancock, J. & J. Kushlan, 1984. The Herons Handbook. Harper and Row, New York, 288 pp.

Heinz, G. H., 1996. Selenium in birds. In Beyer, W. N., G. H. Heinz & A. W. Redmon-Norwood (eds), Interpreting Environmental Contaminants in Animal Tissues. Lewis Publishers, Boca Raton, FL., 453–464.

Heinz, G. H., 1979. Methylmercury: reproductive and behavioral effects on three generations of mallard ducks. Journal of Wildlife Management 43: 394–401.

Henny, C. J., 1997. DDE still high in white-faced ibis eggs from Carson Lake, Nevada. Colonial Waterbirds 20: 478–484.

Henny, C. J. & G. B. Herron, 1989. DDE, selenium, mercury and white-faced ibis reproduction at Carson Lake, Nevada. Journal of Wildlife Management 53: 1032–1045.

Henny, C. J., L. J. Blus & C. S. Hulse, 1985. Trends and effects of organochlorine residues in Oregon and Nevada wading birds, 1979–1983. Colonial Waterbirds 8: 117–128.

Henny, C. J., L. J. Blus, A. J. Krynitsky & C. M. Bunck, 1984. Current impact of DDE on black-crowned night-herons in the intermountain west. Journal of Wildlife Management 48: 1–13.

Henny, C. J., J. L. Kaiser, R. A. Grove, V. R. Bentley & J. E. Elliott, 2003. Biomagnification factors (fish to osprey eggs from Willamette River, Oregon, U.S.A.) for PCDDs, PCDFs and OC pesticides. Environmental Monitoring and Assessment 84: 275–315.

Hickey, J. J. & D. W. Anderson, 1968. Chlorinated hydrocarbons and eggshell changes in raptorial and fish-eating birds. Science 162(3850): 271–273.

Hurlbert, A. H., T. Anderson, K. K. Sturm & S. H. Hurlbert, in press. Fish and fish-eating birds at the Salton Sea: a century of boom and bust. Lake and Reservoir Management.

Lincer, J. L., 1975. DDE-induced eggshell thinning in the American kestrel: a comparison of the field situation with laboratory results. Journal of Applied Ecology 12: 781–793.

Molina, K. C. & K. K. Sturm, 2004. Annual colony site occupation and patterns of abundance of breeding cormorants, herons and ibis at the Salton Sea. Studies in Avian Biology 27: 42–51.

Moreau, M. F., J. Surico-Bennett, M. Vicario-Fisher, D. Crane, R. Gerads, R. M. Gersberg & S. H. Hurlbert, 2007. Contaminants in tilapia (*Oreochromis mossambicus*) from the Salton Sea, California, in relation to human health, piscivorous birds and fish meal production. Hydrobiologia 576: 127–165.

Newton, I. & M. B. Haas, 1988. Pollutants in merlin eggs and their effects on breeding. British Birds 81: 258–269.

NIWQP, 1998. Guidelines for interpretation of the biological effects of selected constituents in biota, water and sediment. National Irrigation Water Quality Program, Information Report No. 3, Denver, CO, 212 pp.

Ohlendorf, H. M., 2003. Ecotoxicology of selenium. In Hoffman, D. J., B. A. Rattner, G.A. Burton Jr. & J. Cairns Jr. (eds), Handbook of ecotoxicology, 2nd ed. Lewis Publishers, Boca Raton, FL, 466–500.

Ohlendorf, H. M. & K. C. Marois, 1990. Organochlorines and selenium in California night-heron and egret eggs. Environmental Monitoring and Assessment 15: 91–104.

Ohlendorf, H. M., J. P. Skorupa, M. K. Saiki & D. A. Barnum, 1993. Food-chain transfer of trace elements to wildlife. In Allen, R. G. & C. M. U. Neale (eds), Management of Irrigation and Drainage Systems: Integrated Perspectives. American Society of Civil Engineers, New York, 593–603.

Patten, M. A., G. McCaskie & P. Unitt, 2003. Birds of the Salton Sea: Status, Biogeography and Ecology. University of California Press, Berkeley, CA, 363 pp.

Riedel, R., L. Caskey & B. A. Costa-Pierce, 2002. Fish biology and fisheries ecology of the Salton Sea, California. Hydrobiologia 473: 229–244.

Roberts, C. A., 2000. Environmental contaminants in piscivorous birds at the Salton Sea, 1992-1993. Final Report. U.S. Fish and Wildlife Service, Carlsbad, CA.

Setmire, J. G., R. A. Schroeder, J. N. Densmore, S. L. Goodbred, D. J. Audet & W. R. Radke, 1993. Detailed study of water quality, bottom sediment, and biota associated with irrigation drainage in the Salton Sea area, California,

1988–1990. U.S. Geological Survey, Water-Resources Investigations Report 93-4014, Sacramento, CA.

Skorupa, J. P., 1998. Selenium poisoning of fish and wildlife in nature: lessons from twelve real-world examples. In Frankenberger, W. T. Jr. & R. A. Engberg (eds), Environmental Chemistry of Selenium. Marcel Dekker, Inc., New York, 315–354.

Skorupa, J. P. & H. M. Ohlendorf, 1991. Contaminants in drainage water and avian risk thresholds. In Dinar, A. & D. Zimmerman, (eds), The Economics and Management of Water and Drainage in Agriculture. Kluwer Academic Publishers, Boston, MA, 345–368.

State of California, 2003. An act to amend Section 2081.7 of the Fish and Game Code and to amend Section 1013 of the Water Code, relating to the resources, and making an appropriation therefore. Senate Bill No. 317, Chapter 612, Sacramento, 8 pp.

Stickel, L. F., S. N. Wiemeyer & L. J. Blus, 1973. Pesticide residues in eggs of wild birds: adjustment for loss of moisture and lipid. Bulletin of Environmental Contamination and Toxicology 9: 193–196.

Thompson, D. R., 1996. Mercury in birds and terrestrial mammals. In Beyer, W. N., G. H. Heinz & A. W. Redmon (eds), Interpreting Environmental Contaminants in Animal Tissues. Lewis Publishers, Boca Raton, FL, 341–356.

Hydrobiologia (2008) 604:151–172
DOI 10.1007/s10750-008-9316-1

SALTON SEA

Occurrence, distribution and transport of pesticides into the Salton Sea Basin, California, 2001–2002

Lawrence A. LeBlanc · Kathryn M. Kuivila

Abstract The Salton Sea is a hypersaline lake located in southeastern California. Concerns over the ecological impacts of sediment quality and potential human exposure to dust emissions from exposed lakebed sediments resulting from anticipated shrinking of shoreline led to a study of pesticide distribution and transport within the Salton Sea Basin, California, in 2001–2002. Three sampling stations—upriver, river mouth, and offshore—were established along each of the three major rivers that discharge into the Salton Sea. Large-volume water samples were collected for analysis of pesticides in water and suspended sediments at the nine sampling stations. Samples of the bottom sediment were also collected at each site for pesticide analysis. Sampling occurred in October 2001, March–April 2002, and October 2002, coinciding with the regional fall and spring peaks in

pesticide use in the heavily agricultural watershed. Fourteen current-use pesticides were detected in water and the majority of dissolved concentrations ranged from the limits of detection to 151 ng/l. Diazinon, EPTC and malathion were detected at much higher concentrations (940–3,830 ng/l) at the New and Alamo River upriver and near-shore stations. Concentrations of carbaryl, dacthal, diazinon, and EPTC were higher in the two fall sampling periods, whereas concentrations of atrazine, carbofuran, and trifluralin were higher during the spring, which matched seasonal use patterns of these pesticides. Current-use pesticides were also detected on suspended and bed sediments in concentrations ranging from detection limits to 106 ng/g. Chlorpyrifos, dacthal, EPTC, trifluralin, and DDE were the most frequently detected pesticides on sediments from all three rivers. The number of detections and concentrations of suspended sediment-associated pesticides were often similar for the river upriver and near-shore sites, consistent with downstream transport of pesticides via suspended sediment. While detectable suspended sediment pesticide concentrations were more sporadic than detected aqueous concentrations, seasonal trends were similar to those for dissolved concentrations. Generally, the pesticides detected on suspended sediments were the same as those on the bed sediments, and concentrations were similar, especially at the Alamo River upriver site. With a few exceptions, pesticides were not detected in suspended or bed sediments from the off-shore sites. The partitioning of pesticides

Guest editor: S. H. Hurlbert
The Salton Sea Centennial Symposium. Proceedings of a Symposium Celebrating a Century of Symbiosis Among Agriculture, Wildlife and People, 1905–2005, held in San Diego, California, USA, March 2005

L. A. LeBlanc (✉)
School of Marine Sciences, University of Maine, 5741 Libby Hall, Room 215, Orono, ME 04469-5741, USA
e-mail: Lawrence.leblanc@umit.maine.edu

K. M. Kuivila
California District Office, US Geological Survey Water Resources Division, 6000 J Street, Sacramento, CA 95819, USA

 Springer

between water and sediment was not predictable from solely the physical–chemical properties of individual pesticide compounds, but appear to be a complicated function of the quantity of pesticide applied in the watershed, residence time of sediments in the water, and compound solubility and hydrophobicity. Sediment concentrations of most pesticides were found to be 100–1,000 times lower than the low-effects levels determined in human health risk assessment studies. However, maximum concentrations of chlorpyrifos on suspended sediments were approximately half the low-effects level, suggesting the need for further sediment characterization of lake sediments proximate to riverine inputs.

Keywords   Alamo river · New River · Saline lake · Chlorpyrifos · Dacthal · EPTC · Trifluralin · DDE

Introduction

The Salton Sea is a saline lake located in the desert of southeastern California. Bounded by the Imperial and Coachella Valleys, it is proximate to areas of intense agricultural production. Approximately 143,259 hectares of land was devoted to growing crops in 2005 (Imperial Irrigation District, 2006). Located in an arid desert climate, water loss from evaporation is large, and is more or less balanced by inputs from agricultural irrigation. Continuous inputs from irrigation drainage (approximately 1.7 km^3/year), coupled with high rates of evapotranspiration (approximately 1.8 m/year) and no outlet or groundwater infiltration has led to an increasing salinity in the Salton Sea (Schroeder et al., 2002). The largest rivers that flow into the sea, in order of their annual discharge, are the Alamo, New, and Whitewater rivers. Forty-two year median daily flows ranged from 12.8 to 32.3 m^3/s for the Alamo River, 12.9 to 21.5 m^3/s for the New River and 2.3 to 3.4 m^3/s for the Whitewater River (LeBlanc et al., 2004). The Alamo and New Rivers flow into the southern end of the basin and the Whitewater River flows into the northern end. Current salinity in the Salton Sea measured as specific conductance is approximately 55,000 µS/cm; this is roughly equivalent to 45 g/l (Setmire and Schroeder, 1998; Schroeder et al., 2002).

Concerns about toxic chemicals entering the lake through agricultural runoff have led to several pesticide-related studies over the years. Concerns over the declining environmental quality arise from the fact that the Salton Sea provides important over-wintering habitat for several species of migrating birds. The Sonny Bono National Wildlife Refuge, located at the southern end of the lake, was created to preserve critical wetland habitat, which has disappeared at an alarming rate from southern California. In addition to ecological considerations, there are also human health concerns over toxicants in the Salton Sea. Changing water use patterns may ultimately lead to a smaller lake, exposing previously submerged lake bed. Potential human and animal exposure to toxicant-laden dust arising from exposed lake bed, has led to increasing worry about sediment-associated toxicants including pesticides, based on experience elsewhere. For example, there have been numerous studies on dust emissions in the dry lakebed of Owens Lake in Owens Valley, California (e.g., Tyler et al., 1997; Rya et al., 2002; Gillette et al., 2004). Water diverted from the watershed to Los Angeles via the California aqueduct resulted in the drying of Lake Owens, and the dust from its bed is considered to be a human health hazard, due to the presence of toxic trace metals (Gill et al., 2002). These types of concerns have been noted in other highly saline lakes, such as the Aral Sea in Uzbekistan (Kunii et al., 2003; Wiggs et al., 2003).

Approximately 1.4 million kg of active ingredients of pesticides and agricultural products are applied to agricultural fields in the Salton Sea watershed area each year (De Vlaming et al., 2000). The Alamo and New rivers in the south and the Whitewater River in the north are made up primarily of water used to irrigate surrounding fields and then drained, known as "agricultural return water" (Schroeder et al., 2002). De Vlaming et al. (2000, 2004) have found river water collected from the Alamo River to consistently exhibit toxicity to aquatic test organisms over a 10-year sampling program.

Several studies conducted since the late 1960s have identified elevated concentrations of pesticides in waters that drain into the Salton Sea area in southeastern California, including agricultural drains (Eccles, 1979; Setmire, 1979; Spencer et al., 1985; Setmire et al., 1990, 1993; Schroeder et al., 1993) and the New and Alamo Rivers (Irwin, 1971; Setmire, 1979, 1984; Crepeau et al., 2002).

Organochlorine pesticides were reported in bottom material collected from irrigation drainage canals (Setmire et al., 1993). More recently, Crepeau et al. (2002) reported elevated concentrations of a number of current-use pesticides in water collected in 1996 and 1997 from the Alamo River and within the Salton Sea. Active ingredient compounds included carbaryl, chlorpyrifos, cycloate, dacthal, diazinon, and EPTC, all found to be highest in fall and atrazine, carbofuran and malathion found to be highest in late winter/early spring. The elevated pesticide concentrations in the Alamo River corresponded with acute toxicity measured during standardized 96-h aquatic toxicity tests done by De Vlaming et al. (2000). Sapoznihkova et al. (2004) reported detectable levels (ng/g) of trifluralin, diazinon, and dacthal, along with DDT metabolites in sediments and in fish tissues collected from the Salton Sea Basin in 2000–2001.

Requirements for the Regional Water Quality Control Board (RWQCB), Colorado River Region, to develop total maximum daily loads (TMDLs) for contaminants, including pesticides, pointed to a need for a more complete understanding of the occurrence and transport of pesticides within the Salton Sea Basin. A TMDL is the maximum amount of a pollutant that can be discharged daily into a water body from all sources and still maintain water quality standards (Riverside County Flood Control and Water Conservation District, 2006). In addition, activities such as water transfers to the San Diego County Water Authority begun in 2003 will shrink the size of the Salton Sea, exposing more of the shoreline which may contain trace metal and organic contaminants. A recent report by the Pacific Institute released in 2006 estimated that 134 km^2 of lakebed will eventually become exposed by the year 2030, in the absence of any remediation efforts, as a result of water transfers to cities in coastal California. Various remediation plans to maintain the level of lake at a given level have also been proposed. The plan favored by the Salton Sea Authority, which involves the construction of a dike between the north and south basins, to create a deepwater lake in the North basin, and a mosaic of wetlands and brine ponds in the South basin will result in less than 2,833 ha (7,000 ac)—less than 8% of the 36,422 ha (90,000 ac) in the South Basin—being exposed and subject to wind erosion (Salton Sea Authority, 2006). The Cascades Plan, proposed by a consortium of interested parties, proposes to create a mosaic of lakes of increasing salinity (from exterior to interior) and a mosaic of wetland habitats surrounding two central brine ponds, may result in even less dry lake bed exposed (The Salton Sea Restoration Consortium, 2004). In order to better understand the transport dynamics of agricultural pesticides to the Salton Sea, and assess the importance of suspended solids as a medium for transport of pesticides into the basin, the US Geological Survey (USGS) conducted sampling and analysis of water, suspended and bed sediments for current-use pesticides, from fall 2001 up to fall 2002. The primary objective of this study was to determine the occurrence and transport of historical and current-use pesticides into the Salton Sea Basin. A major focus was to examine the distribution of pesticides in water, suspended sediment, and bed sediment in the three major rivers that drain into the lake basin.

Materials and methods

Sample collection

Sampling occurred during October 20–29, 2001 (fall 2001), March 14–22, April 16–18, 2002 (spring 2002), and September 16–25, 2002 (fall 2002). Sampling sites were established along the flowpath of the three major rivers that discharge into the Salton Sea basin (three sites per river) to examine the transport of pesticides (Fig. 1). The upriver sites were established approximately 1.6 km upstream of the mouths of the New and Alamo Rivers and 6 km upstream of the mouth of the Whitewater River. The Whitewater River upriver site had a greater upstream distance (6 km) because this was the closest site to the actual point of discharge that could be sampled safely. The near-shore sites were established in the Salton Sea as close as possible to the river discharge point (at approximately 1 m depth), and the off-shore sites were established in water 4.6 m (15 ft) deep just offshore from the river mouths. The three sites associated with a given river were generally sampled within 7 days of one another. Concentrations at each station represent an instantaneous evaluation of pesticide content, and so transects may or may not represent a continuous transport series. Latitude, longitude, and approximate distance from the river mouth are listed for each site in Table 1.

Fig. 1 Location of
sediment sampling sites
within the Salton Sea basin

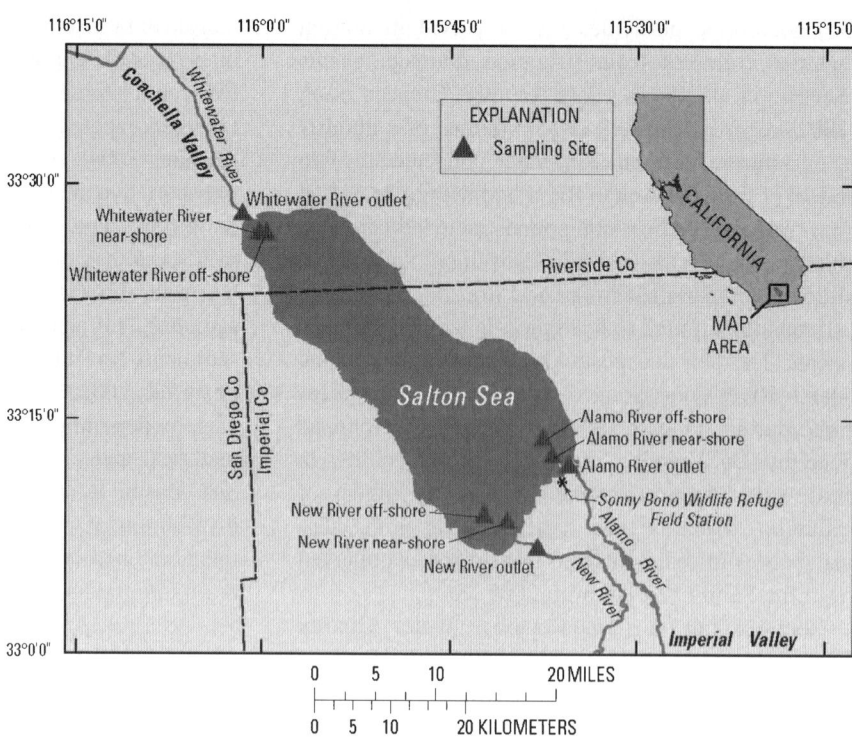

Table 1 List of sampling sites and their coordinates

Station name	Latitude	Longitude	Description[a]
Alamo River upriver	33°11′56″	115°35′46″	Bridge at Garst Road, 1.6 km upstream from river mouth
			River width: 25 m, river depth 3.5 m, total discharge, 903 ft³/s
Alamo River near-shore	33°12′42″	115°37′14″	Within Salton Sea, 250 m from river mouth
			Approx depth: 1.5 m
Alamo River off-shore	33°14′00″	115°38′00″	Within Salton Sea, 2.9 km from river mouth
			Approx depth: 5 m
New River upriver	33°05′59″	115°38′56″	Bridge at Lack Road, 1.6 km upstream from river mouth
			River width: 26 m, river depth ∼1 m, mean total discharge, 606 ft³/s
New River near-shore	33°08′03″	115°41′40″	Within Salton Sea, 100 m from river mouth
			Depth: ∼1 m
New River off-shore	33°08′35″	115°43′45″	Within Salton Sea, 3.5 km from river mouth
			Approx depth: 5 m
Whitewater River upriver	33°31′29″	116°04′44″	Culvert at Lincoln Road, 6 km upstream from river mouth
			River width: 20 m, river depth ∼5 m, mean total discharge, 67 ft³/s
Whitewater River near-shore	33°30′06″	116°03′15″	Within Salton Sea approximately 500 m from river mouth
			Approx depth: 1.5 m
Whitewater River off-shore	33°29′58″	116°02′35″	Within Salton Sea approximately 1.5 km from river mouth
			Approx depth: 5 m

[a] Distances from river mouth are approximations, discharge measurements from US Geological Survey (2001)

Large volumes of water (300–900 l) for the determination of aqueous and suspended sediment-associated pesticide concentrations were collected at the nine sampling sites using a large peristaltic pump powered by a portable generator and equipped with a stainless steel and Teflon inlet hose. At the upriver

sites the samples were collected from a single point at the center of the river channel at a depth of 0.5 m. At the three near-shore sites, the hose was positioned such that the sampling occurred at the mixing zone between the fresh river water and the saltier bottom water, which was between 0.1 and 0.5 m depth for all sites. Prior to sampling, the depth of the mixing zone (easily visible from the boat) was determined by a hand-held conductivity meter. The hose was positioned at the halocline and the up and down rocking of the boat caused the hose to continuously sample through the mixing zone. At the off-shore sites water again sampled at 0.5 m below the surface. Samples for pesticides were pumped directly into pre-cleaned 20 l stainless steel containers. Water collection from the off-shore sites required multiple boat trips, but sampling always was completed within a 24-h period. Duplicates from each site were processed for October 2001 samples. Due to budgetary constraints, single composite samples were analyzed for the following spring (March/April 2002) and fall (September 2002).

Bed sediments were collected concurrently with the water samples at each of the nine sampling sites. Samples taken at the upriver sites were collected using either a 23-cm Ekman grab sampler or a 5-cm diameter, Teflon-barreled hand corer. Samples were collected only from the top 2 cm of undisturbed sediment collected during each grab. At the river sites, multiple core samples (5–7) were composited to make a sample. At the near-shore and off-shore sites, composites of 2–3 grabs were taken using the Eckman dredge. Sediment was scooped into cleaned, 0.5-l glass jars using a stainless steel spoon.

Processing of large volume water samples

Suspended sediments were isolated by pumping the water through a flow-through centrifuge (Westphalia model KA-2, Westfalia Corporation, Odele, Federal Republic of Germany). Water (300–900 l) was pumped through the centrifuge, which spun at 9,800 revolutions per minute (rpm). The centrifuge was operated at a centrifugal force of $9,500g$ and a flow rate of 2 l/min which has been shown to be optimal for efficient capture of a wide range of grain sizes (2 μm and larger) at a wide range of suspended sediment concentrations (Horowitz et al., 1989). Details of this procedure can be found in LeBlanc

et al. (2004). Sediment and sediment-water slurry was carefully removed from each of the concentric centrifuge bowls and further dewatered by centrifuging in 200-ml stainless steel centrifuge bottles for 20 min at 10,000 rpm in a high-speed refrigerated centrifuge (Sorvall RC-5B centrifuge, Du Pont Company, Wilmington, Delaware) in the USGS California District Organic Chemistry Laboratory in Sacramento. Sediment was stored frozen in pre-cleaned glass jars (at $-20°C$) until analysis.

Samples for the determination of aqueous pesticide concentrations were collected in 1 l amber glass bottles from the upriver of the flow-through centrifuge during the beginning, middle, and end of the centrifuging process.

Water extraction

Pesticides were extracted from water samples using solid-phase extraction (SPE) cartridges. This method is described in detail in Crepeau et al. (2000). Water was pumped through pre-cleaned and conditioned C8 SPE cartridges (Varian Bond-Elut, 500 mg, 300 cm³ size barrel, Varian Analytical Corporation, Walnut Creek, California) using 12-volt ceramic-piston metering pumps (at a rate of 20 to 25 ml/min), after the addition of 200 ng of the terbuthylazine surrogate compound. Cartridges were labeled, refrigerated; and within 3 days of collection sent to the USGS California District Organic Chemistry Laboratory in Sacramento for storage. Upon receipt of the samples in Sacramento, the cartridges were dried using carbon dioxide for at least 1 h and stored frozen (at $-20°C$) until analysis, which did not exceed 2 months from the time of collection (Crepeau et al., 2000).

Samples were eluted from the SPE cartridges using 9 ml of ethyl acetate and reduced in volume to 500 μl by evaporation using a stream of nitrogen gas (N-evap, Organomation Associates, Kansas City, Missouri.). The deuterated polyclic aromatic hydrocarbon (PAH) compounds d_{10}-acenaphthene, d_{10}-phenanthrene, and d_{10}-pyrene were added as internal standards. Extracts were brought to a final volume of 200 μl and analyzed using gas chromatography/mass spectrometry (GC/MS).

Detection limits for dissolved pesticides ranged from 0.6 to 7.2 ng/l (Table 2). The recoveries (mean ± standard deviation) for pesticides spiked

Table 2 Method detection limits for pesticides in water and sediment samples, and key physical–chemical descriptors[a] of pesticides detected in water, suspended sediments and bed sediments in the Salton Sea Basin, October, 2001

Pesticides	Water MDL (ng/l)	Sediment MDL (ng/g)	Matrix[b]	Molecular weight (g/mol)	Water solubility (mg/l)	Log K_{oc}[c]	Henry's law (atm*m³/mol)
Alachlor[d]	2.1	1.1					
Atrazine[e]	4.2	0.6	W	215.68	30	2.2	2.7×10^{-9}
Bifenthrin	ND[f]	0.9	BS	422.9	0.1	5.11–5.48	na
Butylate[d,e]	1.8	0.5	SS, BS	217.38	45	2.73	5.5×10^{-8}
Carbaryl[e]	4.2	1.2	W, SS, BS	201.22	104	2.5	1.3×10^{-8}
Carbofuran	3.3	3	W, SS, BS	221.3	1,350, 700	2.1	5.8×10^{-9}
Chlorpyrifos	4.2	1.5	W, SS, BS	350.6	0.3, 1.1	3.78	7.8×10^{-5}
Cycloate[e]	1.5	1.8	W	215.4	75–95	2.6	2.3×10^{-5}
Cyfluthrin	ND	7.9	SS, BS	434.3	0.0012–0.003	4–5.13	na
Cypermethrin	ND	5.6					
Dacthal	1.2	0.6	W, SS, BS	332	0.5	3.78	2.2×10^{-6}
Diazinon	3.6	1.5	W, SS, BS	304.36	40	2.76	1.13×10^{-7}
Diethatyl-ethyl[d,e]	3.6	0.7	SS, BS	311.8	105	3.15	1.28×10^{-8}
EPTC	4.5	0.7	W, SS, BS	189.31	365	2.38	1×10^{-5}
Esfenvalerate	ND	1.4	SS	419.9	0.002	3.72	na
Ethalfluralin[d,e]	2.4	1.9					
Fonfos4	2.4	2.5	SS, BS	283.8	5.30×10^2	2.3, 2.46	9.2×10^{-9}
Hexazinone[d,e]	5.7	3.2	W	330.36	145	3.25	2×10^{-8}
λ-cyhalothrin	ND	0.5	BS	449.9	5.00×10^{-3}	na[4]	na
Malathion[e]	2.1	1.5					
Methidathion4	5.4	3.4	W	271.4	73	2.48–2.85	2×10^{-8}
Methyl parathion[d,e]	4.2	1.6	W, SS, BS	281.3	0.3	1.48–3.70	2×10^{-8}
Metolachlor[d]	3.3	1					
Molinate4,5	2.7	2					
Napropamide[e]	7.2	1.6					
Oxyfluorfen	4.2	6.1	W, SS, BS	361.7	0.1	2.63–2.80	9.25×10^{-6}
Pebulate[d,e]	0.6	0.8					
Pendimethalin	2.4	4					
Permethrin	ND	1.4	SS, BS	391.3	1.53×10^{-5}	4.8	na
Phosmet[d,e]	4.2	0.8					
Prometryn	ND	1.8					
Simazine[e]	6.9	2.1	W	201.67	5	2.14	3.4×10^{-9}
Sulfotep[d,e]	1.2	1.1					
Thiobencarb[d,e]	3.9	4.4					
Trifluralin	3	1.4	W, SS, BS	335.5	.3, variable	3.63	3.95×10^{-5}
p, p′-DDD	ND	2.9					
p, p′-DDE	ND	3.7					
p, p′-DDT	ND	3.9					

[a] Multiple values, as reported in the literature, are listed here; [b]Matrix refers to the environmental compartment where a compound was detected. W, water; BS, bed sediment; SS, suspended sediment; [c]Koc = the organic carbon–water partition coefficient; [d]Pesticide not detected in any aqueous samples; [e]Pesticide not detected in any sediment samples; [f]ND, not determined

to the water matrix spike samples were 85 ± 20%, but generally were greater than 80%. Replicate samples showed excellent agreement, and concentrations differed by no more than 10%. Further details on quality assurance/quality control can be found in LeBlanc et al. (2004).

Sediment extraction

Extractions from suspended sediment and bottom sediment samples were performed using either soxhlet or microwave-assisted solvent extraction (MASE). The MASE system was an MSP 1000 (CEM Corporation, Matthews, North Carolina.). Wet sediment was used for extraction, and approximately 5 g (dry weight) was used per sample. Complete details of the method can be found in LeBlanc et al. (2004) and in LeBlanc et al. (in preparation).

Prior to extraction, a surrogate solution containing 400 ng each of $^{13}C_6$-α-benzenehexachloride (α-BHC), d_{10}-chlorpyrifos and $^{13}C_3$-simazine (Cambridge Isotope Laboratories, Inc, Andover, Massachusetts) was added to each sample. The sediments were extracted using a mixture of methylene chloride and acetone (50:50 volume/volume). Comparisons of MASE and soxhlet extraction techniques for the extraction of a wide variety of hydrocarbons, including current-use pesticides on soils and sediments, have yielded similar results (Pastor et al., 1979; Blanco et al., 2000; Camel, 2000). Methylene chloride was isolated by aqueous back extraction and then dried over sodium sulfate and reduced to 1 ml through rotary evaporation. These extracts, many of which were dark in color, were cleaned by passage through 500 mg of activated carbon in an SPE cartridge (6-cm³-sized barrel, Restek Corporation, Bellefonte, Pennsylvania), followed by passage through a gel permeation/high pressure liquid chromatography (GPC/HPLC) system consisting of a Perkin Elmer HPLC (PE 410 4-stage pump, LC-95 UV fixed wavelength detector, Perkin Elmer Corporation, Norwalk, Connecticut) and a polydivylbenzene GPC column (300 × 7.5 mm, 50-Å pore size, Polymer Laboratories, Amherst, Massachusetts). GPC/HPLC provided additional matrix cleanup, including the separation and elimination of elemental sulfur, a major interferent. Extracts were reduced to final volume of 500 μl, after the addition

of internal standards (the same as for the water samples).

The recoveries (mean ± standard deviation) of d_6-α-HCH, d_3-simazine, and d_{10}-chlorpyrifos surrogate compounds were 66 ± 16, 91 ± 35, and 78 ± 19%, respectively. Matrix interference negatively affected surrogate recoveries in some sediment samples collected from off-shore, which consisted primarily of algal detritus.

Average recoveries for five sediment matrix spikes were between 51 and 109% for all compounds, with an overall mean ± standard deviation recovery of 72 ± 22% for the pesticides that were detected in suspended and bed sediments. Method detection limits ranged from 0.6 to 7.9 ng/g (Table 2). A detailed description of QA/QC samples can be found in LeBlanc et al. (2004).

Instrumental analysis

Water and sediment extracts were analyzed for pesticides using a Saturn 2000 GC/MS ion trap system (Varian, Inc., Walnut Creek, California). Run conditions are listed in Table 3. During portions of the run, selected ion storage (SIS, equivalent to selected ion monitoring in a quadrupole instrument) was used to optimize instrument sensitivity to select analytes. Calibration of instrument response to each pesticide was made using an eight-point standard curve that spanned the range of sample

Table 3 Run conditions for the Saturn 2000 GC/MS system

Injection conditions	Splitless injection, pressure pulse of 50 psi[a] for 1.5 min.
Injection temp	275°C
Oven program	60°C, hold for 0.5 min, 80–120°C at 10 dpm[b]
	120–200°C at 3 dmp, hold for 5 min.
	200–219°C, hold for 5 min.
	219–300°C at 10 dpm, hold for 10 min.
GC/MS conditions	Range SIS, collecting 60–450 m/z[c]
Analytical column	CPSIL 8-MS (Varian Corp.),
	30 m × 0.25 mm
	0.5 mm film thickness

[a] psi, pounds per square inch

[b] dpm, degrees per minute

[c] *m/z*, mass to charge

concentrations. Standards were purchased from Supelco Inc. (Bellefonte, Pennsylvania). In addition to the standard curve, a mid-level standard was run every six injections to verify that the response was within 10% of the standard curve. Each sample was analyzed twice, and replicate injections with greater than 25% variability were reanalyzed.

Organic carbon content in sediments was determined using a Perkin Elmer CHNS/O analyzer (Perkin Elmer Corporation, Norwalk, Connecticut). Sediments were weighed into 5 × 9 mm silver boats (Costech Analytical Technologies, Valencia, California) after being exposed to concentrated hydrochloric acid fumes 24 h to remove carbonate minerals.

Statistical analysis

Statistical analysis was performed using Sigma Stat statistical software (SPSS Inc., Chicago, IL). Correlation analyses using Pearson Product Moment Correlation were performed to test the strength of associations among pesticide use patterns, frequency of detection, and mean pesticide concentrations for aqueous, suspended sediment and bed sediment samples. For calculating mean pesticide concentrations and standard deviations around the mean, non-detect values in the dataset were replaced with ½ the detection limit. The data were then log-transformed and tested for normality using the Kolmogrorv-Smirnov Goodness of Fit Test and the Shapiro-Wilks W test. The geometric mean and standard deviation were obtained by back-transforming (i.e., raising to the 10 power) means and standard deviations calculated from the log-transformed data. Errors around the mean were obtained by multiplying the standard deviation by the geometric mean, after these quantities were back-transformed.

Records of pesticide use in the Salton Sea watershed for 2001–2002 were extracted from the California Department of Pesticide Regulation Pesticide Use Database (California Department of Pesticide Regulation, 2002) and the data compiled to quantify pesticide use within the watershed. For the correlation analyses, each sampling date (October, 2001; March–April 2002 and September–October 2002) was treated independently of one another, and geometric mean pesticide concentrations from upriver stations were calculated at each

sampling date ($n = 3$). Pesticide use data was compiled for the major application periods which occurred prior to and during the sampling events, which were fall 2001 (pesticide use records for April 2001–Oct 2001), spring 2002 (pesticide use records for November 2001–March 2002) and fall 2002 (pesticide use records for April 2002–October 2002, Fig. 4a). In addition, total pesticide use was compared to pesticide concentrations from upriver sites averaged over all sampling dates and rivers. Total pesticide use was determined from the same records, as kilograms of active ingredient applied between April, 2001 and October, 2002.

Results

Aqueous pesticide concentrations

Aqueous pesticide concentrations ranged from below detection limits to 3,800 ng/l, with the highest concentrations detected at the upriver and the near-shore sites (LeBlanc et al., 2004 contains individual concentration results for each sample). Generally, the compounds detected most frequently and with the highest mean aqueous concentrations were those pesticides applied in the largest quantities, and included atrazine, chlorpyrifos, dacthal, diazinon, EPTC, malathion, pendimethalin, and trifluralin (Fig. 2a–c).

The highest aqueous concentrations were found most often in the Alamo River, although the New River also had high concentrations of pesticides (Fig. 2a). EPTC was detected at the highest concentration of all the pesticides (3,490 and 3,830 ng/l) at the New River upriver site and at the Alamo River near-shore site, respectively, during fall 2001. The concentration of malathion was elevated at the Alamo River upriver site, especially during spring 2002, with a concentration of 1,100 ng/l (Fig. 2a). Diazinon concentrations were elevated at the Alamo River upriver sites in both fall 2001 and fall 2002, with concentrations of 789 ng/l and 970 ng/l, respectively. The Whitewater River had the lowest overall aqueous pesticide concentrations (Fig. 2a, c).

Seasonal patterns in both frequency of detection and geometric mean aqueous concentrations at upriver sites paralleled pesticide-use patterns (Fig. 3 a–c), consistent with observations made by Eccles

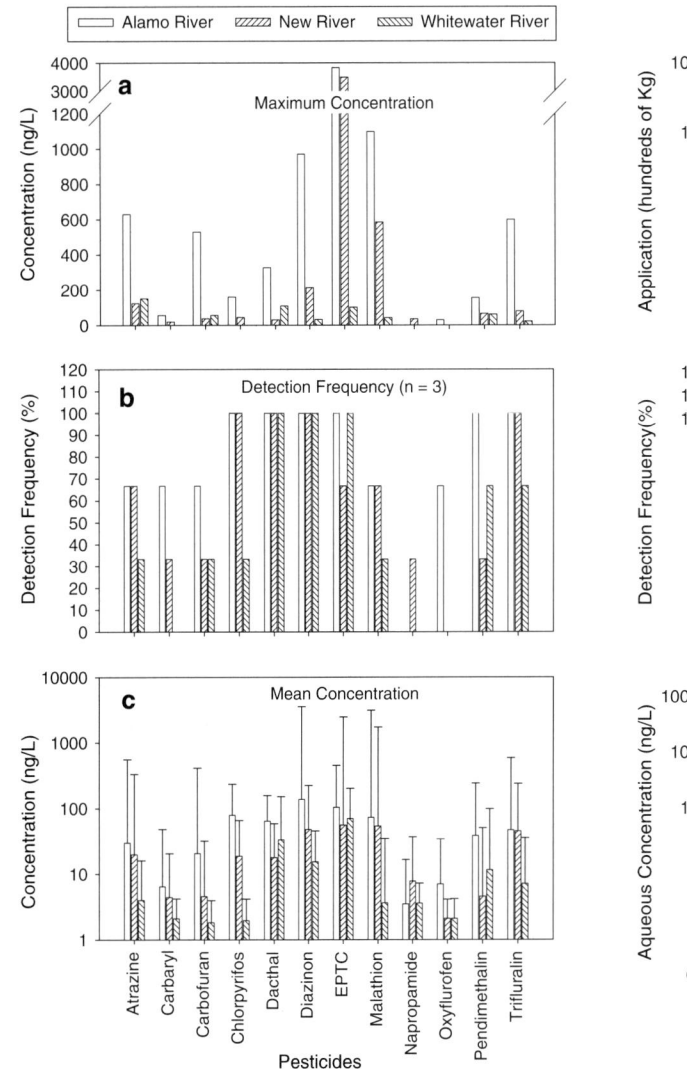

Fig. 2 Comparison of (**a**) maximum aqueous pesticide concentrations detected from upriver sites over all sampling events to (**b**) to the number of times these compounds were detected in the aqueous phase at the upriver sites over all times, to (**c**) geometric mean aqueous concentrations of pesticides detected at the upriver sites from all sampling events between the Alamo, New and Whitewater Rivers. Error bars are one standard deviation of the mean of the log concentrations, back-transformed and multiplied by the geometric mean

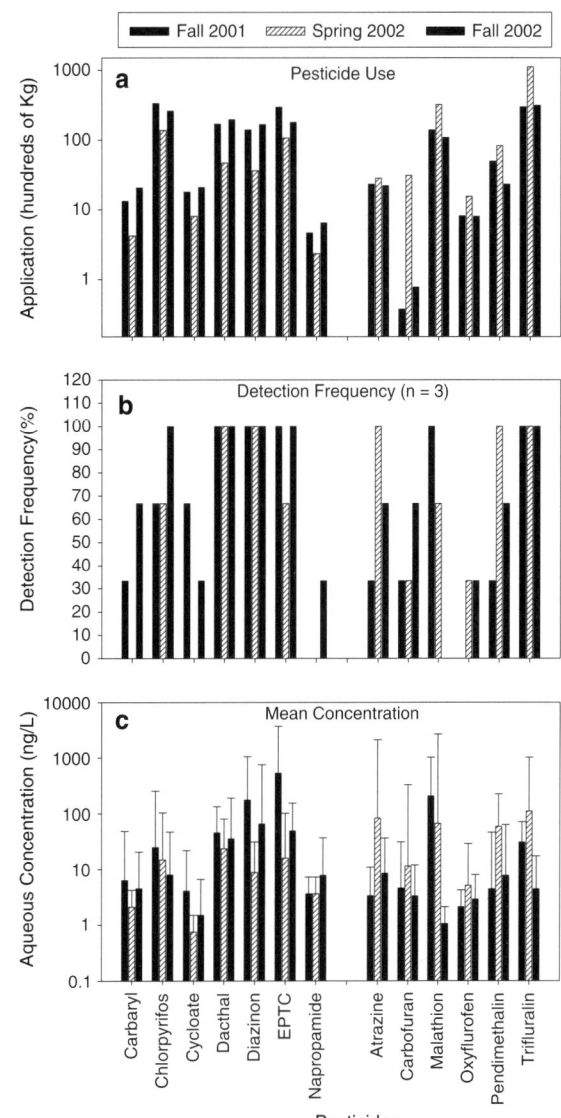

Fig. 3 Comparison of (**a**) the amount of individual pesticide compounds applied within the Salton Sea watershed to (**b**) the frequency at which pesticides were detected at upriver stations and (**c**) the geometric mean aqueous pesticide concentrations at the upriver stations. Error bars are one standard deviation of the mean of the log concentrations, back-transformed and multiplied by the geometric mean

(1979), De Vlaming et al. (2000) and Crepeau et al. (2002). Those pesticides elevated in concentration during the fall-sampling periods included chlorpyrifos, dacthal, diazinon, and EPTC. Those elevated in the spring included atrazine, carbofuran, malathion, and trifluralin (Fig. 3b, c).

The frequency of detection and mean concentrations of individual pesticides were correlated to

pesticide application patterns. Total pesticide use was positively and significantly ($P < 0.05$) correlated to the number of detections of pesticides from upriver sites summed over all sampling times. Mean aqueous pesticide concentrations at upriver sites averaged over all rivers and sampling times however was not significantly correlated with use (Table 4). The same

Table 4 Pearson Product Moment correlation coefficients and *P* values for comparisons between pesticide use and concentration, and pesticide use and detection frequency at upriver sites

Pesticide use vs. Detection frequency

	Aqueous		Suspended sediment		Bed sediment	
	r^e	P^f	r	P	r	P
Total[a]	**0.67**	**0.01**	**0.80**	**0.001**	**0.69**	**0.006**
F01[b]	**0.72**	**0.0005**	**0.79**	**0.007**	**0.55**	**0.04**
S02[c]	0.37	0.21	0.39	0.17	0.34	0.24
F02[d]	**0.66**	**0.01**	**0.86**	**0.0001**	**0.74**	**0.002**

Pesticide use vs. Geometric mean concentration

	Aqueous		Suspended sediment		Bed sediment	
	r	P	r	P	r	P
Total	0.34	0.26	**0.81**	**0.0004**	0.09	0.76
F01	0.51	0.08	**0.68**	**0.007**	−0.17	0.56
S02	**0.74**	**0.004**	**0.81**	**0.0004**	0.22	0.46
F02	0.38	0.20	0.15	0.61	0.33	0.25

[a] Total, the sum of pesticide active ingredient applied within the Salton Sea watershed from April, 2000–October, 2002)

[b] F01, total pesticide use during the months preceding the fall 2000 sampling event (April, 2000–October, 2001)

[c] S02, pesticide use during the months preceding the spring 2002 sampling event (November, 2001 March, 2002)

[d] F02, pesticide use during the months preceding the fall 2002 sampling event (April, 2002–October, 2002)

[e] r, value of the nonparametric Spearman Rank correlation coefficient

[f] P, the probability associated with the Spearman Rank correlation; values that are significant ($P \leq 0.05$) are in bold

correlations were performed separately for each sampling event. A statistically significant correlation between pesticide use and mean aqueous concentration was found in spring 2002 (Table 4).

Concentration patterns of dissolved pesticides along the river transects had maxima, most often, although not exclusively at either upriver or near-shore sites suggesting that pesticides travel in pulses down the rivers, into the lake basin. Spatial trends for all three sampling events are presented in Fig. 4a–i, for the Alamo (Fig. 4a, d, g), New (Fig. 4b, e, h) and Whitewater Rivers (Fig. 4c, f, i) during the fall 2001. Detectable concentrations of a few pesticides at offshore sites (e.g., atrazine, dacthal, and malathion) demonstrated transport into the lake basin of dissolved pesticides.

Organic carbon in suspended and bed sediments

Organic carbon (OC) concentrations are presented in Table 5 for both suspended and bed sediments. Concentrations ranged from 1.1 to 1.7% for upriver sites in the New and Alamo Rivers. OC in the Whitewater River was higher and ranged between 4 and 5.4% for upriver sites. Many offshore suspended sediments had extremely high OC concentrations, ranging from 27.1 to 41.9%. Suspended sediments from the near-shore sites were intermediate between these two extremes, and ranged from 1.0 to 33.6%. Bed sediments were more uniform in organic carbon concentrations, and ranged from 0.4 to 6.2%.

Pesticides associated with suspended and bed sediments

Nineteen pesticides were detected on suspended sediments and concentrations ranged from the limits of detection to 106 ng/g. The current-use pesticides most frequently detected were chlorpyrifos, dacthal, EPTC and trifluralin, which were also among those used in the highest amounts prior to and during the sampling periods (Fig. 5a). Of these, trifluralin and chlorpyrifos had the highest maximum and average concentrations (Fig. 5c, e). Diazinon and permethrin were also detected, but in lower concentrations and less frequently. Other pesticides, such as carbaryl, carbofuran, oxyflurofen, and pendimethalin were detected more sporadically. P,p′-DDE, a metabolite of DDT, was consistently found on suspended sediments in concentrations ranging from 5.8 to 101 ng/g. The greatest number and highest concentrations of pesticides were detected at the upriver and near-shore sites, primarily from the Alamo and New Rivers. In offshore suspended sediments, only carbaryl was detected at the Alamo River offshore site in spring 2002 (43.6 ng/g) and in the New River offshore site in fall 2002 (89.5 ng/g, from LeBlanc et al., 2004).

Differences in the concentration and occurrence of suspended sediment-associated pesticides by season were not as pronounounced as the differences seen with the water samples, but still reflected seasonal use trends (Fig. 6a–e). Chlorpyrifos and diazinon, which had high aqueous concentrations in fall 2001, had slightly higher mean concentrations on suspended

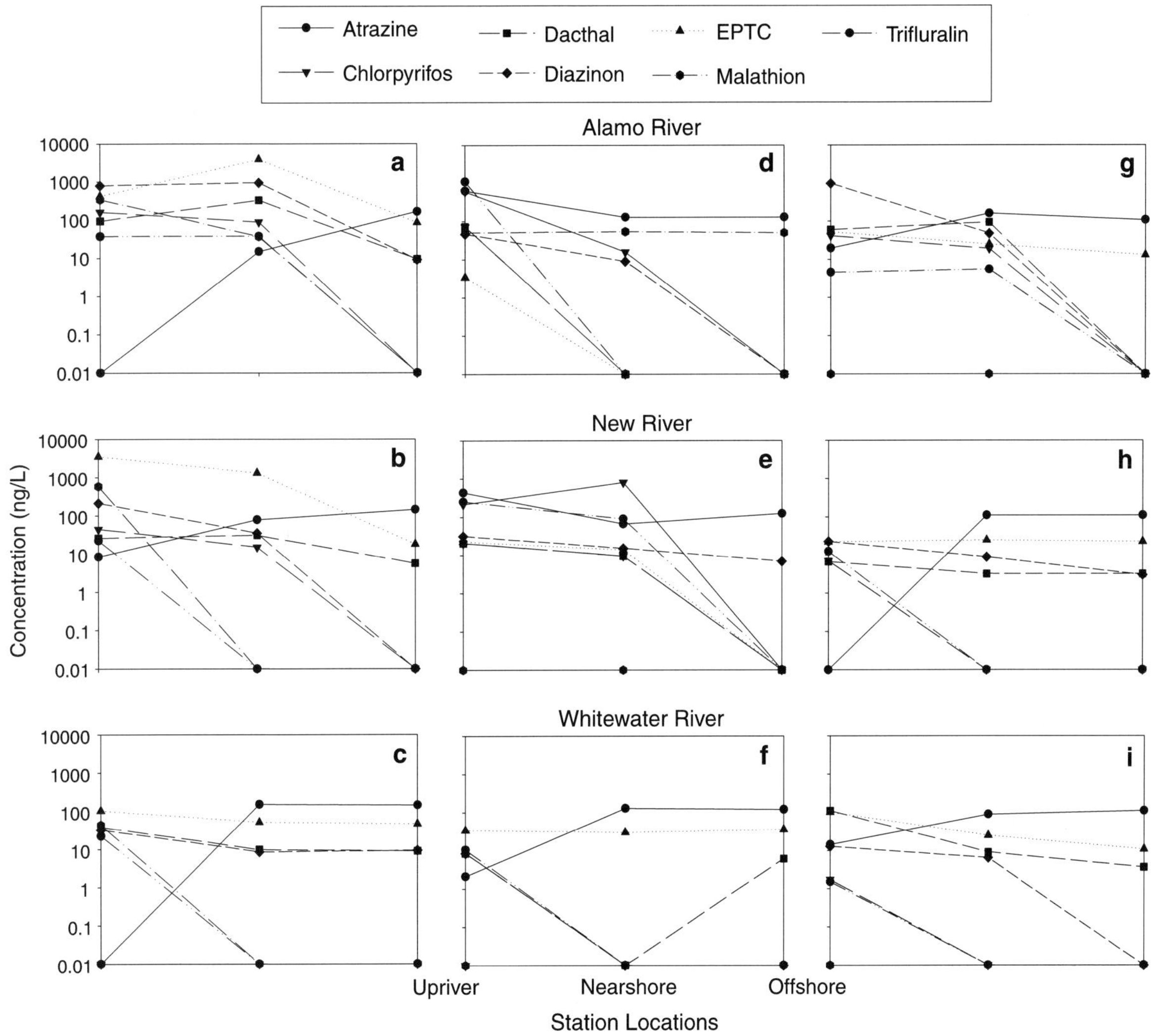

Fig. 4 Aqueous concentrations of representative current-use pesticides along the river transects for the Alamo, New, and Whitewater Rivers during the fall, 2001 (**a–c**), spring 2002 (**d–f**) and fall 2002 (**g–i**) sampling events. For graphing purposes, non-detected values given a value of 0.01 ng/l

sediment when compared to spring 2002 (Fig. 6d). Similarly, trifluralin was used to a greater extent in the time period preceeding the spring 2002 sampling, and highest in aqueous concentration at this time also had higher mean suspended sediment concentrations at the upriver sites compared to concentrations from fall 2001 and from fall 2002 (Fig. 6d).

Concentration patterns of suspended sediments-associated pesticides along the river transects had maxima at nearshore sites in addition to upriver sites, demonstrating transport of sediment-associated pesticides into the Salton Sea. Spatial trends for all three sampling events are presented in Fig. 7a–i. No

pesticides were detected from suspended sediments collected at offshore sites.

Eighteen pesticides were detected in bed sediments, primarily at upriver sites, although nearshore sites had detectable concentrations of a few pesticides, notably DDE, chlorpyrifos, dacthal, permethrin, and trifluralin (Figs. 5f and 8a–i). The compounds were generally the same pesticides detected in the suspended sediments, with concentrations ranging from detection limits to 89.5 ng/g (LeBlanc et al., 2004). As with the suspended sediments, the pesticides detected with the greatest frequency were among those applied in large amounts

Table 5 Percent organic carbon in suspended sediments and bed sediments sampled during fall 2001, spring 2002, and fall 2002

Sampling sites	Percent organic carbon		
	Fall 2001	Spring 2002	Fall 2002
Suspended sediments			
Alamo River upriver	0.93	1.55	1.15
Alamo River near-shore	1.2	11.1	3.49
Alamo River off-shore	28.4	35.9	27.1
New River upriver	1.74	1.9	1.33
New River near-shore	2.71	4.41	5.31
New River off-shore	40.2	na	35.1
Whitewater River upriver	3.99	5.45	4.78
Whitewater River near-shore	42.7	28.8	33.6
Whitewater River off-shore	43.7	36.6	41.9
Bed sediments			
Alamo River upriver	0.56	1.17	0.49
Alamo River near-shore	1.04	1.26	1.39
Alamo River off-shore	0.38	0.36	0.6
New River upriver	0.37	0.47	0.25
New River near-shore	0.94	0.25	1.83
New River off-shore	0.78	1.39	0.83
Whitewater River upriver	0.46	1.13	0.46
Whitewater River near-shore	na	1.65	0.94
Whitewater River off-shore	4.25	3.25	6.23

na, Not analyzed

in the watershed area and included chlorpyrifos, dacthal, diazinon, EPTC, permethrin, and trifluralin as well as p, p′-DDE (Fig. 5b). Chlorpyrifos, DDE, and trifluralin, which had the highest number of detections, also had the among highest detected and mean concentrations in bed sediments (Fig. 5d, f).

Correlations between the total amount of pesticides applied prior to and during the three sampling events and pesticide detection frequency at all rivers and times (upriver sites only) were significant for both suspended and bed sediments (Table 4). Total use was also significantly correlated with geometric mean pesticide concentrations in suspended sediments from upriver sites, averaged over river and season. Correlations between seasonal pesticide application rates and detection frequency for the corresponding sampling event were significant in fall, 2001 and fall, 2002 for suspended and bed sediments.

Correlations between pesticide use and geometric mean pesticide concentration were significant for suspended sediments in fall 2001 and spring 2002.

Discussion

Dissolved pesticides

Patterns of abundance and concentration of dissolved pesticides reflected the use patterns reported for the Salton Sea watershed. Distinct differences were seen in the occurrence and concentrations of dissolved pesticides during the spring and fall sampling events, which closely matched patterns of pesticide application for this watershed. Seasonal differences in fall 2002 were not as dramatic as seen in fall 2001 and spring 2002, but also matched pesticide application records. Elevated aqueous concentrations of diazinon, seen during both fall sampling events in the Alamo River, is one of the major contributors to aquatic toxicity as determined in standard toxicity tests (De Valming et al., 2000, 2004). These observations are supported by positive correlations between pesticide use and frequency of detection. When the data were broken out by season however, not all correlations had significance at the 95% level, due most likely to high variability in the data. Also mean aqueous concentrations were not strongly correlated with pesticide use, due to high variability. Sampling only once per season only gives a "snapshot" of concentrations and is not truly representative of seasonal concentration patterns.

Spatial patterns of dissolved pesticides along the river transects, showed maxima and minima at the various sites, as opposed to being a uniform concentration throughout. This suggests that pesticides enter the rivers in pulses. With only a snaphsot view of concentrations (i.e., one sampling per sampling event) it is not possible to completely characterize the nature of the inputs in terms of observing and quantifying true maximum concentrations as these pulses move down the rivers. The fact that dissolved concentrations are detectable within the lake basin, provides evidence for transport of pesticides into the lake from rivers via the dissolved phase.

Concentrations of dissolved pesticides in the Alamo River were within an order of magnitude of those reported in Crepeau et al. (2002), who sampled

Fig. 5 (**a**, **b**) Comparisons of pesticide detection frequency at upriver sites between the Alamo, New and Whitewater Rivers for (**a**) suspended sediments, and (**b**) bed sediments (**c**, **d**) Comparison of maximum concentrations of pesticides from upriver sites at all sampling events between the Alamo, New and Whitewater Rivers for (**c**) suspended sediments and (**d**) bed sediments. (**e**, **f**) Comparisons of geometric mean concentrations of pesticides from upriver sites at all sampling events between the Alamo, New and Whitewater Rivers for (**e**) suspended sediments and (**f**) bed sediments. Error bars are one standard deviation of the mean log concentrations, back-transformed and multiplied by the geometric mean

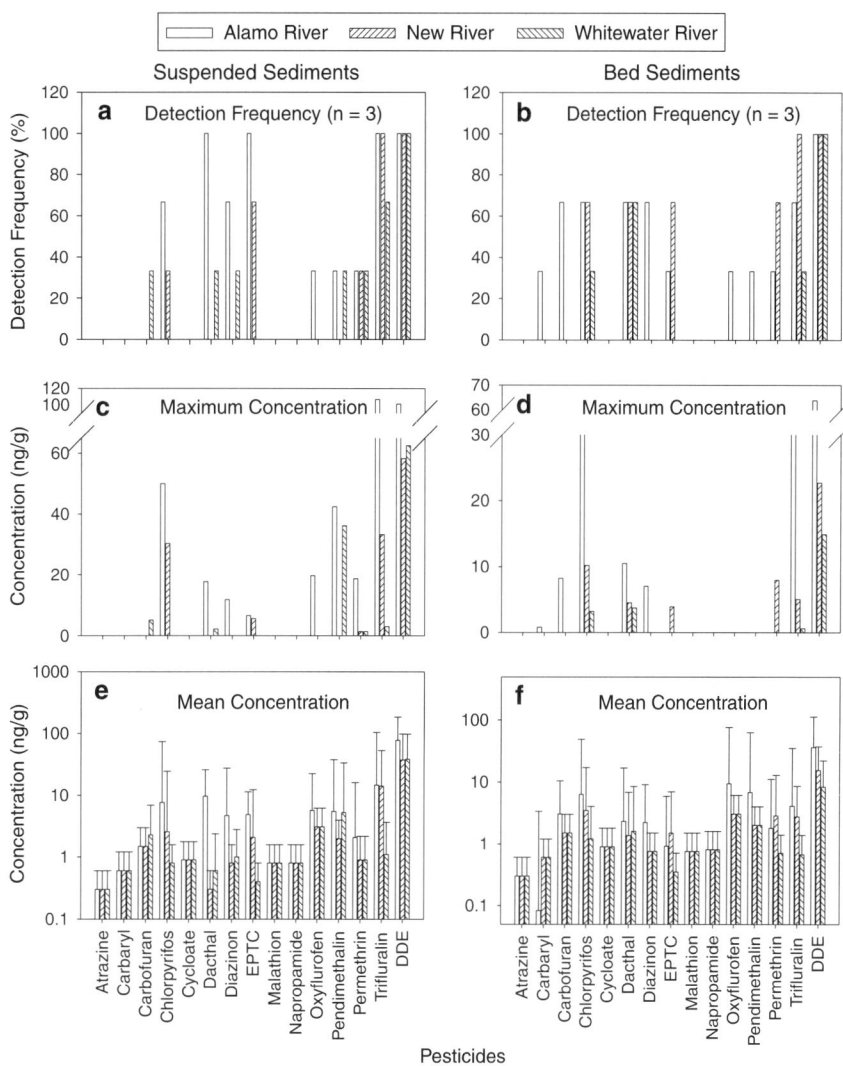

monthly during 1996. However, concentrations of EPTC, which in 1996 were reported to be as high as 13,000 ng/l and often above 1,000 ng/l, were comparable only to the fall 2001 concentrations (3,800 ng/l). During the spring and fall of 2002, aqueous concentrations of EPTC were only 13 to 100 ng/l. It should be noted however that sampling by Crepeau et al. (2002) was done much more frequently (monthly) and so the likelihood of sampling of a maximum or a "peak" concentration was more likely.

Sediment-associated pesticides

While the occurrence and distribution of sediment-associated pesticides was much more sporadic than for dissolved pesticides, leading to high variability in the dataset, trends similar to those observed for aqueous pesticide concentrations were still apparent. The concentration and abundance of current-use pesticides associated with suspended sediments showed a degree of correspondance with pesticide use patterns. Pearson Product Moment correlation coefficients were significant at the 95% level when total pesticide use over the year was compared with frequency of occurrence and mean concentrations for suspended sediments. Statistically significant correlations were also observed for fall 2001 and spring 2002 (pesticide use vs. detection frequency) and fall 2001 and fall 2002 (pesticide use vs. geometric mean concentration). This is consistent with a source of suspended sediment from surrounding agricultural

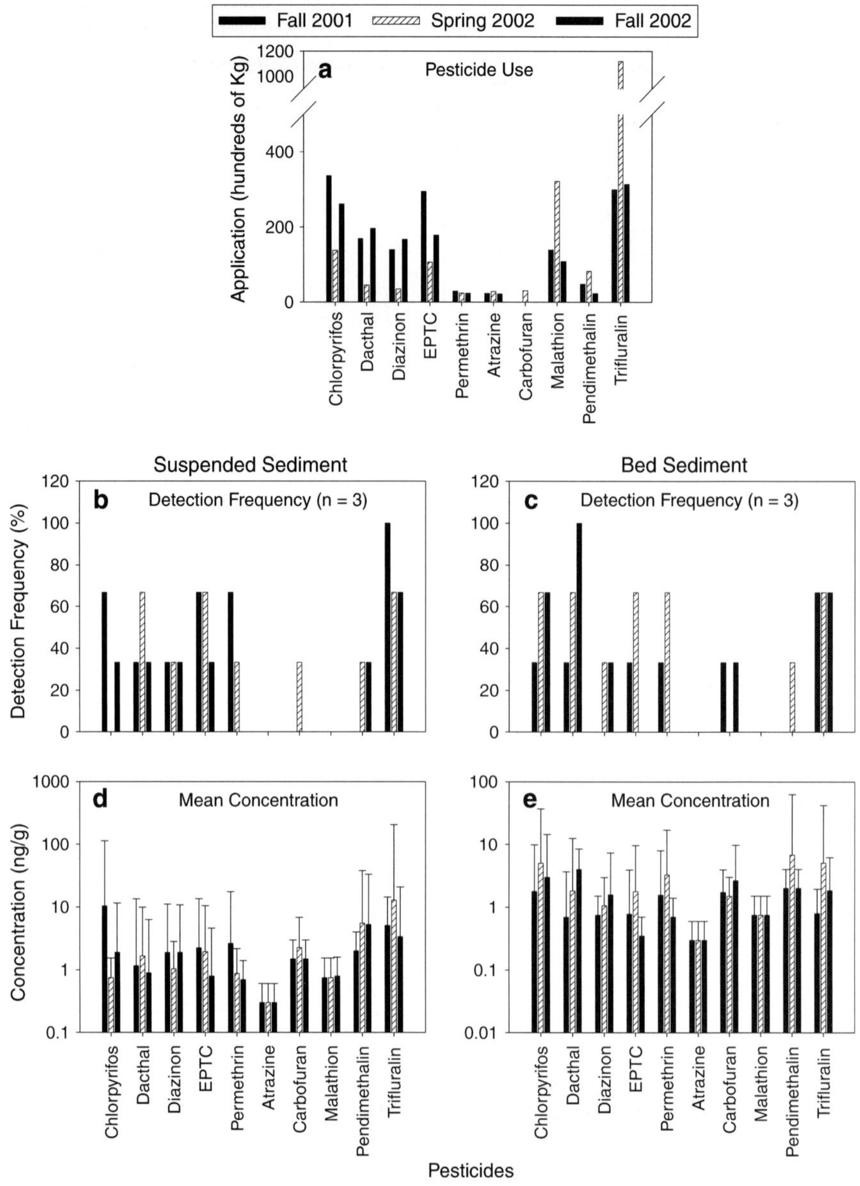

Fig. 6 Comparison of (**a**) the amount of individual pesticide compounds applied within the Salton Sea watershed prior to fall 2001, spring 2002 and fall 2002 sampling events, to the frequency of detection of sediment-associated pesticides at upriver stations for the time period following the pesticide application for (**b**) suspended and (**c**) bed sediments, and to geometric mean concentration of sediment-associated pesticides detected at the upriver sites collected during the fall 2001, spring 2002, and fall 2002 sampling events for (**d**) suspended and (**e**) bed sediments. Each bar represents three samples (*n* = 3) for a total of nine upriver sites (three per sampling event). Error bars are one standard deviation of the mean of the log concentrations, back-transformed and multiplied by the geometric mean

fields. That seasonal trends were not as apparent in bottom sediments is not surprising. The high velocity of current flow resulted in a primarily sandy river bottom at the upriver sites. In order to sample fine-grained sediments (which are known to have higher concentrations of contaminants), it was necessary to seek out quiescent areas within the rivers, such as behind bridge abutments, or in sediment-trap-like structures (such as tractor tires) located in the river. These settling areas may change and be subject to sediment resuspension events, and so may not be useful as a record of seasonal variability.

Evidence for downriver transport of pesticides via suspended solids was provided by the fact that pesticides were detected at upriver sites, which are close to the point of discharge into the Salton Sea, as well as at near-shore sites (Fig. 7). Bottom sediments from near-shore sites in the New and Alamo Rivers for fall 2001 also had pesticide concentrations that were sometimes higher than those at the upriver sites, consistent with the settling of river-derived particles at the near-shore sites (LeBlanc et al., 2004). Pesticides were rarely detected in off-shore suspended or bed sediments, which may be due to dilution of river-

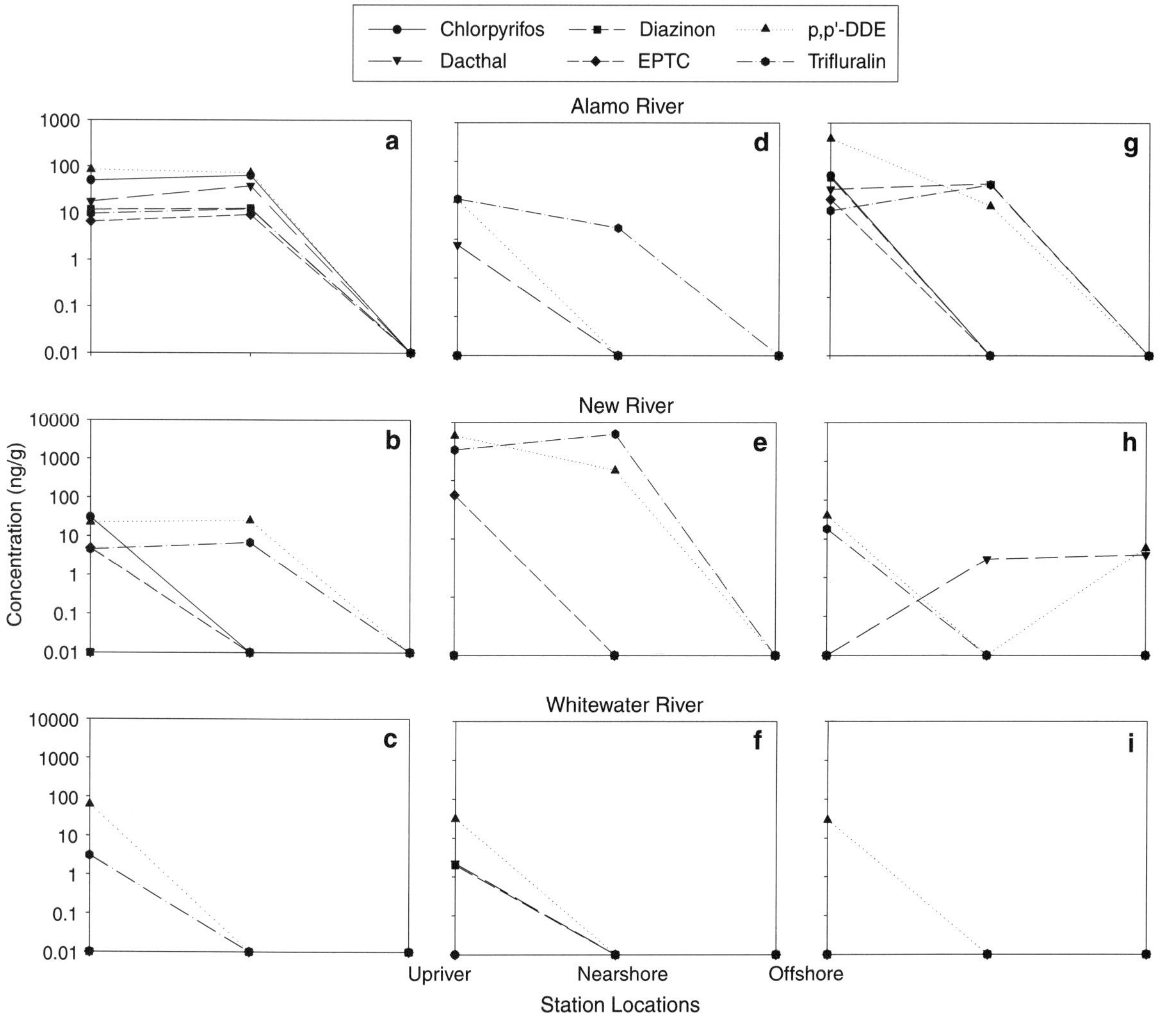

Fig. 7 Concentrations of representative current-use pesticides in suspended sediments along the river transects for the Alamo, New, and Whitewater Rivers during the fall, 2001 (**a–c**), spring 2002 (**d–f**) and fall 2002 (**g–i**) sampling events. For graphing purposes, non-detects values were given a value of 0.01 ng/g

derived suspended sediment by algal material formed within the Salton Sea. However, it is also possible that large amounts of algae in the offshore sediments caused matrix interference in the offshore samples, leading to low recoveries of spiked surrogates in these samples and an understimation of pesticide loads in offshore suspended sediments. A recent study by Vogl and Henry (2002), also did not find evidence for pesticides in lake bed sediments. While most of the sampling sites in the aforementioned study were distributed throughout the lakebed, a few sites were located close to the river mouths, as well as 1.65 km up the Alamo, New and Whitewater Rivers. Why pesticides were not found in these samples is

not known. In our study, we performed extensive cleanup of sediment extracts, as described above, to remove as much interfering matrix as possible. Also, special effort was made at the upriver sites to locate depositional areas, such as behind bridge abutments. These sites contained a higher percentage of fine-grained material, and perhaps higher concentrations of sediment-associated pesticides than the more common sandy sediments found on the river bottoms.

Comparisons of pesticide concentrations in suspended and bottom sediments suggested that suspended sediments come from more than one source, and were not completely derived from resuspension of bed sediments. In several instances,

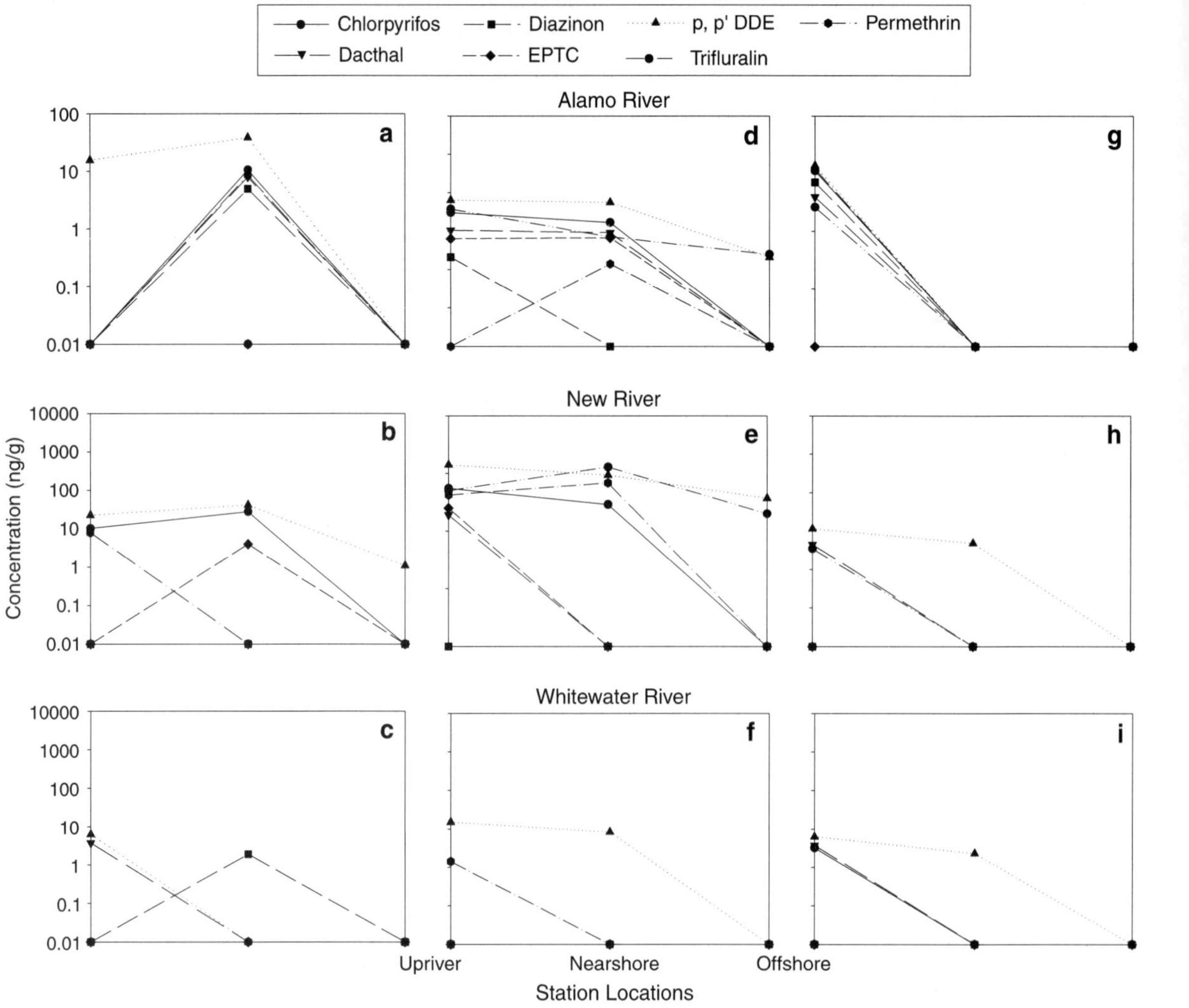

| ● Chlorpyrifos | ■ Diazinon | ▲ p, p' DDE | ● Permethrin |
| ▼ Dacthal | ◆ EPTC | ● Trifluralin | |

Alamo River

Fig. 8 Concentrations of representative current-use pesticides in bed sediments along the river transects for the Alamo, New, and Whitewater Rivers during the fall, 2001 (**a–c**), spring 2002 (**d–f**) and fall 2002 (**g–i**) sampling events. For graphing purposes, non-detects were given a value of 0.01 ng/g

concentrations of pesticides (for example, dacthal, EPTC, and trifluralin in the Alamo River in spring and fall 2002) were higher in the suspended sediments and lower or not detected in the bed sediments (LeBlanc et al., 2004). However, in other instances suspended and bed sediment concentrations were similar, suggesting that pesticides in the suspended solids pool may at times come from resusension of bottom sediments.

The metabolite p,p'-DDE was consistently detected in bed sediments from the upriver and the near-shore sites of all three rivers, with concentrations up to 64 ng/g at the upriver sites. Also detected was p,p'-DDD, although not as frequently.

Concentrations of p,p'-DDE and p,p'-DDD at the upriver sites of all three rivers were similar to those reported in Setmire et al. (1990).

Distribution of pesticides in water, suspended solids, and bed sediments

There was a high degree of overlap in the detection of pesticides among water, suspended solids, and bed sediment samples. Carbaryl, carbofuran, chlorpyrifos, dacthal, diazinon, EPTC, and trifluralin, among the pesticides applied most heavily in the Salton Sea watershed, were often found in all three matrices, especially at the upriver and near-shore sampling sites. Pesticides found only in sediments included lambda-cyhalothrin, metolachlor, and permethrin. Those detected only in water included atrazine, cycloate, malathion, napropamide, and simazine (Table 2).

In order to better understand the distribution of current-use pesticides among these three compartments, pesticide concentrations were compared, taking into account compound physical–chemical properties which are known to affect partitioning between aqueous and particulate phases in the environment. Key physical–chemical properties of detected pesticides are listed in Table 2. If physical chemistry controls the distribution and partitioning of pesticides between sediments and water, one would expect more hydrophobic compounds (i.e., compounds with lower water solubility and higher organic carbon–water partition coefficients[log K_{oc}]) to have a greater tendency to associate with sediments compared to water.

Predictions of dissolved concentrations based on sediment pesticide concentrations and partitioning between sediment and water at equilibrium were made for the fall 2001 samples, where there was the greatest number of pesticide co-occurrences between water and sediment. The following equation

(derived in Schwarzenbach et al., 1993) was used: $C_w = \frac{C_s}{K_d} = \frac{f_{oc} \cdot C_{oc}}{K_d}$.

An example calculation and a definitition of the terms are provided in Table 6.

Predicted aqueous values determined using suspended-sediment concentrations were at times above the measured aqueous values, illustrating that suspended-sediment concentrations were not always at equilibrium with the surrounding water (Table 7). Trends were not consistent however, and did not follow expectations based upon compound hydrophobicity. Carbaryl (relatively hydrophilic) and chlorpyrifos (relatively hydrophobic) had aqueous concentrations that were below the predicted aqueous concentrations. Aqueous concentrations of EPTC, (higher water solubility and a lower log K_{oc} than carbaryl) were at times lower than (predicted) equilibrium concentrations (e.g., Alamo River upriver site), yet at other times were close to predicted values (e.g., Alamo River near-shore site). Aqueous concentrations of trifluralin (hydrophobic, similar to chlorpyrifos in aqueous solubility and log K_{oc} values) were below the predictions in the Alamo River upriver and near-shore sites, yet were similar (within a factor of 2) at the New River upriver and near-shore sites, and the Whitewater River upriver site (Table 7). Bed sediments also did not show consistent trends, although predicted

Table 6 Example calculation of an aqueous concentration of chlorpyrifos based on equilibrium partitioning and measured values of suspended sediment organic carbon and chlorpyrifos concentration from the Alamo River upriver site in the Salton Sea Basin, California, fall 2001

Example calculation: $C_W = (f_{OC} \cdot C_{OC})/K_d$

Alamo River suspended sediment, fall 2001

Compound	Log K_{OC}[a]	K_{OC}[b]	f_{OC}[c] (g OC/g sed)	K_d[d]	C_s[e] (ng/g dry weight)	C_{OC}[f] (ng/g OC)	C_{OC}[g] (mg/kg OC)	C_W[h] (mg/l)	C_W[i] (ng/l)
CHLORPYRIFOS	3.78	6025.6	0.0093	56.0	50	5376	5.38	0.0009	892

Definition of terms

[a] Organic carbon:water partition coefficient (Mackay et al., 1997)

[b] $K_{OC} = 10^{(LogKOC)}$

[c] f_{OC}, fraction of sediment organic carbon (measured by CHN analysis)

[d] K_d, sediment:water partition coefficient for chlorpyrifos $= f_{OC} \times K_{OC}$

[e] C_s, sediment concentration of compound (chlorpyrifos) in ng/g sediment dry weight

[f] C_{OC}, sediment concentration of compound (chlorpyrifos) in ng/g sediment organic carbon; ng/g sediment OC $=$ (ng/g sediment dry weight)/(f_{OC})

[g] Units conversion: mg/kg $=$ (ng/g)/1,000

[h] C_w $=$ Predicted aqueous concentration (mg/l)

[i] Units conversion: ng/l $=$ (mg/l)*1,000,000

Table 7 Predicted and measured aqueous concentrations derived from suspended sediments and bed sediments from the upriver and near-shore sites of the Alamo, New and Whitewater rivers in the Salton Sea Basin, California, fall 2001

Pesticides	Suspended sediment		Bed sediment	
	Predicted (ng/l)	Measured[a] (ng/l)	Predicted (ng/l)	Measured (ng/l)
Alamo River upriver				
Carbaryl	nd	nd	452	56.4
Chlorpyrifos	892	161	nd	nd
EPTC (eptam)	2,958	418	nd	nd
Trifluralin	245	37.4	nd	nd
Alamo River near-shore				
Chlorpyrifos	884	87.7	168	87.7
Dacthal	528	328	126	328
Diazinon	1,810	936	819	936
EPTC (eptam)	3,196	3,830	922	3,830
Trifluralin	238	37.6	16	37.6
New River upriver				
Chlorpyrifos	290	44.3	458	44.3
EPTC (eptam)	1,174	3,490	4,394	3,490
Trifluralin	62.0	35.0	76.0	35.0
New River near-shore				
Carbaryl	nd	nd	1,850	<4
Chlorpyrifos	nd	nd	489	15.0
EPTC (eptam)	nd	nd	1,774	1,300
Trifluralin	58.8	27.3	47.4	27.3
Whitewater River upriver				
Dacthal	nd	nd	133	38.1
Trifluralin	17.6	22.5	31	22.5
Whitewater River near-shore				
Diazinon	nd	nd	200	8.5
Trifluralin	nd	nd	7.1	<3

[a] Measured aqueous concentrations are from the same samples in both the suspended sediment and bed sediment columns

aqueous concentrations of EPTC and trifluralin (which differ markedly in physical–chemical properties) were often within a factor of two or less of equilibrium partitioning predictions.

When measured aqueous concentrations are lower than predictions based upon equilibrium partitioning, one interpretation is that compounds sorbed onto the solid phase have not desorbed sufficiently to reach equilibrium with surrounding waters. There are a

number of factors in the environment of the Salton Sea that could contribute to this. High salinity is known to decrease the aqueous solubility of organic compounds, and increase the partitioning onto solids (Means, 1995; Xe et al., 1997; Turner and Rawling, 2001). Given the extremely high salinity of Salton Sea water and the high conductivity of the river water, which is comprised primarily of irrigation return water, one may expect to see the "salting out" phenomenon observed in the afore-mentioned studies. However, if salinity were the sole controlling factor, one may expect the amount of underprediction to be much higher at the nearshore sites, within the highly saline Salton Sea, compared to the riverine (upriver) sites. Also, measured aqueous values were not always below predictions based upon equilibrium, as mentioned above.

Loss of dissolved concentration through evaporation is another mechanism by which water concentrations could be lower than equilibrium predictions. Despite its high solubility, EPTC also has a relatively high Henry's Law Constant (Table 2), a partition coefficient that describes the distribution of an organic compound between water and air at equilibrium at a specified temperature (usually 20°C). Having a higher Henry's law constant indicates a greater tendency to partition into the atmosphere relative to the aqueous phase. Water temperatures during the fall sampling were higher than 20°C (between 28 and 29°C) for the New and Alamo Rivers, which also would increase the tendency to partition into the atmosphere. Chlorpyrifos and trifluralin have Henry's Law constants of the same order of magnitude as EPTC. By contrast, carbaryl, which also whose aqueous concentrations were also observed to be below equilibrium predictions, has a Henry's Law Constant which is 1,000-times lower than the other pesticides mentioned. Therefore, compound volatility does not completely explain these results.

Another factor affecting eqilibrium distributions between sediment-sorbed pesticides and surrounding waters is whether the sediment has been in contact with water long enough for sufficient desorption to occur. Bergamaschi et al. (2001) noted a disequilibrium between suspended sediment and water pesticide concentrations in the Sacramento and San Joaquin Rivers in northern California. They postulated that elevated concentrations of pesticides associated with suspended sediments were due to

fresh soil entering the rivers from surrounding agricultural fields during storm runoff, with a relatively short residence time in the river. Domagalski and Kuivila (1993) also noted this phenomenon for diazinon associated with suspended sediments in San Francisco Bay. It is highly and likely that soils are washed into the Salton Sea watershed on a more continous basis compared to the storm-driven runoff events cited above for northern California rivers. Suspended sediments (and bed sediments) collected in this study may have been in contact with surrounding waters for longer time periods than in the aforementioned studies, where attempts were made to capture suspended solids washed into rivers immediately following rain events.

Although the distribution of pesticides between water and sediments was not quantitatively predictable based on physical–chemical properties, concentration patterns appeared to be due to a combination of pesticide physical–chemical properties and pesticide use patterns. Trifluralin, (relatively hydrophobic) was detected in water and sediment samples from all three sampling periods. Use of trifluralin was highest in the watershed in spring 2002, which matched concentration patterns in the Alamo and New Rivers. Chlorpyrifos (also relatively hydrophobic), with higher use during the fall application periods, was detected in water and sediment samples most frequently during the fall. Dacthal, with a hydrophobic nature similar to chlorpyrifos, was detected consistently and at the highest concentrations during both fall sampling periods (the time of highest use) in the Alamo River, despite much lower quantities applied relative to chlorpyrifos (LeBlanc et al., 2004). EPTC, despite being very water soluble, had high use in the watershed and was detected frequently in sediment as well as water samples, although sediment concentrations were low. Atrazine, with a lower aqueous solubility but a similar log K_{oc} to EPTC and lower use, was detected only in water. Permethrin, which is extremely hydrophobic and used in lower quantities than the other pesticides mentioned, was detected in sediments but not in water during fall 2001 and spring 2002.

Thus, it would appear that the amount of pesticide applied is equally as important as compound physicochemical properties in determining whether a pesticide will be detected in sediment samples. The distribution of pesticides among water, suspended sediments, and bed sediments appears to be a complicated function of pesticide use, physical–chemical properties, and perhaps the residence time of sediments in the watershed.

Pesticide concentrations and human health

Maximum concentrations of dissolved pesticides were as high as 3.8 parts-per-billion (ppb) for EPTC, and were found in the hundreds of parts-per trillion (ppt) range in several samples. Maximum concentrations in sediments were at approximately 100 ppb. It was shown by De Vlaming et al. (2000) that dissolved pesticide concentrations are sufficient in concentration to induce toxicity in aquatic test organisms. Studies relating pesticide exposure to human health have been compiled and evaluated by health scientists at the Environmental Protection Agency Office of Research and Development and the National Center for Environmental assessment and listed on the Integrated Risk Information website (EPA, 2007). These studies relate deleterious effects seen from oral doses of pesticides given to test animals (primarily mice and rats) over time periods of several days–weeks. Deleterious effects, such as lowered cholinesterase activity in blood plasma, begin to manifest themselves at concentrations known as the low effects level, which range from 0.1 mg/kg for chlorpyrifos, (equal to 100 parts-per-billion), 0.9 to 1.5 mg/kg for DDE (equal to 900–1,500 parts-per-billion), 10 mg/kg for dacthal, 150 parts-per million for trifluralin, and 214 parts-per-million for diazinon. Sediment concentrations of chlorpyrifos in suspended sediments were found to be as high as 64 parts-per-billion, which is more than half the low-effects level resulting from chronic oral exposure. Chlorpyrifos was detected in bed sediments at 30 ppb in the Alamo upriver site and 10–16 ppb at the Alamo River nearshore sites, which is 3–10-times lower than the low effect concentration. Maximum sediment concentrations of p, p′-DDE were 101 ppb for suspended sediments and 63.8 ppb for bed sediments, which is approximately 10–20 times lower than the low effects level. Maximum concentrations of dacthal (38 ng/g) and trifluralin (106 ng/g) on suspended sediments were 100 times or more lower than the low effects level. Diazinon and pendimethalin were found to be at levels which are

1,000-times lower than the low effects levels. While these concentrations are not directly comparable to effects seen from chronic oral intake (effects due to exposure from inhalation would more comparable), they demonstrate that sediment concentrations of a few pesticides in the Salton Sea are not insignificant in terms of potential exposure and human health, and warrant futher characterization, especially in near-shore areas adjacent to river inputs.

Summary and conclusions

Current-use pesticides were detected in water, suspended sediments, and bed sediments collected from the Salton Sea Basin, California, during the fall 2001, spring 2002, and fall 2002. Dissolved concentrations were similar to concentrations reported in previous studies, which were shown to cause toxicity to aquatic organisms. Differences in pesticide concentrations between spring and fall most likely reflected pesticide use during the application periods. Aqueous pesticide concentrations along the transects (river upriver, near-shore delta, and off-shore sites) had maxima at either upriver sites, near-shore or off-shore sites, consistent with input pulses from surrounding agricultural fields, which travel downstream. The number of pesticides detected and their concentrations generally were higher in samples from the New and Alamo Rivers than those in the samples from the Whitewater River. Overall, aqueous concentrations were higher in fall 2001 than in spring 2002 and fall 2002.

Differences in sediment pesticide concentrations were seen among rivers. More detections and higher concentrations were seen in suspended and bed sediment samples from the New and Alamo Rivers compared to the Whitewater River. An exception was p, p′-DDE, which was found in similar concentrations in bed sediment samples from all three rivers. Seasonal trends for pesticides associated with suspended sediments were not as pronounced as those for the dissolved concentrations, although they were similar for a few pesticides, notably chlorpyrifos, which had higher sediment concentrations in fall 2001, and trifluralin, which had higher concentrations in spring 2002. In the Alamo River, suspended sediment-associated pesticide concentrations were nearly as high or higher at the near-shore site compared with concentrations at the upriver site

consistent with downstream transport of sediment-associated pesticides. There was only sporadic detection of pesticides in suspended and bed sediments from the off-shore sites, which also had the highest degree of matrix interference. Concentrations detected on sediments in upriver and nearshore sites were well below levels at which deleterious effects manifest themselves in mice and rat feeding studies, although one pesticides (chlorpyrifos) was found at a level that was only half of the low effects level from chronic oral intake studies. Further characterization of nearshore lake sediments for current-use and legacy pesticides, especially near the major riverine inputs is needed to better assess human health risks due to exposed lake bed sediments.

As with water, the pesticides detected in sediments were the pesticides most heavily used in the watershed; thus, the differences in concentration between seasons were at least partly related to pesticide-use patterns. Chlorpyrifos, dacthal, and EPTC, used heavily during fall 2001, were detected in suspended sediments. Trifluralin, used heavily during spring 2002, was detected at high concentrations at this time. Concentrations of pesticides in water were sometimes below predictions based upon equilibrium partitioning with suspended and bed sediments, and at other times were close to equilibrium. Distribution of current-use pesticides between sediments and water is probably controlled by a number of factors, including use and the residence time of the sediments in the watershed.

Acknowledgments The authors thank James Orlando, Roy Schroeder and G. Edward Moon for help with field sampling and boat operation. Thanks to Dr. Roy Schroeder for help in directing field operations, enlightening discussions and generous sharing of his knowledge of the Salton Sea. Thanks to Jim Orlando for retrieval and summary of pesticide use data, and creation of the Salton Sea area map. Thanks to Robert Gale and Bill Foreman for helpful discussions concerning the sediment multi-residue method. Finally, thanks to Charles Pelizza and the staff of the Sonny Bono National Wildlife Refuge for the generous use of their facilities and equipment, including the Boston Whaler used for sampling.

References

Bergamaschi, B. A., K. M. Kuivila & M. S. Fram, 2001. Pesticides associated with suspended sediments entering San Francisco Bay following the first major storm of water year 1996. Estuaries 24: 368–380.

Blanco, E. V., P. L. Mahia, M. S. Lorenzo, D. P. Rodriguez & E. F. Fernandez, 2000. Optimization of microwave-assisted extraction of hydrocarbons in marine sediments: comparison with the Soxhlet extraction method. Analytical and Bioanalytical Chemistry 366: 283–288.

California Department of Pesticide Regulation, 2002. Pesticide use data for 2001–2002 [digital data]. California Department of Pesticide Regulation, Sacramento, California.

Camel, V., 2000. Microwave-assisted solvent extraction of environmental samples. Trends in Analytical Chemistry 19: 229–248.

Crepeau, K. L., L. M. Baker & K. M. Kuivila, 2000. Method of analysis and quality-assurance practices for determination of pesticides in water by solid-phase extraction and capillary-column gas chromatography/mass spectrometry at the US Geological Survey California District organic chemistry laboratory, 1996–1999. US Geological Survey Open-File Report 2000-229, 19 pp.

Crepeau, K. L., K. M. Kuivila & B. Bergamaschi, 2002. Dissolved pesticides in the Alamo River and the Salton Sea, California, 1996–97. US Geological Survey Open-File Report 2002-232, 7 pp.

De Vlaming, V., V. Connor, C. DiGiorgio, H. C. Bailey, L. A. Deanovic & D. E. Hinton, 2000. Application of whole effluent toxicity test procedures to ambient water quality assessment. Environmental Toxicology and Chemistry 19: 42–62.

De Vlaming, V., C. DiGiorgio, S. Fong, L. A. Deanovic, M. D. Carpio-Obeso, J. L. Miller, M. J. Miller & N. J. Richard, 2004. Irrigation runoff insecticide pollution of rivers in the Imperial Valley, California (USA). Environmental Pollution 132: 213–229.

Domagalski, J. L. & K. M. Kuivila, 1993. Distributions of pesticides and organic contaminants between water and suspended sediment, San Francisco Bay, California. Estuaries 16: 416–426.

Eccles, L. A., 1979. Pesticide residues in agricultural drains, southeastern desert area, California. US Geological Survey Water-Resources Investigations Report 79-16, 60 pp.

Environmental Protection Agency, 2007. Accessed from http://www.epa.gov/IRIS/index.htmlT on 12/19/06.

Gill, T. E., D. A. Gillette, T. Niemeyer & R. T. Winn, 2002. Elemental geochemistry of wind-erodible playa sediments, Owens Lake, California. Nuclear Instruments and Methods in Physics Research Section B–Beam Interactions With Materials and Atoms 189: 209–213.

Gillette, D., D. Ono & K. Richmond, K., 2004. A combined modeling and measurement technique for estimating windblown dust emissions at Owens (dry) Lake, California. Journal of Geophysical Research-Earth Surface 109 (F1): Art. No. F01003.

Horowitz, A. J., K. A. Elrick & R. C. Hooper, 1989. A comparison of instrumental dewatering methods for the separation and concentration of suspended sediment for subsequent trace element analysis. Hydrological Processes 2: 163–184.

Imperial Irrigation District, 2006. Accessed from www.iid.com/media/cropAc-2006_01_P10.pdf on October 6, 2006.

Irwin, G. A., 1971. Water-quality data for selected sites tributary to the Salton Sea, California, August 1969–June 1970. US Geological Survey Open-File Report, 12 pp.

Kunii, O., M. Hashizume, M. Chiba, S. Sasaki, T. Shimoda, W. Caypil & D. Dauletbaev, 2003. Respiratory symptoms and pulmonary function among school-age children in the Aral Sea region. Archives of Environmental Heath 58: 676–682.

LeBlanc, L. A., R. A. Schroeder, J. L. Orlando & K. M. Kuivila, 2004. Occurrence, distribution and transport of pesticides, trace elements and selected inorganic constituents into the Salton Sea Basin, California, 2001–2002. US Geological Survey Scientific Investigations Report 2004-5117, 40 pp.

LeBlanc, L. A. & K. M. Kuivila (in preparation) A multiresidue method for the detection of current-use pesticides in suspended and bottom sediments.

Mackay, D., W. Y. Shiu & K. C. Ma, 1997. Illustrated handbook of physical-chemical properties and environmental fate for organic chemicals: Volume V—Pesticide Chemicals. Lewis Publishers, Boca Raton, Fl.: 812.

Means, J. C., 1995. Influence of salinity upon sediment-water partitioning of aromatic hydrocarbons. Marine Chemistry 51(1): 3–16

Pastor, A., E. Vasquez, R. Ciscar & M. de la Guardia, 1979. Efficiency of the microwave-assisted extraction of hydrocarbons and pesticides from sediments. Analytica Chimica Acta 344: 241–249.

Riverside County Flood Control and Water Conservation District, 2006. Accessed from http://www.floodcontrol.co.riverside.ca.us/districtsite/content/glossary.htm#T on 10/19/06.

Rya, J. H., S. D. Gao, R. A. Dahlgran & R. A. Zierenberg, 2002. Arsenic distribution, speciation and solubility in shallow groundwater of Owens Dry Lake, California. Geochimica et Cosmochimica Acta 66: 2981–2994.

Salton Sea Authority, 2006. Salton Sea Authority plan for multipurpose project. accessed from http://www.saltonsea.ca.gov/SSA-Plan-Board-Review-Draft_7-20-06.pdf on January 19, 2007, 68 pp.

Sapoznikhova, Y., O. Bawardi & D. Schlenk, 2004. Pesticides and PCBs in sediments and fish from the Salton Sea, California, USA. Chemosphere 55: 797–809.

Schroeder, R. A., A. M. Rivera, B. J. Redfield, J. N. Densmore, R. L. Michel, D. R. Norton, D. J. Audet, J. G. Setmire & S. L. Goodbred, 1993. Physical, chemical and biological data for detailed study of irrigation drainage in the Salton Sea area, California, 1988–1990. US Geological Survey Open-File Report 93–83, 179 pp.

Schroeder, R. A., W. H. Orem & Y. K. Kharaka, 2002. Chemical evolution of the Salton Sea, California: nutrient and selenium dynamics. Hydrobiologia 473: 23–45.

Schwarzenbach, R. P., P. M. Gschwend & D. M. Imboden, 1993. Environmental Organic Chemistry. John Wiley and Sons, Inc., New York, 681.

Setmire, J. G., 1979. Water-quality conditions in the New River, Imperial County, California. US Geological Survey Water Resources Investigation Report 79–86, 63 pp.

Setmire, J. G., 1984. Water quality in the New River from Calexico to the Salton Sea Imperial County, California. US Geological Survey Water-Supply Paper 2212: 42 pp.

Setmire, J. G. & R. A. Schroeder, 1998. Selenium and salinity concerns in the Salton Sea area of California in Frankenberger. In Jr, W. T., Engberg R. A. (eds.), Environmental Chemistry of Selenium: Marcel Dekkar Inc., New York, p. 205–221.

Setmire, J. G., R. A. Schroeder, J. N. Densmore, S. L. Good-bred, D. J. Audet & W. R. Radke, 1993. Detailed study of water quality, bottom sediment, and biota associated with irrigation drainage in Salton Sea area, California, 1988–90. US Geological Survey Water-Resources Investigations Report 93-4014, 102 pp.

Setmire, J. G., J. C. Wolfe & R. K. Stroud, 1990. Reconnaissance investigation of water quality, bottom sediment, and biota associated with irrigation drainage in the Salton Sea area, California, 1986–87. US Geological Survey Water-Resources Investigations Report 89-4102, 68 pp.

Spencer, W. F., M. M. Cliath, J. W. Blair & R. A. LeMert, 1985. Transport of pesticides from irrigated fields in surface runoff and tile drain waters. US Department of Agriculture Conservation Research Report 31, 71 pp.

The Salton Sea Restoration Consortium, 2004. Salton Sea Restoration–The Cascade Alternative, accessed from www.saltonsea.water.ca.gov/documents on January 19, 2007, 55 pp.

Turner, A. & M. C. Rawling, 2001. The influence of salting out on the sorption of neutral organic compounds in estuaries. Water Research 35: 4379–4389.

Tyler, W. W., S. Kranz, M. B. Parlange, J. Albertson, G. G. Katul, G. F. Cochran, B. A. Lyles & G. Holder, 1997. Estimation of groundwater evaporation and salt flux from Owens lake, California, USA. Journal of Hydrology 200: 110–135.

U S Geological Survey, 2001. National Water Information System (NWIS Web), US Geological Survey, accessed March 15, 2003 at http://wateredata.usgs.gov/nwis.

Vogl, R. A. & R. H. Henry, 2002. Characteristics and contaminants of the Salton Sea sediments. Hydrobiologia 473: 47–54.

Wiggs, G. F. S., S. L. O'Hara, J. Wegerdt, J. Van der Meer, I. Small & R. Hubbard, 2003. The dynamics and characteristics of aeolian dust in dryland Central Asia: possible impacts on human exposure and respiratory health in the Aral Sea basin. Geographical Journal 169, 142–157 Part 2.

Xie, W.-H., W.-Y. Shiu & D. Mackay, 1997. A review of the effect of salts on the solubility of organic compounds in seawater. Marine Environmental Research 44: 429–44.

Hydrobiologia (2008) 604:173–179
DOI 10.1007/s10750-008-9318-z

Evaluation of potential impacts of perchlorate in the Colorado River on the Salton Sea, California

G. Chris Holdren · Kevin Kelly · Paul Weghorst

© U.S. Bureau of Reclamation (Department of Interior) 2008

Abstract Ammonium perchlorate, a component of rocket fuel, entered Lake Mead through drainage and shallow groundwater in the Las Vegas Valley, Nevada, and is now found in the lower Colorado River from Lake Mead to the international boundary with Mexico. Perchlorate is a threat to human health through reduction of thyroid hormone production. Perchlorate has been found in water throughout the lower Colorado system and in crops in the California's Imperial Valley, as well as in several other states, but it has not previously been included in investigations of the Salton Sea. Because perchlorate behaves conservatively in the Colorado River, it was postulated that it could be accumulating at high levels along with other salts in the Salton Sea. Results show that perchlorate is not accumulating in the Sea, although it is present in tributaries to the Sea at levels similar to those found in the Colorado River. Bacterial reduction of perchlorate is the most likely explanation for the observed results.

Keywords Saline lakes · Lake Mead · Water supply · Bacterial reduction · Alamo River · New River · Whitewater River

Guest editor: S. H. Hurlbert
The Salton Sea Centennial Symposium. Proceedings of a Symposium Celebrating a Century of Symbiosis Among Agriculture, Wildlife and People, 1905–2005, held in San Diego, California, USA, March 2005

G. C. Holdren (✉) · K. Kelly · P. Weghorst
Bureau of Reclamation, P.O. Box 25007 (86-68220),
Denver, CO 80225, USA
e-mail: choldren@do.usbr.gov

Introduction

Perchlorate was found by the Metropolitan Water District in southern California water supplies in July 1997, when advances in analytical techniques lowered the detection limit for perchlorate from approximately 400 mg m^{-3} to 4 mg m^{-3}. In addition to its presence in local groundwater supplies, perchlorate was found in water from the Colorado River, where the source was traced upstream to Lake Mead. Further investigations indicated that perchlorate was entering Lake Mead through Las Vegas Wash, a tributary to Lake Mead, and through shallow groundwater along the Wash. Las Vegas Wash drains the entire Las Vegas Valley, including the Henderson, Nevada, area, where perchlorate was manufactured. Figure 1 shows the relative locations of Lake Mead, the Colorado River, and the Salton Sea. Perchlorate is now found throughout the lower Colorado River as a result of these past releases.

Ammonium perchlorate was manufactured at two plants in Henderson, Nevada, which produced the

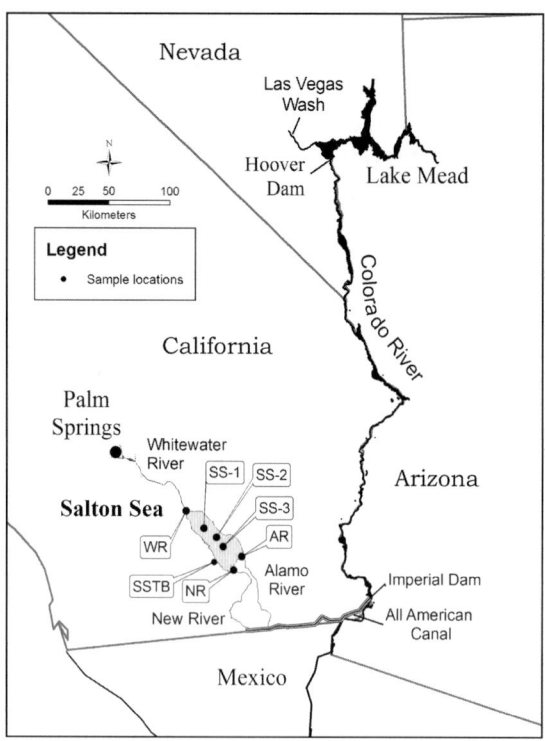

Fig. 1 Relative locations of Lake Mead, the Colorado River, and the Salton Sea, and locations of sampling stations

entire supply of ammonium perchlorate for the U.S. Department of Defense and the National Aeronautics and Space Administration from approximately 1951 through 1998 (Sellers et al., 2007). The two facilities also manufactured sodium, potassium, and magnesium perchlorate during that period. Production ceased at one plant in 1988 following an explosion, but the second plant continued production until 1998. Perchlorate was released to the environment through disposal of wastes into unlined ponds and through other leaks from the facilities and is now found in the lower Colorado River from Lake Mead to the international boundary with Mexico.

Perchlorate is an oxyanion with very low chemical reactivity in aquatic solutions. It is extremely soluble in water and is not expected to precipitate from solution or be removed by adsorption reactions. As a result, treatment options are limited (Urbansky, 1998; Sellers et al., 2007). By volume, more perchlorate is used as a component of solid rocket fuel than all its other uses combined (Susarla et al., 1999).

The natural occurrence of perchlorate appears to be limited mainly to soils or potash deposits in arid climates derived from ancient marine seabeds now located in arid climates, although some perchlorate may also be produced during lightning storms. Only a small number of contaminated sites, associated with industrial perchlorate production (Henderson, Nevada), rocket testing (San Gabriel, California), or military munitions (Karnack, Texas; Indian Head, Maryland; Cape Cod, Massachusetts), have been identified (Sellers et al., 2007).

Perchlorate is a threat to human health. It reduces thyroid hormone production by competitively inhibiting the uptake of iodide, which can impair the development of the gland (Clark, 2000; Greer et al., 2002; Clewell et al., 2003; Dohan et al., 2003; Kirk, 2006). Susceptible sub-population groups include pregnant women, children, and individuals with compromised thyroid function.

There were no regulatory criteria for perchlorate concentrations in drinking water at the time it was found in the Colorado River, although the U.S. Environmental Protection Agency (EPA) had established a provisional reference dose for perchlorate of 4 mg m^{-3} (parts per billion) in 1992. After its discovery in Colorado River water, perchlorate was added to the U.S. EPA Contaminant Candidate List (CCL) and the Unregulated Contaminant Monitoring Rule (UCMR) in 1998 (Federal Register, 1998; U.S. EPA, 1998). Following the recent release of a National Research Council recommendation, the U.S. EPA set its safe daily dose for perchlorate at 0.7 µg kg^{-1} day^{-1} in February 2005 (Hogue, 2005; Renner, 2005). This would correspond to a drinking water standard of 24.5 mg m^{-3} for a 70-kg male adult drinking 2 l of water per day, assuming 100% of the total daily dosage of perchlorate is received through drinking water. While a final EPA regulatory criterion for perchlorate may require several more years to develop, the State of California adopted a Public Health Goal of 6 mg m^{-3} for perchlorate in 2004 (Tikkanen, 2006).

At the time of its discovery in the lower Colorado River in 1997, perchlorate concentrations were above 4 mg m^{-3} in all reaches of the river. This water is used for drinking water, irrigation, and recreation by millions of people in Nevada, California, and Arizona (Potius, 2000), as well as Mexico. In the Imperial Valley and surrounding areas, Colorado River water is used for irrigating over 70% of the U.S. winter lettuce crop from October to March (Logan, 1998). Subsequent investigations by the U.S. Food and Drug Administration detected perchlorate in lettuce grown

in California and four other states (Sellers et al., 2007).

Although perchlorate was found in water and crops in the Imperial Valley, it has not previously been investigated at the Salton Sea. Because perchlorate behaves conservatively in the Colorado River and because the Salton Sea serves as a sink for wastewaters derived from Colorado River water transported to agriculture and municipalities in the Coachella, Imperial, and Mexicali valleys, it was postulated that perchlorate could be accumulating at high levels along with other salts in the Salton Sea. This assessment was conducted to evaluate that possibility because of the potential importance of high levels of perchlorate on management alternatives for the Sea and for operation and decommissioning of experimental evaporation ponds after their use for research on salinity control methodologies for the Salton Sea.

Methods

Sample sites and collection

Water samples were collected from the Salton Sea, from the three major rivers discharging to the Sea, and from evaporation ponds at the Salton Sea Test Base from June 24–26, 2003. Locations of the sampling stations are also included in Fig. 1.

The Sea and riverine samples were collected from the same locations used by Holdren & Montaño (2002) during their 1999 limnological investigation of the Sea. The Sea locations are North Basin (SS-1), Mid-Lake (SS-2), and South Basin (SS-3). Samples were collected at the surface and near the bottom at each of the Sea locations.

The Alamo River (AR) site is located at the Garst Road bridge just downstream of USGS Station Number 10254730. Water in the Alamo River includes both surface runoff and agricultural drainage and is composed almost entirely of water derived from the Colorado River. The average discharge for the Alamo River was 23.6 $m^{-3} s^{-1}$ in 2003, which is about 2% higher than the average for the period of record (Agajanian et al., 2005).

The New River (NR) site is located west of Lack Road at USGS Station Number 10255550. The New River also contains runoff and drainage derived from Colorado River water and wastewater discharges

from treatment facilities in both the United States and Mexico, as well as some groundwater from Mexico. Average discharge in the New River during 2003 was 16.0 $m^{-3} s^{-1}$, about 9% above the long-term average at that site (Agajanian et al., op. cit.).

The Whitewater River contains surface runoff from the surrounding partially alpine watershed, groundwater from the Coachella Valley, and some agricultural runoff derived from Colorado River water brought into the valley by the Coachella Canal. The Whitewater River (WR) sampling site is located at the Lincoln Street crossing west of USGS Station 10259540 (Fig. 1). Flows in the Whitewater River are much lower than the other two rivers, averaging 1.75 $m^{-3} s^{-1}$ in 2003. Long-term averages for this site are not meaningful because flows greater than 5.66 $m^{-3} s^{-1}$ have not been computed since 1992 (Agajanian et al., op. cit.).

Brine samples were collected from a disposal pond at the Salton Sea Test Base (SSTB), located southwest of the Salton Sea. This facility was used by the Bureau of Reclamation and the Salton Sea Authority to conduct research on salinity control measures and salt disposal processes that could be applicable to Salton Sea restoration alternatives. The facility contained 11 cells with a total surface area of 7.3 ha. Four samples were collected from different areas in a 0.8 ha waste disposal pond containing the concentrated waste from a series of seven ponds that progressively evaporated Salton Sea water. A description of the facility and results of experiments conducted at the Salton Sea Test Base were summarized by Weghorst (2004). An additional sample from this facility was collected from an experiment which continuously evaporated Colorado River water, transported to the Salton Sea Test Base via the Coachella Canal, in a Class A stainless steel evaporation pan.

Grab samples were collected from the Sea surface, the river stations, and the waste disposal pond at the Salton Sea test base. A Van Dorn sampler was used to collect bottom samples from the Sea stations. Samples were collected in both 60-ml and 250-ml plastic bottles at each site and shipped to the participating laboratories.

Sample analysis

Replicate samples from the Salton Sea and river samples were sent to four laboratories that routinely perform perchlorate analyses, although none of the laboratories had specific experience with the analysis

of perchlorate in brines. Recent studies (Stetson et al., 2006) indicate that perchlorate can degrade with time in unfiltered surface water samples, but perchlorate concentrations in filtered samples and groundwater samples were stable for well over 300 days. All perchlorate analyses for this study were conducted within two weeks of collection to ensure that sample degradation would not be a problem.

Three of the participating laboratories, the Denver Environmental Chemistry Laboratory (DECL) of the U.S. Bureau of Reclamation; the Southern Nevada Water System Laboratory in Boulder City, Nevada (SNWS); and MWH Laboratories in Monrovia, California (MWH), used ion chromatography—conductivity detection (IC) based on EPA Method 314.0 for perchlorate analyses (U.S. EPA, 2000). This IC method is capable of detecting perchlorate with a detection limit of about 0.5 mg m^{-3}, although due to the non-selective nature of detection, perchlorate cannot be unambiguously identified by this method.

The IC method is also vulnerable to sensitivity loss caused by high sample ionic strength. With brine samples, laboratories can use either dilution or ion exchange cleanup to alleviate the problems caused by high ionic strength.

The fourth laboratory, Exygen Research in State College, Pennsylvania (Exygen), used an in-house liquid chromatography-mass spectroscopy-mass spectroscopy (LC/MS/MS) method. The LC/MS/MS method can detect and quantify perchlorate concentrations by monitoring atomic mass transitions ($99 \rightarrow 83$ and $101 \rightarrow 85$ atomic mass units) that are unique to the perchlorate ion. As a result, this method not only avoids most of the interferences expected from brine samples, but also provides confirmation of the presence of perchlorate.

Analyses of total dissolved solids (TDS) were also provided for Salton Sea and river samples by DECL. EPA Method 160.1, evaporation at 180°C (U.S. EPA, 1979), was used for the gravimetric analyses of TDS concentrations. Relationships between specific gravity and dissolved solids developed by Weghorst (2004) were used to estimate TDS concentrations in brine samples.

Results

Perchlorate results are presented in Table 1. Table 1 also includes TDS concentrations for study samples.

Samples with perchlorate concentrations lower than the laboratory detection limits are listed as less than (<) the detection limit for that sample. Detection limits reported in Table 1 varied widely among the participating laboratories. For high-salinity samples, including the Salton Sea and brine samples in this study, the detection limit for EPA Method 314.0 is largely determined by the effort put into removing potential contaminants. DECL used cleanup procedures to specifically remove sulfate, chloride, and carbonate to achieve lower detection limits. SNWS and MWH used their typical analytical procedures for drinking water, in combination with sample dilution, to reduce interferences. This resulted in higher detection limits for samples sent to the SNWS and MWH laboratories. The LC/MS/MS method used by Exygen is capable of very low detection limits in low-salinity waters. Exygen used barium and silver cartridges for pretreatment, but they also diluted the samples, which elevated their detection limits for the high-salinity samples.

No perchlorate was found at concentrations above the detection limits for samples from the Salton Sea. Although detection limits for some of the laboratories were very high, the results do indicate that perchlorate is not accumulating in the Salton Sea.

Perchlorate was found by three of the four laboratories in the Alamo River at levels (3.85–4.8 mg m^{-3}) near those previously reported for the lower Colorado River. Exygen found a similar level (3.56 mg m^{-3}) in the New River, and both DECL and Exygen reported perchlorate at low levels (1.22 and 1.8 mg m^{-3}, respectively) in the Whitewater River; lower detection limits were possible for the Whitewater River samples because of the lower salinity at that location compared to other study sites.

No perchlorate was found in any of the evaporation ponds at the Salton Sea Test Base, although detection limits for these highly concentrated brine samples were all high. Exygen did find perchlorate in the evaporation pan sample from the Salton Sea Test at a concentration of 30.8 mg m^{-3}. This sample of highly concentrated Colorado River water had the highest salt concentration of any sample available for this study (Table 1).

Discussion

Boralessa & Batista (2000) analyzed archived water samples provided by the Southern Nevada Water

Table 1 Perchlorate and total dissolved solids' concentrations in test samples

Sampling location	Laboratory				
	DECL		Exygen	MWH	SNWS
	TDS (g m^{-3})	ClO$_4^-$ (mg m^{-3})	ClO$_4^-$ (mg m^{-3})	ClO$_4^-$ (mg m^{-3})	ClO$_4^-$ (mg m^{-3})
Salton Sea and River samples					
SS1—S	43,480	<2.5	<12.5	<800	<80
SS1—B	45,060	<2.5	<12.5	<800	<80
SS2—S	43,800	<2.5	<12.5	<40	<80
SS2—B	44,840	<2.5	<12.5	<40	<80
SS3—S	45,020	<2.5	<12.5	<800	<80
SS3—B	45,080	<2.5	<12.5	<40	<80
Alamo River	1,890	4.8	3.85	4.3	<80
New River	2,520	<2.5	3.56	<4	<80
Whitewater River	1,010	1.8	1.33	<4	<80
Salton Sea test base samples					
Disposal pond #1	404,000	<100	<12.5	<800	<80
Disposal pond #2	380,000	<100	<12.5	<800	<80
Disposal pond #3	406,000	<100	<12.5	<800	<80
Disposal pond #4	420,000	<250	<12.5	<800	<80
Evaporation pan	480,000		30.8		

Perchlorate detection limits: DECL—2.5 mg m^{-3} for Sea, 100–250 mg m^{-3} for evaporation ponds; Exygen 12.5 mg m^{-3} for Sea and evaporation ponds, 0.25 mg m^{-3} for river samples; MWH-40–800 mg m^{-3} for Sea and evaporation ponds, 4 mg m^{-3} for river samples; SNWS—all samples below the detection limit of 80 mg m^{-3}

Authority to show that perchlorate concentrations in Las Vegas Wash and Lake Mead were relatively unchanged between 1990 and 2000. Those concentrations resulted in perchlorate levels of 4–10 mg m^{-3} in the Colorado River.

Loading calculations developed for nutrients in the Salton Sea by Holdren & Montaño (2002) were applied to perchlorate concentration in influent rivers measured during this study. The average annual inflow of 1.66×10^9 m^3 to the Salton Sea (Salton Sea Authority/USBR, 2000) was allocated to the three major rivers based on measured inflows, and a weighted average perchlorate concentration was calculated from measurable perchlorate concentrations in Table 1. This calculation resulted in a loading of approximately 65 metric tons of perchlorate to the Salton Sea during 1990–2000, the period covered by the Boralessa & Batista (2000) study. This would have resulted in an in-Sea concentration of 7 g m^{-3} if perchlorate had not been lost from the system. Depending upon how long it took for

contaminated water from the manufacturing facilities in Henderson, Nevada, to reach Lake Mead, it is possible that up to five times this amount of perchlorate actually reached the Salton Sea between the time perchlorate manufacturing began in Henderson in 1951 and the time this study was conducted in 2003.

Recent results on the operation of a bioremediation project conducted under the direction of the Nevada Department of Environmental Protection at the American Pacific facility near Las Vegas Wash may be used to explain the lack of accumulation of perchlorate in the Salton Sea. Results from that project indicate that bacteria are able to successfully reduce perchlorate levels by using the perchlorate ion as an electron acceptor (Sellers et al., 2007).

The reduction of perchlorate is very similar to the denitrification and sulfate reduction processes occurring in the Salton Sea. The reduction reactions for these three processes are:

Perchlorate reduction: $ClO_4^- + 8e^- + 8H^+$

$$= Cl^- + 4H_2O, E^\circ = 1.389,$$

Denitrification: $2NO_3^- + 10e^- + 12H^+$

$$= N_2(g) + 6H_2O, E^\circ = 1.24,$$

and Sulfate reduction: $SO_4^= + 8e^- + 9H^+$

$$= HS^- + 4H_2O, E^\circ = 0.24$$

Thermodynamics clearly favor perchlorate reduction and denitrification, but bacterial mediation is necessary to overcome high activation energies for those reactions. Successful treatment of perchlorate by bacterial reduction has been demonstrated (Sellers et al., 2007) and suggests that sufficient numbers of bacteria could be present in sediments from the Salton Sea, the river beds, and irrigation canal bottoms to carry out perchlorate reduction.

Results presented by Holdren & Montaño (2002) indicate that low dissolved oxygen concentrations and redox potentials are present in the bottom waters of the Salton Sea throughout the year. At times during the summer months, dissolved oxygen was almost completely absent from the water column. As a result, conditions in the Salton Sea clearly favor bacterial perchlorate reduction.

Perchlorate concentrations in the Alamo and New Rivers during this study were at the low end of the 4–10 mg m^{-3} range reported by Boralessa & Batista (2000) for the Colorado River over the past ten years, in spite of expected evaporative concentration in the irrigation return flow to these rivers. This may be an indication that the same bacterial reduction processes apparently responsible for removing perchlorate from the Salton Sea may also be partially removing perchlorate in river sediments and agricultural fields.

A lack of bacterial mediation could also explain the observed concentration of perchlorate in the evaporation pan sample from the Salton Sea Test Base. Since the evaporation of Colorado River water for this sample occurred in a stainless steel pan, the bacteria which are provided by sediments would not have been present. As a result, bacterial perchlorate reduction did not occur and perchlorate was concentrated.

Conclusions

Results show that perchlorate is not accumulating in the Salton Sea in spite of potentially high loadings from the Colorado River. All four laboratories reported that perchlorate concentrations were undetectable in the Salton Sea and in concentrated brine samples at the Salton Sea Test Base. The lowest detection limit of 2.5 mg m^{-3} for Salton Sea samples is more than three orders of magnitude lower than the conservative estimate of 7 g m^{-3} based on loading calculations between 1990 and 2000. Bacterial reduction of perchlorate is the most likely explanation for the observed results.

Acknowledgments Funding for the project was provided to the Technical Service Center by the Bureau of Reclamation's Lower Colorado Regional Office and the Salton Sea Restoration Program/Project (Lower Colorado Region). The Yuma Area Office and the Technical Service Center provided sampling support for the project. Analytical Services were performed by the Reclamation Environmental Chemistry Laboratory, Denver, Colorado; the Southern Nevada Water System Laboratory, Boulder City, Nevada; Exygen Research, State College, Pennsylvania; and MWH Laboratories, Monrovia, California. Administrative support was provided by the Salton Sea Program Office. Stuart Hurlbert and three anonymous reviewers provided helpful comments for revision of the original manuscript.

References

Agajanian, J., L. A. Caldwell, G. L. Rockwell & G. L. Pope, 2005. Water resources data—California, water year 2004. Volume 1—Southern Great Basin from Mexican border to Mono Lake Basin, and Pacific Slope Basins from Tijuana River to Santa Maria River. Water-Data Report CA-04-1. U.S. Department of the Interior, U.S. Geological Survey, Sacramento, California.

Boralessa, R. & J. Batista, 2000. Historical perchlorate levels in the Las Vegas Wash and Lake Mead. Proceedings of the 2000 AWWA Inorganic Contaminants Workshop, Albuquerque, New Mexico, February 27–29, 2000.

Clark, J. J., 2000. Toxicology of perchlorate. In Urbanksy, E. T. (ed.), Perchlorate in the Environment. Kluwer Academic/Plenium Publishers, New York, 15–30.

Clewell, R. A., E. A. Merrill, K. O. Yu, D. A. Mahle, T. R. Sterner, D. R. Mattie, P. J. Robinson, J. W. Fisher & J. M. Gearhart, 2003. Predicting fetal perchlorate dose and inhibition of iodide kinetics during gestation: a physiologically-based pharmacokinetic analysis of perchlorate and iodide kinetics in the rat. Toxicological Sciences 73: 235–255.

Dohan, O., A. De la Vieja, V. Paroder, C. Riedel, M. Artani, M. Reed, C. S. Ginter & N. Carrasco, 2003. The sodium/iodide symporter (NIS): characterization, regulation, and medical significance. Endocrine Reviews 24: 48–77.

Greer, M. A., G. Goodman, R. C. Pleus & S. E. Greer, 2002. Health effects assessment for environmental perchlorate contamination: the dose response for inhibition of thyroidal function. Environmental Health Perspectives 110: 927–937.

Federal Register, 1998. Federal Register 63(40): 10273–10287.

Hogue, C., 2005. Pollution: EPA sets safe dose for perchlorate. Chemical and Engineering News, February 28: 14.

Holdren, G. C. & A. Montaño, 2002. Chemical and physical characteristics of the Salton Sea, California. Hydrobiologia 473: 1–21.

Kirk, A. B., 2006. Environmental perchlorate: why it matters. Analytica Chimica Acta: 567: 4–12.

Logan, B. E., 1998. A review of chlorate- and perchlorate-respiring microorganisms. Bioremediation Journal 2: 69–79.

Potius, F. W., P. Damian & A. Eaton, 2000. Regulation of perchlorate in drinking water. In Urbansky, E. T. (ed.), Perchlorate in the Environment. Kluwer Academic/Plenium Publishers, New York, 31–37.

Renner, R., 2005. Perchlorate report doesn't dispel controversy, Environmental Science & Technology: 39–96A.

Salton Sea Authority/USBR, 2000. Draft Salton Sea restoration project environmental impact statement/environmental impact report. Report prepared for the Salton Sea Authority and the U.S. Department of the Interior, Bureau of Reclamation, by Tetra Tech, Inc.

Sellers, K., W. Alsop, S. Clough, M. Hoyt, B. Pugh, J. Fobb & K. Weeks, 2007. Perchlorate: Environmental Problems and Solutions. CRC Press, Taylor and Francis Group, Boca Raton, Florida, 224.

Stetson, S. J., R. B. Wanty, D. R. Heisel, S. J. Kalkhoff & D. L. Macalday, 2006. Stability of low levels of perchlorate in drinking water and natural water samples. Analytica chimica Acta 567: 108–112.

Susarla, S., T. W. Collette, A. W. Garrison, N. L. Wolfe & S. C. McCutcheon, 1999. Perchlorate identification in fertilizers. Environmental Science & Technology 33(19): 3469.

Tikkanen, M. W., 2006. Development of a drinking water regulation for perchlorate in California. Analytica Chimica Acta 567: 20–25.

Urbansky, E. T., 1998. Perchlorate chemistry: implications for analysis and remediation. Bioremediation Journal 2: 81–95.

U.S. EPA., 1979. Methods for chemical analysis of water and wastes. EPA document no. EPA-600/4-79-020. U.S. Government Printing Office, Washington, DC.

U.S. EPA., 1998. Drinking water contaminants list. EPA Document No. 815-F-98-002. U.S. Government Printing Office, Washington, DC.

U.S. EPA., 2000. Methods for the determination of organic and inorganic compounds in drinking water. Volume 1. EPA document no. EPA815-R-00-014. U.S. Government Printing Office, Washington, DC.

Weghorst, P.A., 2004. Salton Sea salinity control research project. U.S. Bureau of Reclamation, Lower Colorado Region, Boulder City, Nevada, and Salton Sea Authority, La Quinta, California. Available at: http://www.usbr.gov/lc/region/saltnsea/pdf_files/salincntrl/report.pdf.

Hydrobiologia (2008) 604:181–195
DOI 10.1007/s10750-008-9317-0

SALTON SEA

Fundamentals of estimating the net benefits of ecosystem preservation: the case of the Salton Sea

**Kurt A. Schwabe · Peter W. Schuhmann ·
Kenneth A. Baerenklau · Nermin Nergis**

Abstract This article, both theoretical and methodological in nature, argues the potential merits of using a net benefits' framework as a tool to aid policy makers in their efforts to compare Salton Sea restoration alternatives and inform the public as to the potential magnitude and distribution of trade-offs associated with each alternative. A net benefits' approach can provide a more accurate comparison and evaluation of the potential net returns from public spending on Salton Sea restoration than what would be provided under the suggested criteria of current legislative mandates. Furthermore, a net benefits' framework provides a more lucid and systematic accounting framework by which to enumerate the full array of benefits and costs of each alternative for policy analysis. Finally, net benefits' analysis serves to add transparency to the decision-making process so that the public gains an understanding of how its scarce resources, including both financial and natural capital, are being appropriated. Additionally, we illustrate and emphasize the importance of estimating the non-market values associated with many of the ecosystem services provided by the Salton Sea and describe the major techniques that do so.

Keywords Ecosystem services · Net benefits ·
Non-market valuation · Recreation

Guest editor: S. H. Hurlbert
The Salton Sea Centennial Symposium. Proceedings of a Symposium Celebrating a Century of Symbiosis Among Agriculture, Wildlife and People, 1905–2005, held in San Diego, California, USA, March 2005

K. A. Schwabe (✉) · K. A. Baerenklau
Department of Environmental Sciences, University of California, Riverside, CA 92521, USA
e-mail: kurt.schwabe@ucr.edu

P. W. Schuhmann
Department of Economics and Finance, University of North Carolina, Wilmington, NC 27251, USA

N. Nergis
San Diego, CA 92101, USA

Introduction

The objectives of this article are two-fold. First, we emphasize the potential merits of using a net benefits' framework as a tool to aid policy makers in their efforts to compare Salton Sea restoration alternatives and inform the public as to the potential magnitude and distribution of trade-offs associated with each alternative. A net benefits' framework is a framework that uses the differences between the benefits and costs of a policy or action as a means of comparison. Currently, legislation mandates that the Secretary of the Resources Agency for the State of California establish "suggested criteria for selecting and evaluating alternatives" (Section 2081.7 of the California State Fish and Game Code, part (e)). Two explicitly

mentioned criteria include an evaluation of the construction, operation, and maintenance costs of each alternative, hereafter referred to as engineering costs, and the identification of a cost-effective, technically feasible option. Relative to these suggested criteria, a net benefits' framework can provide a more accurate comparison and evaluation of the potential net returns from public spending on Salton Sea restoration. Furthermore, a net benefits' framework provides a more lucid and systematic accounting framework by which to enumerate the full array of benefits and costs of each alternative. Finally, net benefits' analysis serves to add transparency to the decision-making process so that the public gains an understanding of how its scarce resources, including both financial and natural capital, are being appropriated.

Second, we emphasize the importance of estimating the non-market values associated with many of the ecosystem services provided by the Salton Sea, describe the major techniques that do so, and suggest how these techniques could be applied to the Sea. The Salton Sea is a natural asset that provides many services to society, including unpriced non-market goods and services for bird watching, fishing, boating, and camping. The Sea is home to the endangered desert pupfish (*Cyprinodon macularius* Baird and Girard), as well as over 400 species of migratory and resident birds, approximately fifty of which have garnered special status as threatened, endangered, or species of concern. These sorts of ecosystem services have been shown to be highly valued by society in other regions around the state, nation, and world, and, accordingly, should be treated as such in any objective economic analyses concerning their use. Indeed, as the National Research Council (2004) argued recently, assigning a dollar figure to non-market ecosystem services is essential to accurately weight the trade-offs among environmental policy options. Overlooking these values often results in an implicit value of zero being assigned to them in the economic analyses, which is incorrect and unnecessary because numerous analyses exist that have estimated the monetary value of similar services. Wilson & Carpenter (1999), for example, provide a summary of the economic value of freshwater ecosystem services in the U.S., noting 30 refereed published articles in the scientific literature from 1971 to 1997. This literature is quite extensive and includes values derived for all manner of ecosystems, including wetlands.

We begin with a description of the mandate imposed upon the California Resources Agency in its endeavor to identify a preferred alternative. Shortcomings of the process are identified relative to what might be provided by a net benefits' framework in evaluating alternative restoration plans. A description of how to perform a net benefits' analysis is then provided. Since many of the benefits associated with restoring the Salton Sea reside in the ecosystem services the Sea provides to society, a description of the main non-market valuation techniques used to estimate the value of these services is presented. Summaries of previous studies attempting to value a healthy Salton Sea are provided, including a recent report developing suggestive estimates of the recreation and preservation values, using the results from non-market valuation studies of somewhat similar California habitats. Finally, we summarize our findings.

Background and motivation

The California Resources Agency is mandated to identify a preferred alternative for the restoration of the Salton Sea. As noted by California State Senator Ducheny at a recent conference centered on the Salton Sea, this will be no easy task (Remarks by State Senator Ducheny at "*The Salton Sea Centennial Symposium*," San Diego, CA, April 1, 2005). First, the legislature will be required to operate under a set of preexisting rules and regulations that will limit how much financial and natural capital (i.e., water) it can allocate to solving this problem. California has been mandated to adhere to its Colorado River water entitlement of 4.4 million acre-feet, down from the approximately 5.2 million acre-feet it has grown accustomed to using. Additionally, as part of the Quantification Settlement Agreement (QSA) signed in 2003, the Imperial Irrigation District has agreed to transfer 200,000 acre-feet of water to San Diego (Cohen & Hyun, 2006).

Second, given the complex set of linkages in the region, mostly driven by water, the effectiveness of any particular restoration alternative will be largely dependent on, or place restrictions upon, upstream users of the water, particularly agriculture. Agriculture, whose drainage flows are responsible for nearly 85% of the inflow into the Salton Sea, will be largely responsible for this cutback; as such, less applied irrigation water will lead to less drainage and thus less inflow to the Salton Sea. Less inflow will strain the effectiveness of any particular restoration solution and leave more Salton Sea lakebed exposed as shorelines recede. As the area of exposed lakebed increases, so will the amount of fine windblown dust in this high wind region which is currently not in compliance with state and federal air quality standards and is characterized by the highest rates of hospitalization of children for asthma in the state (Cohen & Hyun, 2006). Alternatively, requirements for agriculture to maintain historic inflow levels will likely affect the economic health of this very poor region that has nearly 19% of its population considered living in poverty (United States Department of Agriculture, 2004). These impacts may directly affect the productivity and profitability of agriculture and, consequently, labor and income associated with agriculture and agricultural-related activities.

Third, across the feasible set of restoration alternatives there are significant differences in habitat configuration, elevation, and both the quantity and quality of inflow assumed; consequently, each alternative will provide a different array of ecosystem services. Hence, the benefits of restoration are likely to differ depending on the alternative chosen, including those benefits associated with recreation and preservation at the local, state, and national level.

As part of its decision-making process, the Resources Agency will perform a "…restoration study to determine a preferred alternative for the restoration of the Salton Sea Ecosystem and the protection of wildlife dependent on that ecosystem" and report the findings to the legislators (California State Fish and Game Code, Section 2930: 93). Elements of this study are to include:

(a) an evaluation of restoration alternatives including consideration of salinity control, habitat creation and restoration, and different shoreline elevations and surface area configurations,
(b) consideration of a range of possible inflow conditions,
(c) suggested criteria for selecting and evaluating alternatives, including, but not limited to, at least one most cost-effective, technically feasible alternative, and
(d) an evaluation of the magnitude and practicality of costs of construction, operation, and maintenance of each alternative evaluated.

These elements, in addition to providing some necessary bounds on the problem, identify factors that can substantially influence the costs and benefits of any particular alternative. For instance, consider item (b) related to inflow conditions. The engineering costs of the final restoration alternatives considered in the Salton Sea Reclamation Act of 1998 varied between $320 million and $1.4 billion depending on the inflow assumption. While none of the alternatives listed in the 1998 legislation was enacted, the cost estimation exercise did highlight how the engineering costs of any particular alternative depend on the values assigned to possible factor inputs.

Consider the remaining elements listed above. Similar to the 1998 legislation, engineering costs of each restoration alternative are to be estimated and compared (d), yet an explicit call for the identification of a cost effective solution is to be included (c). Cost effectiveness typically refers to the least-cost approach to achieving a particular level of environmental quality, or, in this case, ecosystem services. Yet, as emphasized in (a), each restoration alternative likely will provide a different array of ecosystem characteristics.

Hence, while the stated intent of the restoration study is to inform policy makers of the potential trade-offs associated with each alternative and it is acknowledged that any particular restoration strategy can deliver a different stream of benefits to society, there is no discussion of how to evaluate and quantify these benefits. Unlike goods that are bought and sold in the marketplace, the economic benefits of natural resources are not revealed through market transactions. The benefits derived from these resources are thus termed "non-market values." Most of the benefits from restoring the Salton Sea consist of these non-market values (e.g., those values we place on recreation or the preservation of endangered and threatened species).

Implicit in performing a cost-effectiveness analysis or engineering cost comparison is that the benefits are assumed constant across restoration alternatives, in contrast to what is suggested in element (a) above.

This is clearly not the case with the proposed alternatives for restoration of the Salton Sea, either in the alternatives proposed in the 1998 legislation or the eight alternatives listed in the Salton Sea Ecosystem Restoration Draft Programmatic Environmental Impact Report (California State Resources Agency, 2006). Because the California Resources Agency is not required to consider how the returns to each investment will differ relative to the costs, their selection may not allocate resources to their highest valued uses since achieving such efficiency requires a comparison of the costs and benefits of each alternative. The process of itemizing, quantifying, and comparing the costs and benefits is known as benefit-cost analysis. Whether formalized or not, this practice is perhaps the most fundamental tool used in decision making by individuals, private organizations (e.g., firms), and public institutions (e.g., state governments).

Further, as indicated in the 1998 legislation, the attractiveness of any alternative is inextricably linked to assumptions about the inputs (e.g., inflows), outputs (e.g., level of ecosystem services), and scale of analysis. For instance, what might be considered the cost-minimizing engineering solution may not be the cost-effective alternative when the impacts on regional agricultural production, the regional economy, or human health from poorer air quality also are included. Continuing, what might be the regionally efficient solution may not be the efficient solution from the state perspective, and so on. Given that the state government will be involved, it seems reasonable to assume that the California Resource Agency would consider the local, regional, and statewide impacts in their efforts to choose a preferred alternative. This does not suggest that broadening the analysis even further has no value, though. As Ciriacy-Wantrup (1964) noted, consideration of the broad impacts of a policy may be a preliminary step toward broadening the repayment base—a base which is sometimes rather narrow if confined to primary benefits. Enumerating and quantifying the benefits of Salton Sea preservation to a broader population might be a first step toward justifying federal assistance.

Finally, it is important to note that while the popular press has only recently begun extolling the importance of placing a value on non-market environmental goods and services (e.g., The Economist,

2005, April 3rd–29th: 76–78; Business Week, 2004, December 29th; Infocus Magazine, 2005, 4.3; Outside Magazine, March, 2005: 106–123), these values, and the non-market valuation methods used to estimate them, have been given standing in legislative mandates and by state and federal government agencies for decades, including the Comprehensive Environmental Response, Liability, and Compensation Act (CERCLA) of 1980, the Oil Pollution Act (OPA) of 1990, U.S. Water Resources Council, the U.S. Department of Interior, and the U.S. Forest Service. Federal and state agencies also consider non-market values when making natural resource allocation decisions. Since 1979, for example, the U.S. Army Corps of Engineers and Bureau of Reclamation have been required to assess the value of recreation benefits in cases where federal projects impact areas of high visitation (Loomis, 2005). The U.S. Environmental Protection Agency (EPA) is required to conduct benefit-cost analyses of environmental regulations and must include estimates of non-market benefits. CERCLA mandates that lost recreation values and "passive use" values from toxic waste sites and hazardous material spills must be assessed to measure the full value of damaged natural resources. Many states have funded studies measuring non-market values associated with recreation and ecosystem preservation, including the State of California, which sponsored an analysis of the values of protecting Mono Lake as a bird habitat (Loomis, 2005). Hence, the validity of valuing changes in environmental or natural resource quality and its usefulness in guiding resource allocation decisions has been invoked at state and federal levels.

Net benefits: background and conceptual issues

As noted above, a commonly employed litmus test in judging whether a project should be undertaken or not is whether it passes the present value benefit-cost test (i.e., whether the present value benefits are at least as great as the present value costs).The formal use of benefit-cost analysis for large water-related projects can be traced back to Eckstein (1958) in his evaluation of federal water-resource programs. In particular, Eckstein (1958: 2) references the Flood Control Act of 1936, which suggests that only projects where "the benefits, to whomsoever they

may accrue, are in excess of the estimated costs" would be considered. Eckstein described benefit-cost analysis as a very promising approach for evaluating the use of scarce natural and financial capital that can provide a much stronger foundation for policy decisions than what might otherwise be available. This is especially true when many agencies with jurisdictional overlap are involved in the decision-making process, such as in the case of the Salton Sea. In response to the problems associated with multi-agency involvement and overlap, Eckstein stressed the importance of a general set of standards by which projects can be appraised and compared. Such standards, he continued, would also serve a wider interest in informing the public about the merits of a project and what they will be asked to forgo in return.

It should be emphasized that just because the estimated benefits of an alternative are in excess of the estimated costs do not mean that this alternative is the economically efficient alternative. Indeed, there may be more than one alternative that meets this condition. Of course, the alternative that is in the best interest of society from an economic efficiency perspective is that alternative providing the highest net benefits, which are defined as the difference between the total benefits and total costs.

Why there has not been greater focus on using benefit-cost analysis or net benefits' analysis in the context of Salton Sea restoration is puzzling, especially when such an approach has been prominent for more than 30 years at the federal level in consideration of major environmental, health, and safety regulations (Morgenstern, 1997). Under President Clinton's Executive Order 12866, federal agencies were allowed to "include both quantifiable measures and qualitative measures of costs and benefits" and to "select those approaches that maximize net benefits (including potential economic, environmental, public health and safety, and other advantages; distributive impacts, and equity)." Furthermore, numerous real world examples exist of governments incorporating the benefits of preserving natural and environmental resources into their decision making, both in the U.S. and abroad. Such evaluations cover a wide array of resources, including the Glen Canyon Dam (Bishop et al., 1987), Hell's Canyon (Krutilla & Fischer, 1975), Mono Lake (Loomis, 1987), the spotted owl in the Pacific Northwest (Hagen et al., 1992), Kootenai Falls in Montana (Duffield, 1982), and the Kakadu

Conservation Reserve in Australia (Imber et al., 1991), to name a few. In these and other studies, the preservation benefits associated with environmental and natural resources were quantified and given standing in benefit-cost analysis. In each case, the quantification of such benefits either supported an action for preservation or modified an existing development scheme to be more environmentally friendly. In all cases, a large—if not the largest—component of the value of preservation was non-market value.

Before moving on to the various steps involved in estimating the net benefits, it is useful to clarify what economists mean by economic value, especially in the context of environmental and natural resource goods and services. Economic value is defined by what one (or a group) would be willing and able to pay for a good, not by what one has to pay for it—what one has to pay for a good is what it costs and is considered an expenditure. In contrast to the benefits of an action, the costs of achieving a particular objective can be measured by what is forgone to achieve that objective, and include both direct engineering costs as well as *opportunity costs*. The former includes both current and discounted future costs, and the latter represents the value associated with the opportunity to use the forgone resources in another activity.

We must also recognize that economic value, which is meaningful from an anthropocentric perspective only, extends beyond the marketplace to *non-market goods* such as clean air or water, open space, and wildlife preservation. Furthermore, the economic value of these goods comprises both *use* and *non-use* values. The values associated with catching tilapia for consumption and bird watching would be examples of *use* values associated with the Sea, while the value that people derive from knowing that the Salton Sea ecosystem exists for current and future generations would be an example of *non-use* value.

Components of net benefits' estimation

For large projects, efforts to estimate the net benefits may seem insurmountable; thus it is best to have a road map as to what might be the necessary steps to perform a net benefits' analysis. Borrowing upon

previous works (Boardman et al., 1996; Morgenstern, 1997), we present a description of the main steps in performing a net benefits' analysis and identify how each step could be applied to the restoration of the Salton Sea.

Specify the portfolio of alternative projects

Finding the efficient solution requires identifying, investigating, and comparing numerous alternatives with the outcome that would occur if no action were taken, i.e., the baseline. In the Salton Sea Reclamation Act of 1998, over 50 proposals were identified, of which five were given additional scrutiny, but eventually deemed "too costly and too impractical to implement." (California Department of Water Resources, 2003: iii). These analyses did prove valuable in that the CDWR, along with other agencies (e.g., USBR, USFWS, Salton Sea Authority), could now focus on a narrower set of feasible alternatives. Currently there are eight proposals to evaluate relative to two "no-action" alternatives. These alternatives differ in many dimensions including construction and maintenance costs, strategies for salinity control, shoreline elevations, water body size, depth, salinity, and surface area configurations, and wildlife habitat.

Well-defined objectives and criteria will go far in narrowing the possible choice set. The Salton Sea Restoration Act puts forth the following objectives to be considered when evaluating alternatives: sustain avian biodiversity at the Salton Sea without maintaining elevation of the entire Sea, maintain near-current salinity and elevation, and represent the most cost effective technical alternative. Regarding this last objective, care must be taken here not to prejudge alternatives too quickly based on *ex ante* costs alone for risk of defining the choice set too narrowly. Because the proposed alternatives provide a wide array of environmental benefits, the most cost-effective solution may not be that which maximizes net benefits.

Decide whose benefits and costs have standing

The benefits and costs of a particular regulation or action can be realized at the local, regional, state,

national, and even international level. As Boardman et al. (1996) note, national governments typically consider costs and benefits at the national level. It is not uncommon, though, for cities, municipalities, or states to overlook the impacts of their actions on one another in terms of who counts. Political boundaries and the level of administrative unit will often drive who is included in a benefit-cost study.

With respect to the Salton Sea, many different groups will be directly or indirectly impacted by the choice of restoration alternative, extending from recreational users of the Sea, to the localities around the Sea, to the growers in the Imperial and Coachella agricultural regions, to state, federal, and tribal agencies, as well as to those living abroad. A potential difficulty that arises with so many agencies and political boundaries is that what might be efficient at one level of analysis (or political boundary) may not be efficient at another level (or boundary). For instance, a cost effective restoration alternative might not be cost effective when the impacts on agriculture or regional employment and income are considered, or when the impacts on human health from dust particles from the exposed seabed are acknowledged. What might be the efficient solution for Imperial County and residents of the Salton Sea might not be efficient for the state of California. Performing a broad-based net benefits' analysis to determine the extent of the market would provide transparency as to the distribution of benefits and costs among different stakeholders. This will be useful so that criteria other than efficiency, such as equity, are part of the decision-making process.

Catalog the impacts and select measurement indicators

Many types of impacts may result from regulatory and policy actions. What is necessary for benefit-cost analysis is to catalog these impacts as either benefits (positive impacts) or costs (negative impacts) and decide upon a measurement unit for the impact. These impacts can be measured in a variety of ways, including economic, environmental, and health effects. Economic indicators include jobs, time, income, and changes in consumer and producer welfare. Environmental indicators may include quantitative assessments of species viability, ecosystem

productivity, and water and air quality. Indicators of public health might include the avoidance of health care costs or benefits associated with changes in quality or longevity of life.

Such categorization and measurement are certainly suitable for the Salton Sea restoration alternatives. Benefits and costs may accrue to landowners, farmers, local businesses (especially those relying on tourism), recreationists such as bird watchers and anglers, environmental groups, and local and regional governments. While the category of measurement for the engineering costs of restoration will be dollars, each restoration alternative may affect or be affected by upstream activities related to agriculture. That is, the response from agriculture to the reduction in California's take of Colorado River water can affect the inflow volume into the Sea; consequently, additional mitigation activities will be required to offset the reduced volume and surface elevation of the Sea. For instance, as the shoreline recedes due to lower inflow and in lieu of additional mitigation, increased dust and particulates will be generated exacerbating an already exorbitant regional air quality problem.

Alternatively, if the inflow volume is required to remain constant, then presumably agriculture may need to engage in additional water conservation schemes (e.g., reduce applied water rates, more efficient irrigation measures, land fallowing) to reduce their applied water rates sufficiently to provide enough mitigation water to maintain inflow volume requirements. These activities can and should be measured in terms of productivity, additional labor hours, income, and employment. The indirect impacts from the restoration alternatives may be very important and thus, at a minimum, should be acknowledged via a categorization of this type. Insight into possible agricultural-related and regional impacts would be further enhanced by applying a regional agricultural production model for agricultural activities (e.g., Schwabe et al., 2006) and a social accounting matrix model (multiplier analysis) to account for the employment and income effects within the region (e.g., Berck et al., 1991).

Additionally, the impacts of the restoration alternatives on ecosystem services should be considered. These impacts will differ by restoration alternative, and thus will have varying effects on tourism and recreation, such as time and income spent on recreating, and wages and income earned from tourism. Such impacts may extend beyond the immediate area, certainly to the state, and perhaps to the nation in terms of non-use values. Indeed, perhaps the largest benefit associated with preserving and restoring the Salton Sea does not necessarily accrue to current users of the Sea, but rather to people who care about the Sea regardless of whether they tangibly use the Sea currently. People have been observed benefiting from environmental resources, and willing to pay to protect them, just by knowing that the resources exist. For example, Sanders et al. (1990) estimate what people are willing to pay (i.e., their value) for preserving free flowing rivers with no intention of ever visiting them. Alternatively, Olsen et al. (1991) estimate people's willingness to pay (value or benefits) for maintaining salmon migrations, again, without actively engaging in any recreation activities (e.g., fishing, photography) involving these salmon. This sort of value is called a non-use or passive-use value and captures that value people have for resources for possible future use by themselves, future use by future generations, current use by others, or simply because they think it is the right or moral thing to do.

Predict the impacts quantitatively over the life of the project

A comprehensive classification of impacts and their associated costs and benefits is complicated by the extent to which direct impacts transfer across agents, markets, and natural systems, and the degree to which this transfer is measurable. The future time path of changes to health, the economy, and the environment must be estimated in some way. This estimation may rely on an extensive review of existing scientific knowledge and data or may rely on a new analysis.

As part of the restoration plan, the California Resources Agency is to prepare a Program Environmental Impact Report (PEIR) that will analyze the potential environmental impacts of the alternatives included in the Ecosystem Restoration Plan. For the restoration of the Salton Sea, a lengthy time horizon and complex interactions will certainly make the estimation of cause and effects difficult and costly. When constructing and evaluating the Draft PEIR, this cost must be weighed against the importance of accurate estimation or the cost of making the wrong

decision. It may indeed be impossible to directly measure some impacts. When this is the case, a proxy measurement must be constructed to account for the impacts.

Monetize (attach dollar values to) all impacts

Once all impacts have been identified, cataloged, and estimated, their monetary value must be determined. In this way, benefits and costs can be compared in dollars. When the impacts occur through markets (such as costs associated with construction or the benefits of created jobs), monetization is relatively straightforward. These values can be derived using the appropriate demand curve and estimated changes in market prices and quantities. The estimation of non-market (and especially non-use) values presents a challenging problem in measuring the full value of resources, but is facilitated by well-established valuation techniques, the most popular of which are discussed below.

Discount benefits and costs and obtain present value

Many projects related to the environment will have costs and benefits that accrue over time. For restoration projects, it is often the case that costs are borne "up front," or in the present, while benefits do not accrue until sometime in the future. Because dollars or resources consumed today are worth more than the same dollars or resources consumed in the future (due to peoples' preferences to consume now rather than later), values that occur in different time periods need to be converted into a common period equivalent by "discounting" future values to their present value via a social discount rate.

The social discount rate, as noted in Pearce & Turner (1990), should reflect the rate at which society is willing to trade current dollars for future dollars and depends on the degree of risk associated with the future payoff. Since there are a wide variety of opinions as to society's aversion to risk, choice of this rate is a matter of much debate. Mathematically, lower discount rates make the present value of future dollars appear higher and vice versa. Hence, higher

discount rates weaken the case for projects with benefits that occur over long time horizons relative to up-front costs. With this in mind, using a predetermined rate removes the temptation to choose a discount rate to achieve a desired net benefits result. Indeed, many projects funded by the U.S. government use a real (inflation adjusted) discount rate of 7% (Boardman et al., 1996). However, because the discount rate can affect the outcome substantially, a range of discount rates and their corresponding net benefits should be analyzed and presented.

Once future costs and benefits for project or policy alternatives have been discounted, the present value of costs should be subtracted from the present value of benefits to arrive at the net present value (NPV) of each alternative. When deciding among competing alternatives, including the baseline, the project or policy with the highest NPV will yield the highest net gains to society and, thus, is considered the efficient choice.

Perform sensitivity analysis and make a recommendation

Even with the best available scientific information, most projects will involve some degree of uncertainty in predicting impacts, deriving their monetary value, or discounting future values. This uncertainty may be due to unknown parameters, lack of data, or lack of information about future environmental or economic conditions, which are often complex and difficult to predict. Such uncertainties exist with regard to the restoration alternatives for the Sea, including future annual inflows, salinity, habitat and wildlife impacts from construction, and the amount and nature of dust that will be created as the Sea's elevation drops over time. When some degree of certainty can be assigned, the most probable or plausible values of the uncertain parameters should be identified and reported as a "base case" scenario. Examination of a reasonable range of parameter values and probabilities around this base case acknowledges the uncertainty of the estimation and provides a means of examining the sensitivity of results to underlying assumptions. It is critical for the analyst to report on the robustness of the results to underlying assumptions so that policy makers can be fully informed. If net benefits remain consistent over a range of possible values, one can be

more confident in the results. Within a reasonable range of uncertain parameter values, the alternative with the highest NPV should be recommended. With respect to the Salton Sea, a present value net benefits' approach with sensitivity analysis would seem useful for evaluating the proposed restoration alternatives and their trade-offs.

Valuing environmental goods and services

For most goods and services, the starting point for estimating value is the market price. Yet for many environmental and natural resource goods and services, no such market price exists. For goods such as clean air, biodiversity, endangered species, and wildlife habitat, rarely are there market transactions revealing the price and subsequently the value of these goods and services to the society. Consequently, the value of these goods and services is not readily apparent to policy makers in charge of determining how these scarce and often unique resources are to be allocated. As an example of this problem, consider the decision of how to allocate an acre of land in, say, Sequoia National Forest. There is value associated with the timber that could be obtained from these giant trees. Yet, there is also value in preserving the forest in its present state for recreation activities such as hiking, camping, and photography today and in the future. There is value indirectly in the habitat these forests and trees provide for other wildlife resources we enjoy. There is value also in simply knowing that these resources exist for use by others, and possible future use by current and future generations. As such, we define the value of a resource that is not revealed through market transactions as its non-market value. Without knowledge of these non-market values, benefit-cost analysis is limited in its usefulness in aiding policy makers on how to efficiently and equitably allocate these resources.

The objective of non-market valuation is to estimate the economic value of these environmental and natural resources to society. Quantification of the benefits gives these goods and services standing in benefit-cost analysis. In considering the benefits of preservation, the total value of the resource should be considered, where total value is defined as:

$$Total\ Economic\ Value\ =\ Use\ Value \\ +\ Non-use\ Value.$$

Use value relates to the tangible use of the resource presently, and can include both consumptive use (e.g., *catch and keep* fishing) and non-consumptive use (e.g., photography, or *catch and release* fishing). Non-use value, as described in Kopp & Smith (1993: 340), is that "…component of the value of a natural resource that does not derive from the in situ consumption of the resource." There are four general categories for non-use values, including: *option value*—the value that people place on a good or service for future possible use, *altruistic value*—the value someone places on the preservation of a resource for use by others in the current generation, *bequest value*—the value someone places on the preservation of a resource for use by future generations, and *existence value*—the value one places on a resource for its mere existence, possibly for moral or ethical reasons.

In considering the non-market values associated with preservation of the Salton Sea, a variety of stakeholders come to mind. The Sea provides many non-market benefits to the State of California. Thousands of visitors frequent the Sea annually for bird watching, it has been the only tilapia (*Oreochromis mossambicus* Peters; Chchlidae) sports fishing area in the state, and other activities such as camping, boating, and swimming occur throughout the year. Indeed, the Salton Sea has been considered one of the most productive fisheries in the world (Cohn, 2000), especially during the years from 1960 to 2000. In 1987, there were nearly 2.6 million visits by recreators to the Salton Sea, making it a more popular destination than Yosemite National Park (CIC Research, 1989).

The Sea also provides non-market benefits to the nation as a whole. The Salton Sea is ranked as the second highest birding area in the nation. Indeed, 90% of the North American population of eared grebes (*Podiceps nigricollis* Heermann), more than 80% of the entire western U.S. population of white pelicans (*Pelecanus erythrorhynchos* Gmelin), and nearly half of the U.S. population of Yuma clapper rails (*Rallus longirostris yumanensis* Dickey), an endangered subspecies, utilize this habitat. The Sea is one of the two nesting areas in the western U.S. for

gull-billed terns (*Gelochelidon nilotica* Bancroft), a bird proposed for listing as a threatened species. From a fishery perspective, the Sea has supported eight species of fish, including the federally endangered desert pupfish (*C. macularius*).

Non-market valuation techniques

Three of the most popular methods for estimating non-market values for natural resources include the Travel Cost Method, the Random Utility Model, and the Contingent Valuation Method. The first two techniques are *revealed preference* methods—methods which examine decisions that individuals make regarding market goods that are used together with non-market goods to reveal the value of the non-market goods. These methods require that a link be established between changes in the environmental resource and changes in the observed behavior of people. For instance, changes in water depth and salinity in the Salton Sea may result in fewer fish. Anglers may then move to another part of the Sea, move to a different fishing location, or take fewer fishing trips. In establishing this link, it is important to account for any other factors that may be causing behavior to change. With this information, a demand or marginal willingness to pay function can be estimated, which allows one to estimate the value of environmental resource changes. While revealed preference methods allow for estimation of use values, they cannot be used to estimate non-use values. To elicit such values, *stated preference* methods, which ask people directly about the values they place on non-market goods, must be used. The most widely used stated preference method is the Contingent Valuation Method.

Travel cost method

The travel cost method (TCM), one of the most widely used revealed preference valuation techniques, uses information on actual behavior to estimate a trip demand curve from which the value of the resource can be derived. The demand curve is estimated using visitation data, including travel costs and the number of trips taken by each individual to a particular site. Using distance traveled as a proxy for the price of a trip and the number of trips as the quantity, individual or group demand curves can be estimated for a site. The net benefits of a particular site or the value of the resources within each site can then be estimated.

As noted in Loomis & Walsh (1997), the recreational benefits from a well-done TCM analysis should be fairly accurate, partly as a result of over 45 years of investigating and improving upon this technique. This method has been used by both state and federal agencies to value a wide variety of non-market goods and services. For instance, the TCM was used by Beal (1995) to estimate the value of camping at Carnarvon National Park. Results suggested that the annual net present value for camping at this park alone was nearly $40 million. Other recent analyses include valuing hiking in National Forests in Colorado and Montana (Hesseln et al., 2004), canoeing in Canada (Hellerstein, 1991), hunting in California (Creel & Loomis, 1990), salmon sport fishing in Alaska (Layman et al., 1996), and ecotourism and wildlife viewing in Costa Rica and Kenya (Navrud & Mungatana, 1994; Menkhaus & Lober, 1996). This method could similarly be employed to value the flow of recreation services from the Salton Sea. Application would require a survey of recreationists who use the Sea. In addition to a host of demographic information, survey respondents would be queried about the frequency of their participation in recreation activities at the Sea.

Random utility model

While application of the TCM would provide useful information on the value of recreation services from the Sea in its current state, a variation of this method, the Random Utility Model (RUM), may be more applicable to valuing potential changes in the Sea under the various restoration alternatives. The RUM has been used in a variety of applications, most commonly freshwater and saltwater recreational fishing (Bockstael et al., 1987; Schuhmann & Schwabe, 2004). It has also been used to value a wide assortment of activities at unique recreation areas, such as hiking in the Grand Canyon or Yellowstone National Park, rafting in the Middle Fork of the Salmon River in Idaho, and ecotourism and wildlife viewing in Italy (Font, 2000). RUMs are commonly

used to model the choice among a set of qualitatively different recreation sites. By estimating how the choice of alternative sites is dependent upon the characteristics of those sites, the RUM allows the researcher to value changes in the quality or characteristics of those sites.

Given that each restoration alternative is likely to result in a different level of ecosystem services (e.g., expected changes in length of shoreline, elevation, availability of bird habitat, or fish catch rates), which in turn will differentially impact the quality or quantity of recreational activities, the RUM would be a very appropriate method of estimation. Such an application would require identification of substitute sites for each recreation activity at the Sea, a catalog of current measures of quality at each site, measures of the expected changes in quality that would result from the restoration alternatives, and a survey of recreationists at each site. As the RUM relies on information gained from actual choice occasions, this survey could be conducted in person at the alternative recreation sites. This analysis would provide a quantitative assessment of the likely impacts on recreation benefits prior to any restoration action so that the net benefits of each alternative are more completely understood.

Contingent Valuation Method

The Contingent Valuation Method (CVM) is a well-accepted technique for valuing non-market goods, with far greater than 1600 CVM studies to date estimating non-market values in over 40 countries (Carson et al., 1994). The U.S. Department of Interior (DOI) has adopted CVM to measure non-market values associated with damages under CER-CLA 1980; NOAA has endorsed the use of this method for damage assessment under the Oil Pollution Act of 1990; and it is recommended by the Water Resources Council (1979) for use in benefit-cost analysis.

The goal of CVM is to create a realistic, albeit hypothetical, market where peoples' values for a good are expressed. A CVM survey consists of four main elements. The first element is a description of the program the respondent is asked to value or vote upon. This element often involves a description of the baseline services with no action, and an improved level of services with some type of policy action. Identifying the conditions of the "no-action" alternative and other restoration options may require research by physical and biological scientists. The second element of the CVM is specifying a mechanism for eliciting value or choice. There are a variety of options for eliciting value, the most well accepted being a referendum type question that asks each respondent to vote "yes" or "no" to a specified price or prices. A payment vehicle describing the manner in which the hypothetical payments are collected is the third element. Such vehicles have included higher taxes or utility bills, or a payment into a trust fund (Loomis et al., 2000). The fourth element consists of collecting information on respondent characteristics including socioeconomic data and environmental attitudes.

Because non-use values entail no actual observable use of a resource, the ability to measure non-use values reliably has been questioned (Hausman, 1993). To assess the reliability of CVM in measuring non-use values, NOAA convened a panel of prominent social scientists co-chaired by two Nobel Laureate economists. The panel concluded that if CVM practitioners follow a certain set of conditions, the results obtained from CVM are likely to be reliable (Arrow et al., 1993). Subsequent research has discussed issues associated with the conclusions of the NOAA panel and provided additional procedures that ensure CVM reliability (Hanemann, 1994).

Examples of benefits' estimation for preserving the Salton Sea

To date, little has been done in terms of quantifying, in monetary terms, the benefits of preserving the Salton Sea. No studies were found that used the methods discussed above to estimate the possible benefits from the proposed restoration alternatives. This is unfortunate since such information can be extremely useful in informing policy makers of the relative attractiveness of one option over another and justifying, *ex ante* or *ex post*, a particular decision. Two studies that have attempted to estimate the value of the Salton Sea include CIC Research (1989) and the Inland Empire Economic Databank and Forecasting Center (IEEC, 1998); unfortunately, neither study estimates the non-market benefits of preservation.

 Springer

CIC Research (1989), for instance, focused on estimating the expenditures of Salton Sea recreationists and, subsequently, the potential impact those expenditures might have on both the local and regional economy, often referred to as the *secondary market effect*. Based on responses from a telephone survey of Southern Californian residents and an intercept survey at the Salton Sea, approximately 154,600 households engaged in recreation at the Salton Sea in 1987 for a total of 2.6 million recreation days. Household expenditures that could be directly related to recreation at the Salton Sea amounted to $76 million, of which $53 million was spent directly in counties contiguous with the Salton Sea. Using regional and local economic multipliers, it was estimated that the $76 million in direct expenditures generated an additional $221 million in secondary market impacts. Unfortunately, the ability to use these impacts to measure the benefits of preserving the Salton Sea is tenuous since this study measures expenditures, not benefits. Hence, very little in terms of the value of preserving the Salton Sea can be gleaned from the CIC Research study with expenditure information alone.

Alternatively, the IEEC (1998) focused on the economic benefits of cleaning up the Salton Sea. IEEC categorized these benefits into how changes in Salton Sea water quality would affect (i) privately held developable property within a one-half mile of the Salton Sea shore and (ii) public sector revenues generated from taxes on property values and economic activity in the area. In estimating the benefits to private property owners for changes in water quality, IEEC considered changes in property values that would likely accompany an increase in water quality using retail market values from other tourist and recreation markets in the Southwest. Added to these privately held property values, they calculated the expected change in tax revenue that would accompany the changes in both property values and economic activity.

The combined present value benefits of increasing Sea water quality was estimated to be between $2.6 and $3.2 billion, with slightly over half accruing to private property owners and the rest generated from tax revenues. A serious problem with these estimates, from both a qualitative and quantitative perspective, is their treatment of tax flows. While they account for tax dollars earned, they do not account for tax dollars paid—symmetry should apply. Furthermore, tax generation is simply a transfer of wealth from one group to another. If the taxes are paid by agents outside the region, then local governments in the Salton Sea vicinity would experience an increase in tax revenues while governments elsewhere would experience a decrease. More problematic is their estimation of the value of preventing further degradation of the Sea. In particular, they assume that the benefits of prevention can be approximated by the costs of prevention with the justification that if society is observed incurring the cost, it must be that the benefits exceed these costs (IEEC, 1998: 13). Yet, politicians and government agents may make decisions regarding resources on criteria other than economic efficiency.

Neither CIC Research (1989) nor IEEC (1998) nor any other study to date directly estimates the non-market values from preserving the Salton Sea. In cases like this where a primary valuation study for the resource of concern is absent, economists sometimes rely on existing valuation studies for similar resources to obtain a somewhat less accurate but still potentially useful benefits' estimate. The use of previous non-market valuation studies to inform current decisions is known as *benefits' transfer* (Freeman, 2003; Rosenberger & Loomis, 2003).

K2 Economics (2007) recently conducted a simple benefits' transfer study for the Salton Sea that relied primarily on estimated values for two similar natural resources in California: San Joaquin Valley (SJV) wetlands and the Mono Lake ecosystem. Citing several analyses of contingent valuation survey data for SJV wetlands, K2 Economics determined that a conservative estimate of the current annual value of 1,000 acres of SJV wetlands to the residents of California is approximately $50 million. Assuming wetlands at the Salton Sea provide services similar to those provided by wetlands in the SJV, and assuming people value these services similarly, K2 Economics argued that this estimate also can be applied to wetlands at the Salton Sea. Transferring this value to the wetland acreage associated with each of the eight restoration alternatives implies a current state-wide annual value of at least $600 million, yet more likely in the range of $1.9–$4.4 billion for preserving the Sea.

K2 Economics also used results from multiple contingent valuation surveys for the Mono Lake

ecosystem to develop a separate transferable estimate for the value of preserving the Sea to the residents of California. Making conservative assumptions about values expressed for restoration of Mono Lake, K2 Economics determined that the current state-wide annual value of preserving the Sea is around $1.5 billion and possibly higher. Again, this estimate relies on strong assumptions about similarities between both the services provided by Mono Lake and the Salton Sea as well as the populations receiving the benefits. It also involves a relatively large amount of uncertainty compared to a primary valuation study of the Sea. Regardless, after considering both the SJV and Mono Lake benefits transfer results, K2 Economics concluded that a conservative order-of-magnitude estimate of the non-market benefits provided to the residents of California from preserving the Sea would be in the range of $1–$5 billion annually.

Interestingly, the estimated construction costs of the eight restoration alternatives range from $2.3 to $5.9 billion (California Resources Agency, 2006), a large number indeed. Yet, if one were to take the results from K2 Economics (2007) as a conservative order-of-magnitude estimate, the benefits of preserving the Salton Sea to California residents alone would seem to pass the benefit-cost test (i.e., positive net benefits). Furthermore, consider the results of Loomis (2000) who, in evaluating six different resource preservation programs, finds that residents within the states where each resource is located hold only a fraction of the total national value. This suggests that from a national perspective, the $1–$5 billion range is a very conservative estimate of restoration benefits to residents of the U.S.; consequently, the net benefits of preserving the Salton Sea are large.

Conclusions

The Salton Sea is a natural asset that provides many ecosystem services that directly and indirectly impact the quality of life for people at the local, regional, state, and national level. Such services include boating, fishing, hiking, photography, bird watching, and habitat provision for an abundance of birds, including migratory and resident as well as endangered and threatened species. The viability of this ecosystem and its ability to continue to provide such

services will be dependent on the engineering solutions devised for restoration, yet it will also be influenced by regional agricultural activities, and the quantity and quality of the associated drainage water. Any additional impositions on agriculture will likely impact agricultural-related industries and activities and, subsequently, affect regional employment and income.

In consideration of these effects, this article has argued that a net benefits' framework would be the most useful approach in which to evaluate and compare alternative restoration strategies. Relative to a cost-effectiveness or cost-minimizing approach, net benefits' analysis can provide more accurate information regarding potential net returns associated with a particular restoration alternative and present a clearer picture of the magnitude and distribution of benefits and costs at the local, state, and national levels. Certainly, it would seem useful for the legislature to have information on both the potential returns that each restoration alternative provides as well as the magnitude of the resources society is being asked to forgo to provide those returns. If such returns and costs cannot be quantified and monetized, then at the very least an enumeration of the trade-offs associated with each restoration alternative should be provided to inform the discussion. In addition to a description of how the physical characteristics of the Sea will differ across the restoration alternatives, qualitative information on the differences that each alternative presents for recreation opportunities, air quality, wildlife preservation, and other changes that society values should be provided. This enumeration exercise will help to identify what is missing, both qualitatively and quantitatively, in efforts to account for the full array of impacts of any particular alternative. This framework highlights the trade-offs associated with each alternative and exposes the limitations, thereby stimulating the need for additional scientific research to achieve better understanding.

Equally important, we have emphasized the fact that non-market goods, such as many of the ecosystem services provided by the Salton Sea, have been and should be part of any sound economic analysis involving habitat restoration. The recreational and preservation benefits derived from the natural resources of the Salton Sea will be directly dependent upon which restoration alternative is selected. These types of benefits have been given standing by state

and federal legislation and regulatory mandates, and have been shown to be one of the most important arguments in determining the class and scope of preservation that should occur.

While no primary non-market valuation studies have been performed to estimate the value of preserving the Salton Sea, and thus an accurate comparison of the alternative restoration strategies is limited, other research has estimated the value of somewhat similar ecosystems and their services (e.g., Mono Lake or wetlands in the San Joaquin Valley). The results from these other studies seem to suggest that the benefits of preserving the Salton Sea far exceed the costs. This is not surprising considering the research of Loomis and White (1996), who perform a meta-analysis of valuation studies for rare, threatened, and endangered species. The authors find that even for the most costly endangered species preservation efforts, the benefits are likely to exceed the costs. Yet for the particular case at hand, to accurately compare the trade-offs associated with different restoration alternatives, a primary valuation study is necessary and should be couched as one part, albeit a significant part, of a net benefits' analysis.

References

Arrow, K., R. Solow, E. Leamer, P. Portney, R. Radner & H. Schuman, 1993. Report of national oceanic and atmospheric administration panel on the reliability of natural resource damage estimates derived from contingent valuation. Federal Register 58: 4601–4614.

Beal, D., 1995. A travel cost analysis of the value of carnarvon Gorge national park for recreational use. Review of Marketing and Agricultural Economics 63: 292–303.

Berck, P., S. Robinson, & G. Goldman, 1991. The use of computable general equilibrium models to assess water policies. In Dinar, A. & D. Zilbermann (eds), The Economics and Management of Water and Drainage in Agriculture. Kluwer Academic Publishers, Boston, MA: 489–509.

Bishop, R., K. Boyle & K. Welsh, 1987. Glen Canyon Dam Releases and Downstream Recreation: an Analysis of User Preferences and Economic Values. Glen Canyon Environmental Studies Report No. 27/87. Bureau of Reclamation, Washington, D.C.

Boardman, A., D. Greenberg, A. Vining, & D. Weimer, 1996. Cost-Benefit Analysis: Concepts and Practice. Prentice Hall, New Jersey.

Bockstael, N., M. Hanemann, & C. Kling, 1987. Estimating the value of water quality improvements in a recreation demand framework. Water Resources Research 23: 951–960.

California Department of Water Resources, 2003. Salton Sea Reference Information. State of California, Sacramento, California.

California State Fish and Game Code, Sections 2081 and 2930.

California State Resources Agency, 2006. Salton Sea Update: Ecosystem Restoration Program. The Resources Agency, Department of Water Resources. Department of Fish and Game. May. California.

Carson, R., J. Wright, N. Carson, A. Alberini, & N. Flores, 1994. A Bibliography of Contingent Valuation Studies and Papers. NRDA Inc. San Diego, California.

CIC Research, 1989. The Economic Importance of the Salton Sea Sportfishery: A Report to the California Department of Fish and Game. CIC Research INC., San Diego, California.

Ciriacy-Wantrup, S., 1964. Benefit Cost Analysis and Public Resource Development. In Smith, S. C. & E. N. Castle (eds), Economics and Public Policy in Water Resource Development. Iowa State University Press, Ames, Iowa.

Cohen, M. J., & K. H. Hyun, 2006. Hazard: The Future of the Salton Sea with No Restoration Project. May. The Pacific Institute. Oakland, CA.

Cohn, J. P., 2000. Saving the Salton Sea: researchers work to understand its problems and provide possible solutions. Bioscience 50: 295–301.

Creel, M. & J. Loomis, 1990. Theoretical and empirical advantages of truncated count data estimators for analysis of deer hunting in California. American Journal of Agricultural Economics 72: 434–441.

Duffield, J., 1982. The Value of Recreational Use of Kootenai Falls. Chapter in Kootenai Falls Hydroelectric Project Final EIS. Montana Department of Natural Resources and Conservation, Helena.

Eckstein, O., 1958. Water-Resource Development: The Economics of Project Evaluation. Harvard University Press. Cambridge, MA.

Font, A., 2000. Mass Tourism and the Demand for Protected Natural Areas: A Travel Cost Approach. Journal of Environmental Economics and Management 39: 97–116.

Freeman, A., 2003. The Measurement of Environmental and Resource Value: Theory and Methods. 2nd Ed. Resources for the Future, Washington, DC.

Hagen, D., J. W. Vincent & P. G. Welle, 1992. Benefits of preserving old-growth forests and the spotted owl. Contemporary Economic Policy Volume X: 13–26.

Hanemann, W., 1994. Valuing the environment through contingent valuation. Journal of Economic Perspectives 8: 19–43.

Hausman, J., 1993. Contingent Valuation: A Critical Assessment, North Holland, Amsterdam.

Hellerstein, D., 1991. Using count data models in travel cost analysis with aggregate data. American Journal of Agricultural Economics 73: 860–867.

Hesseln, H., J. Loomis & A. Gonzalez-Caban, 2004. Comparing the economic effects of fire on hiking demand in Montana and Colorado. Journal of Forest Economics 10: 21–35.

Imber, D., G. Stevenson & L. Wilks, 1991. A Contingent Valuation Survey of the Kakadu Conservation Zone, Research Paper No. 3: Resource Assessment Commission, AGPS, Canberra.

Inland Empire Economic Databank, Forecasting Center, 1998. An Economic Analysis of the Benefits of Rehabilitating the Salton Sea: Summary of Results. University of California, Riverside, CA.

K2 Economics, 2007. A Preliminary Investigation of the Potential Non-Market Benefits Provided by the Salton Sea. Final report to the Salton Sea Authority. January. Riverside, CA.

Kopp, J. & V. K. Smith, 1993. Valuing Natural Assets: The Economics of Natural Resource Damage Assessment. Resources for the Future. Washington D.C.

Krutilla, J. & A. Fischer, 1975. The Economics of Natural Environments: Studies in Valuation of Commodities and Amenity Resources. The John Hopkins University Press for Resources for the Future, Baltimore, MD.

Layman, R., J. Boyce & K. Criddle, 1996. Economic valuation of the Chinook Salmon sport fishery of the Gulkana River, Alaska, under current and alternate management plans. Land Economics 72: 113–128.

Loomis, J., 1987. Balancing public trust resources of Mono Lake and Los Angeles' water right: an economic approach. Water Resources Research 23: 1449–1456.

Loomis, J. B., 2000. Vertically summing public good demand curves: an empirical comparison of economic versus political jurisdictions. Land Economics 76: 312–321.

Loomis, J., 2005. Economic values without prices: the importance of nonmarket values and valuation for informing public policy debates. Choices 20: 179–182.

Loomis, J., K. Strange, K. Fausch, & A. Covich, 2000. Measuring the total economic value of restoring ecosystem services in an impaired river basin: results from a contingent valuation survey. Ecological Economics 33: 103–117.

Loomis, J. & R. Walsh, 1997. Recreation Economic Decisions: Comparing Benefits and Costs. 2nd Ed. Venture Publishing, State College, PA.

Loomis, J. B. & D. S. White, 1996. Economic benefits of rare and endangered species: summary and meta-analysis. Ecological Economics 18: 197–206.

Menkhaus, S. & D. Lober, 1996. International ecotourism and the valuation of tropical rainforests in Costa Rica. Journal of Environmental Management 47: 10–16.

Morgenstern, R., 1997. The legal and institutional setting for economic analysis at EPA. In Morgenstern, R. (ed), Economic Analyses at EPA: Assessing Regulatory Impact. Resources for the Future, Washington, DC: 5–25.

National Research Council, 2004. Valuing Ecosystem Services: Toward Better Environmental Decision-Making. The National Academies Press, Washington, DC.

Navrud, S. & E. Mungatana, 1994. Environmental valuation in developing countries: the recreational value of wildlife viewing. Ecological Economics 11: 135–151.

Olsen, D., J. Richards & D. Scott, 1991. Existence and sport values for double the size of Columbia River basin salmon and steelhead runs. Rivers 2: 44–56.

Pearce, D. & R. Turner, 1990. Economics of Natural Resources and the Environment. John Hopkins University Press, Baltimore, MA.

Rosenberger, R. S. & J. B. Loomis, 2003. Benefit transfer. In Champ, P. A., K. J. Boyle & T. C. Brown (eds), A Primer on Nonmarket Valuation. Kluwer, Boston.

Sanders, K., R. Walsh & J. Loomis, 1990. Toward and empirical estimation of the total value of protecting rivers. Water Resources Research 26: 1345–1358.

Schwabe, K., I. Kan & K. Knapp, 2006. Drainwater management to reduce salinity problems in irrigated agriculture. American Journal of Agricultural Economics 88: 133–149.

Schuhmann, P. & K. Schwabe, 2004. An analysis of congestion measures and heterogeneous angler preferences in a random utility model of recreational fishing. Environmental and Resource Economics 27: 429–450.

United States Department of Agriculture, 2004. Economic Research Service. 2004 County-Level Poverty Rates for California. http://www.ers.usda.gov/Data/Poverty.

Water Resources Council, 1979. Federal Register: Procedures for evaluation of National Economic Development. Benefits and Costs in Water Resources Planning, Final Rule. Washington DC.

Wilson, M. & S. Carpenter, 1999. Economic valuation of freshwater ecosystem services in the United States, 1977–1997. Ecological Applications 9: 772–783.